Subsurface Remediation

Subsurface Remediation: Vadose Zones and Aquifers presents the state-of-the-art physical treatment methods for remediating contaminated subsurface formations, bridging theoretical foundations with practical applications through pedagogically enriched content. This book guides readers through the complex mechanics of water flow and solute transport in porous media, offering groundbreaking insights into dispersivity estimation methods. It also includes detailed examinations of inward hydraulic gradient sufficiency evaluation, degradation rate constant estimation methods with mass balance considerations, soil venting mechanics, and pump-and-treat technologies. With over 45 worked examples, end-of-chapter problems, and dual unit systems (SI and Imperial), this resource equips environmental engineers, hydrogeologists, and students with the essential knowledge to effectively remediate contaminated ground water systems either by removing pollutants or by converting them into harmless products. It addresses contamination from industrial spills, agricultural chemicals, urban runoff, and landfill leachates.

- Provides in-depth coverage of widely used physical treatment methods including soil vapor extraction, bioventing, and pump-and-treat systems with real-world implementation guidance for environmental professionals.
- Contains the most current research and methodologies for estimating longitudinal, transverse horizontal, and transverse vertical dispersivity values.
- Includes end-of-chapter problems as well as a comprehensive solution manual for instructors.

Subsurface Remediation

Vadose Zones and Aquifers

Vedat Batu

CRC Press
Taylor & Francis Group
Boca Raton London New York

CRC Press is an imprint of the
Taylor & Francis Group, an **informa** business

Designed cover image: Getty Images

First edition published 2026
by CRC Press
2385 NW Executive Center Drive, Suite 320, Boca Raton FL 33431

and by CRC Press
4 Park Square, Milton Park, Abingdon, Oxon, OX14 4RN

CRC Press is an imprint of Taylor & Francis Group, LLC

ISBN: 978-1-041-23922-2 (hbk)
ISBN: 978-1-041-23921-5 (pbk)
ISBN: 978-1-003-73912-8 (ebk)

DOI: 10.1201/9781003739128

Typeset in Palatino LT Std
by codeMantra

Access the Instructor and Student Resources: www.routledge.com/9781041239222

Contents

Preface....................xi
About the Author....................xiii

1. Definitions and the Presentation Framework of the Book....................1
1.1 Introduction....................1
1.2 Remediation Methods for Vadose Zones and Aquifers....................1
1.2.1 Soil Venting Methods for Vadose Zones....................1
1.2.2 P&T Method for Aquifers....................2
1.3 Presentation of the Framework of the Book....................2
1.3.1 Chapter 1....................2
1.3.2 Chapter 2....................2
1.3.3 Chapter 3....................2
1.3.4 Chapter 4....................2
1.3.5 Chapter 5....................3
1.3.6 Chapter 6....................3
1.3.7 Chapter 7....................4
1.3.8 Chapter 8....................4

2. Physical Characteristics of Fluids....................5
2.1 Introduction....................5
2.2 Dimensions and Units....................5
2.2.1 SI System....................5
2.2.2 English FSS System....................5
2.3 Density, Specific Weight, Specific Volume, and Specific Gravity....................6
2.3.1 Density and Specific Weight of Fluids....................6
2.3.2 Specific Volume....................7
2.3.3 Specific Gravity (*s.g.*) of Fluids....................9
2.3.4 Specific Weight and Density of a Perfect Gas....................9
2.3.5 Units of the *R* Specific Gas Constant and Its Values for Dry Air....................9
2.4 Compressibility of Fluids....................12
2.4.1 Compressible and Incompressible Fluids....................12
2.4.2 Compressibility and Elasticity of Fluids....................12
2.5 Viscosity of Fluids....................14
2.5.1 Definition of Viscosity....................14
2.5.2 Newton's Viscosity Law for Laminar Flows....................14
2.6 Surface Tension and Capillarity....................18
2.6.1 Surface Tension....................18
2.6.2 Capillarity....................18
Problems....................21

3. Fundamental Principles of Water Flow in Porous Media....................23
3.1 Introduction....................23
3.1.1 Porous Materials....................23

3.1.2 The Purpose of This Chapter 23
3.2 Natural Porous Media and Their Characteristics 23
3.2.1 Definitions for Natural Porous Media 23
3.2.2 Types of Natural Water-Bearing Formations 24
3.3 Hydraulic Head and Drawdown 25
3.3.1 Hydraulic Head 25
3.3.2 Drawdown 25
3.4 Porosity and Effective Porosity 25
3.5 Discharge Velocity and Ground water Velocity 26
3.6 Darcy's Law 27
3.6.1 Description of Darcy Experiment 27
3.6.2 Bernoulli's Equation for Porous Media Flow 28
3.6.3 Darcy Experiment 29
3.6.4 Validity of Darcy's Law 30
3.6.5 Hydraulic Conductivity, Permeability, and Transmissivity 30
3.6.6 Darcy and Flow Rate Equations in Terms of Pressure 33
Problem 36

4. Fundamental Principles of Solute Transport in Porous Media 37
4.1 Introduction 37
4.2 Concentration Units 37
4.2.1 Solute and Solvent 37
4.2.2 Relative Concentration Units 38
4.2.3 Dilute Mass Concentration Units 39
4.2.4 Concentration Units Based on Moles 41
4.3 Type of Concentrations 44
4.3.1 Introduction 44
4.3.2 Volume-Averaged Concentration 45
4.3.3 Flux-Averaged Concentration 45
4.4 Physical Mechanism of Solute Transport in Saturated Porous Media without Sorption and Decay: One-Dimensional Case 46
4.4.1 Solute Flux Equation Due to Diffusive Solute Transport 46
4.4.2 Quantification of Molecular Diffusion: Fick's First Law 46
4.4.3 Solute Flux Equation Due to Advective Transport 48
4.4.4 Solute Flux Equation Due to Dispersive Transport 49
4.4.5 Solute Flux Equation under Hydrodynamic Dispersion Conditions 50
4.4.6 Solute Flux Equation in Unsaturated Porous Media for Constant Water Content under Hydrodynamic Dispersion Conditions 50
4.5 Three-Dimensional Advective–Dispersive Solute Flux Equations for Saturated Porous Media—General Case 50
4.6 Three-Dimensional Advective–Dispersive Solute Flux Equations for Saturated Porous Media—Special Cases 51
4.6.1 Solute Flux Equations for Homogeneous and Anisotropic Porous Media 51
4.6.2 Solute Flux Equations for Homogeneous and Isotropic Porous Media 53
4.7 Solute Transport in Saturated Porous Media without Sorption and Decay 54
4.7.1 General Case: Homogeneous and Anisotropic Dispersion 54
4.7.2 Homogeneous and Isotropic Dispersion with Three Constant Velocity Components 57

4.7.3 Homogeneous and Isotropic Dispersion with Uniform Flow 57
4.8 Solute Transport in Saturated Porous Media with Sorption and Decay Conditions 58
4.8.1 Solute Transport Equations under Sorption Conditions 58
4.8.2 Solute Transport Equations under Degradation Conditions 63
4.8.3 Generalized Solute Transport Differential Equations in Saturated Porous Media 64
4.9 Mechanical Dispersion Coefficients for Saturated Porous Media 66
4.9.1 Mechanical Dispersion Equations Based on Laboratory Experiments 67
4.9.2 Generalized Mechanical Dispersion Coefficient Equations Based on Theory and Postulation 69
4.10 Generalized Hydrodynamic Dispersion Coefficient Equations with Three Different Dispersivities: α_L, α_{TH}, and α_{TV} 70
4.11 Hydrodynamic Dispersion Coefficient Equations under Uniform Flow Conditions 71
4.12 Estimation Methods of Dispersivities in Aquifers 72
4.12.1 Introduction 72
4.12.2 Longitudinal Dispersivity Estimation Methods Based on Scale 73
4.12.3 Longitudinal Dispersivity Estimation Method of Zech, Attinger, Bellin, Cvetkovic, Dagan, Dietrich, Fiori, and Teutsch [Zech et al. (2023) Method] Based on the Log Hydraulic Conductivity Variance (σ_f^2) and Longitudinal Correlation Scale (λ_f) 76
4.12.4 Comparative Evaluation of the Methods of Estimation for the Longitudinal Dispersivity (α_L) Based on Pickens and Grisak (1981), Neuman (1990), Xu and Eckstein (1995), and Zech et al. (2023) for Some Selected Aquifers α_L Data 85
4.12.5 Estimation of Transverse Dispersivities 90
4.12.6 Concluding Remarks for Longitudinal Dispersivity Estimation 92
4.13 Some One-Dimensional Analytical Solutions for Solute Transport in Porous Media 93
4.13.1 Introduction 93
4.13.2 One-Dimensional First-Type Continuous Source Solutions 93
4.13.3 One-Dimensional Third-Type Continuous Source Solutions 96
Problems 99

5. Sufficiency Evaluation of the Inward Hydraulic Gradients in Contaminated Aquifers 101
5.1 Introduction 101
5.2 Flow and Dispersion in Opposite Directions 101
5.2.1 Governing Differential Equation 101
5.2.2 Status of the Dispersion Coefficient for Flow against Dispersion 101
5.2.3 Description of the Problem 102
5.2.4 Initial and Boundary Conditions 102
5.3 Transient Solutions for Concentration and Comparison with Numerical Solutions 103
5.3.1 Solution Method 103
5.3.2 Solution for $C(x=0,t)=0$ Boundary Condition 104
5.3.3 Solution for $\frac{\partial C(0,t)}{\partial x}=\frac{U}{2D}C(0,t)$ Boundary Condition 111

5.3.4 Comparison with Numerical Solutions 120
5.4 Steady-State Solutions for Concentration 120
5.4.1 Steady-State Solution for the Boundary Condition $C(x=0,t=\infty)=0$ 120
5.4.2 Steady-State Solution for the Boundary Condition $\frac{\partial C(0,t=\infty)}{\partial x}=\frac{U}{2D}C(0,t=\infty)$ 121
5.4.3 Comparison with Numerical Solutions 122
5.5 Potential Applications to Remediation Projects 122
5.5.1 Evaluation of the Steady-State Concentration Expressions 122
5.5.2 Application of the Steady-State Concentration Expressions 122
Problems 130

6. Estimation of Degradation Rates of Solute Contaminants in Porous Media 132
6.1 Introduction 132
6.1.1 Terminology for Mass Loss during the Transport Process 132
6.1.2 Literature for Degradation Rate Estimates 132
6.2 One-Dimensional Steady-State First-Type Source Solution 133
6.2.1 Problem Description 133
6.2.2 Governing Differential Equation and Boundary Conditions 133
6.2.3 Solution for Flux-Averaged Concentration 134
6.2.4 Advective–Dispersive Flux for Flux-Averaged Concentration 134
6.3 One-Dimensional Steady-State Third-Type Source Solution 135
6.3.1 Problem Description 135
6.3.2 Governing Differential Equation and Boundary Conditions 135
6.3.3 Solution for Volume-Averaged Concentration Variation 136
6.3.4 Advective–Dispersive Flux for Volume-Averaged Concentrations 137
6.4 The Relationship between the First- and Third-Type Source Analytical Solutions 137
6.5 Comparative Evaluation of the First- and Third-Type Source Solutions for Mass Balance at the Inlet 138
6.6 Evaluation of the First-Type Source Solution-Based Method for Estimating Degradation Parameter 142
6.6.1 Literature for Degradation Rate Estimates Based on the First-Type Source Solution 142
6.6.2 Buscheck and Alcantar Method Degradation Rate Estimation Method Based on the First-Type Source Solution 143
6.6.3 Evaluation of the First-Type Source-Solution Based Method 145
6.6.4 Comparative Evaluation of the First-Type Source Condition with an Example 146
6.7 Degradation Rate Estimation Method by Curve-Matching 148
6.7.1 Introduction 148
6.7.2 Curve-Matching Method for Degradation Rate Estimation 148
Problems 151

7. Fluid Dynamics of Soil Venting and Its Associated Remediation Methods for Vadose Zones 152
7.1 Introduction 152
7.2 Governing Equations of Soil Venting 152

7.2.1 Introduction 152
7.2.2 Equation of Continuity for Gases 153
7.2.3 Equation of Continuity for Vapor Flow 156
7.2.4 Linearized Equation of Continuity for Vapor Flow 157
7.3 Transient Analytical Solutions for Soil Gas Flow to Wells and Determination of Unsaturated Zone Properties – Fully-Penetrating Wells 158
7.3.1 Introduction 158
7.3.2 Fully-Penetrating Wells in Unsaturated Zones and Adaptation of the Theis Solution 159
7.3.3 Adaptation of the Theis Solution to the Gas Flow in the Unsaturated Zone 161
7.3.4 Determination of Air Permeability and Air-Filled Porosity 168
7.4 Transient Analytical Solutions for Soil Gas Flow for Partially Penetrating Wells to Determine Unsaturated Zone Properties 172
7.4.1 Introduction 172
7.4.2 Partially Penetrating Well in an Unsaturated Zone Below the Ground Surface without a Leaky Confining Layer 172
7.4.3 Partially Penetrating Well in an Unsaturated Zone Below the Ground Surface with a Leaky Confining Layer 175
7.4.4 Special Solutions for Steady-State Cases 177
7.4.5 Evaluation of the Solutions and Air Permeability Determination from Field Data 178
7.5 Steady-State Analytical Solutions for Soil Gas Flow to Wells and Determination of the Unsaturated Zone Properties 178
7.5.1 Introduction 178
7.5.2 Estimation of Steady-State Reaching Time in a Typical Sand Formation 179
7.5.3 Steady-State Soil Gas Flow to a Fully-Penetrating Well in the Unsaturated Zone 182
Problems 197

8. Pump-and-Treat Method 198
8.1 Introduction 198
8.2 General Description of a Pump-and-Treat System 198
8.2.1 Sources of Ground water Contamination 199
8.2.2 Options for Ground water Cleanup 199
8.2.3 A Typical Pump-and-Treat System 199
8.2.4 Components of a Typical P&T System 200
8.2.5 Main Limitations of P&T Systems 200
8.3 Data Requirements of a P&T System 200
8.3.1 Hydrogeologic Data 200
8.3.2 Contaminant Data 201
8.3.3 Data Collection 204
8.3.4 Data Interpretation 207
8.4 Conceptual Design 208
8.4.1 Aquifer Contamination Cleanup 208
8.4.2 Soil Contamination Cleanup 209
8.4.3 Calculating the Estimated Cleanup Time for Nonreactive Solutes 209
8.4.4 Calculating the Estimated Cleanup Time for Reactive Solutes 211

8.4.5 Description of Tailing and Rebound ... 212
8.4.6 Determination of Critical Inward Hydraulic Gradient ... 213
8.5 Performance Evaluations of P&T Systems ... 213
8.5.1 Remedial Action Objectives ... 213
8.5.2 Monitoring ... 214
8.6 Limitations of Pump-and-Treat (P&T) Systems ... 216
8.6.1 Limitations Due to Nonaqueous Phase Liquids (NAPLs) ... 216
8.6.2 Limitations Due to Contaminant Sorption ... 219
8.6.3 Limitations Due to Solids Dissolution ... 221
8.6.4 Limitations Due to Hydraulic Conductivity and Ground water Velocity Variations ... 222
8.6.5 Limitations Due to the Effects of Matrix Diffusion Contamination ... 223
Problems ... 228

References ... 230

Index ... 237

Preface

As the title *Subsurface Remediation: Vadose Zones and Aquifers* indicates, the main purpose of this book is to present the state-of-the-art remediation methods to clean up contaminated water-bearing formations in subsurface systems below the ground surface with numerous enriched examples based on the rules of the science of learning called "pedagogy." The subsurface formations are composed of the vadose zone (or the unsaturated zone) between the ground surface and the water table as well as the zone below the water table. In a broader sense, the remediation of these formations is the process used to treat polluted water in these formations by removing pollutants or converting them into harmless products. Contaminants in aquifers and vadose zones potentially can be inorganic chemicals, organic chemicals, or radioactive materials that result mostly from human activities as well as from natural sources. Among them are accidental spills from industrial operations, overapplication of fertilizers or pesticides, infiltration from urban runoff, leaking from landfills, and others.

Remediation methods for aquifers and vadose zones are generally based on physical, chemical, and biological treatment techniques as well as their combination. The main focus of this book is to remediate aquifers and vadose zones using physical treatment methods, which require familiarity with the flow and transport mechanisms in aquifers and vadose zones. The well-known physical treatment methods are soil venting for vadose zones (or unsaturated zones) and the pump-and-treat method for aquifers.

This book has eight chapters, which are described briefly below.

In Chapter 1, definitions regarding aquifer remediation methods and the presentation framework are presented.

In Chapter 2, the physical characteristics of fluids along with their units are presented.

In Chapter 3, the water flow mechanics through saturated natural porous materials, along with the related physical laws, principles, and governing differential equations, are presented.

In Chapter 4, the fundamental principles of solute transport in porous media along with its mathematical modeling foundations with details of the governing differential equations that are crucial to be understood in order to use various numerical and analytical mathematical models with their initial and boundary conditions for effective usage of the remediation methods are presented. The most important feature of Chapter 4 is to provide the state-of-the-art status regarding estimation methods of longitudinal dispersivity (α_L), transverse horizontal dispersivity (α_{TH}), and transverse vertical dispersivity (α_{TV}) values based on various methods till the 2020s.

In Chapter 5, the theoretical developments and their practical features regarding sufficiency evaluation of the inward hydraulic gradients in contaminated aquifers are presented. One of the most widely used remediation methods for a contaminant plume in an aquifer is to take control of existing contaminant plumes in aquifers. In such cases, a situation often can occur such that control of contaminant migration in an aquifer by the ground water flow toward the region of higher concentration is required. In other words, the goal is to generate inward flow gradients toward the center of the contaminant plume in order to reduce the concentrations below the imposed concentrations by regulatory requirements. The question is: Are the inward hydraulic gradients, generated by extraction wells or other means, sufficient to take control of the plume? The purpose of Chapter 5 is to provide answers to this question.

During the past half-century, some methods have been developed to estimate the degradation rate constant for various contaminant constituents under laboratory and field conditions using measured dissolved solute concentrations (Buscheck and Alcantar, 1995; Chapelle et al., 1995; Zhang and Heathcote, 2003; Batu, 2010; Batu et al., 2013). These methods are based on the steady-state, one-dimensional solutions of the advective–dispersive solute transport differential equation with degradation and sorption effects, using the first-type (Dirichlet-type) source condition with the exception of Batu (2010) and Batu et al. (2013) for which the third-type (Cauchy-type) source condition is used. The first-type source solutions do not satisfy the mass balance at the inlet, whereas the third-type source solution satisfies the mass balance at the inlet exactly. As a result, methods based on the first-type source solutions may yield errors in estimating degradation rates depending on the longitudinal dispersivity and retardation factor values. Chapter 6 describes these methods along with their theoretical foundations with examples. It has also been proven that the first-type source solutions can generate degradation rates that are several times higher depending on the values of dispersion coefficient (D_x) and retardation factor (R_d).

In Chapter 7, the mechanics of soil venting is presented in detail based on scientific advancements during the past decades. The method of *soil venting* is to remove volatile and semivolatile contaminants from soil. A positive or negative air pressure is applied either passively or actively to soil to remove vapors that are appropriately treated. It is well known that petroleum products and organic solvents are serious contaminant constituents, especially in industrialized countries. Soil vapor extraction (SVE) and bioventing (BV) systems are commonly designed to remediate especially unsaturated zones.

In Chapter 8, the pump-and-treat (P&T) method is described, in which contaminated ground water is extracted from an aquifer, cleaned at the ground surface, and then discharged back to the same aquifer. The P&T is one of the most widely used ground water remediation technologies for dissolved contaminants in ground water. However, under complex site conditions, it may be necessary to combine several remedial actions for a given site.

As emphasized previously, a focus on the rules of pedagogy is one of the key characteristics of this book. To fulfill these features, this book includes 45 worked examples. Beyond this, there are problems at the end of each chapter for which a solution manual accompanies this book for instructors.

The vast majority of the countries of the world have been using the metric system (*Système International d'Unités, SI*). However, in the USA, the English FSS (foot-slug-second) system is still more often used. Therefore, in this book, both systems are used as appropriate.

I wish to express my gratitude to Yılmaz Akyıldız (Emeritus Professor, University of Boğaziçi, Istanbul, Türkiye) for his help regarding some mathematics involved in some chapters; Shlomo P. Neuman (Emeritus Professor, University of Arizona, Tucson, Arizona, USA) and Gedeon Dagan (Emeritus Professor, Tel Aviv University, Tel Aviv, Israel) for their help regarding recommendations and critics for the state-of-the-art dispersivity estimation methods; Alraune Zech (Professor, Utrecht University, Utrecht, The Netherlands) for the papers and data she provided; and John F. Delvin (Professor, University of Kansas, Lawrence, Kansas, USA) and Aldo Fiori (Professor, Università Roma Tre, Rome, Italy) for reviewing several chapters.

I am grateful to Taylor & Francis CRC Press for undertaking to publish this book. Special thanks are due to Joseph Clements.

Vedat Batu
Naperville, Illinois, USA
September, 2025

About the Author

Vedat Batu graduated from the Civil Engineering Faculty of Istanbul Technical University (Istanbul, Türkiye) in 1969 with B.S. and M.S. degrees (combined) in the area of Civil Engineering. In the same year, he was appointed as an instructor in the Hydraulic Engineering Division of the Civil Engineering Department of Karadeniz Technical University (Trabzon, Türkiye). He received his Ph.D. in the area of Hydraulic Engineering from Istanbul Technical University in 1974. He prepared his Associate Professorship thesis in the Hydraulic Engineering area for a 1-year time period in 1976–1977 at the University of Wisconsin, Madison, Wisconsin, USA. After submittal of his thesis and passing all required examinations, he was appointed as Associate Professor (Doçent) in the Department of Civil Engineering of Karadeniz Technical University in June 1979, and he worked as a faculty member until the end of 1982. Lastly, he was the Chairman of the Civil Engineering Department. At the end of 1982, he voluntarily resigned from his position and went to the USA. He also served as a visiting professor in several universities in the USA and in the UK and then he joined the private sector in the USA and worked mostly on environmental engineering-related projects. In Türkiye and the USA, he taught many courses at both undergraduate and graduate levels including hydraulic engineering, open channel hydraulics, hydrology, and ground water hydraulics for over 10 years. His industrial work covers more than 30 years. Since he received his Ph.D. degree in 1974, Dr. Batu has been involved in many scientific projects and published a number of authored and coauthored papers in some internationally recognized journals including *Ground Water, Journal of Hydrology, Journal of Hydraulic Engineering, Journal of Irrigation and Drainage Engineering, Soil Science Society of America Journal,* and *Water Resources Research.* Dr. Batu has developed new concepts and approaches, including many analytical and numerical models in the areas of aquifer hydraulics, vadose zone hydrology, and solute transport in aquifers. Besides this book, he is the author of three volumes entitled *Aquifer Hydraulics: A Comprehensive Guide to Hydrogeologic Data Analysis; Applied Flow and Solute Transport Modeling in Aquifers: Fundamental Principles and Analytical and Numerical Methods* (Taylor & Francis CRC Press, 2006); and *Fluid Mechanics and Hydraulics: Illustrative Worked Examples of Surface and Subsurface Flows* (Taylor & Francis CRC Press, 2024). The contents of these books, as well as the views of many practitioners and academicians, can be seen at www.amazon.com by searching for Dr. Batu's full name. His original papers and books are cited and used by many scientists and practitioners around the world, and these can be seen at scholar.google.com by searching for his name. Apart from practitioners in various engineering areas, these books are being used as textbooks in various universities around the world, including North America, Europe, and the Far East. Dr. Batu was appointed as a full professor in August 2008 in the Department of Civil Engineering of Yeditepe University (Istanbul, Türkiye). Currently, he is with Argonne National Laboratory, Lemont, Illinois, USA.

1

Definitions and the Presentation Framework of the Book

1.1 Introduction

There are generally two types of actions regarding subsurface remediation. The first one is to clean up polluted aquifers and vadose zones. The second one is to protect subsurface water resources from contamination. When these environments become contaminated, it is difficult and expensive to clean them up, depending on their characteristics. Contaminants can be removed from these environments by applying various techniques to bring the water to a standard that is commensurate with various intended uses.

To protect subsurface water resources from contamination or remediate them, the interrelation of surface water bodies and the aquifers below the vadose zone must be well understood. If there are water supply wells near a source of contamination, those wells potentially will be under risk of contamination depending on the distances of the source as well as the hydrogeologic characteristics of the aquifers.

1.2 Remediation Methods for Vadose Zones and Aquifers

Remediation methods for vadose zones and aquifers are generally based on physical treatment, chemical treatment, and biological techniques, as well as their combination. The main focus of this book is to remediate vadose zones and aquifers using physical treatment methods, which require familiarity with the flow and transport mechanisms in vadose zones and aquifers. The well-known physical treatment methods are soil venting for vadose zones (or unsaturated zones) and the pump-and-treat (P&T) method for aquifers. These techniques are briefly described in the following sections.

1.2.1 Soil Venting Methods for Vadose Zones

The method of *soil venting* is to remove volatile and semivolatile contaminants from soil in vadose zones. A positive or negative air pressure is applied either passively or actively to soil to remove vapors that are appropriately treated. It is well known that petroleum products and organic solvents are serious contaminant constituents, especially in industrialized countries. Soil vapor extraction (SVE) and bioventing (BV) systems are being designed to remediate especially unsaturated zones.

The SVE method is one of the most effective and cost-efficient methods of removing volatile organic compounds (VOCs) from unsaturated soils. The BV method is a remediation technology that uses indigenous microorganisms to biodegrade organic constituents

DOI: 10.1201/9781003739128-1

adsorbed to soils in the unsaturated zone. Soils in the capillary fringe and the saturated zone are not affected. When extraction wells are used for BV, the process is similar to SVE. However, while SVE removes constituents primarily through volatilization, BV systems promote biodegradation of constituents and minimize volatilization generally by using lower air flow rates than for SVE.

1.2.2 P&T Method for Aquifers

The P&T method is one of the most widely used aquifer remediation technologies for dissolved contaminants in aquifers. However, under complex site conditions, it may be necessary to combine several remedial actions for a given site. In a P&T system, contaminated ground water is extracted to the ground surface, and after the cleaning process is completed, it is injected back into the aquifer. Selection of a remedial technology is a function of the physical characteristics, reactivity, and mobility of the dissolved contaminant, and characteristics of the aquifer. If the aquifer is very tight, extraction of water at a reasonable rate may cause the well to go dry, creating a narrow and deep capture zone. For such conditions, an intercept drain may be more appropriate.

1.3 Presentation of the Framework of the Book

This book has a total of eight chapters, which are described below.

1.3.1 Chapter 1

In Chapter 1, definitions and the presentation framework are presented.

1.3.2 Chapter 2

In Chapter 2, the physical characteristics of fluids, along with their units, are presented.

1.3.3 Chapter 3

In Chapter 3, the water flow mechanics through saturated natural porous materials, along with the related physical laws, principles, and governing differential equations, are presented. After presenting the concepts of discharge velocity and ground water velocity, all the details of Darcy's law are presented as well.

1.3.4 Chapter 4

In Chapter 4, the fundamental principles of solute transport in porous media, along with their mathematical modeling foundations, with details of the governing differential equations that are crucial to be understood in order to use various numerical and analytical mathematical models with their initial and boundary conditions for effective usage of the remediation methods, are presented. Also, in Chapter 4, only some well-known one-dimensional analytical solute transport solutions are presented, because these solutions are used in the subsequent several chapters, especially in determining the degradation rates of contaminant constituents.

The most important feature of Chapter 4 is to provide the state-of-the-art status regarding estimation methods of longitudinal dispersivity (α_L), transverse horizontal dispersivity (α_{TH}), and transverse vertical dispersivity (α_{TV}) values based on various methods till the 2020s. Among these studies, Pickens and Grisak (1981) recommended a simple "1/10" plume length formula for estimation of the longitudinal dispersivity (α_L), i.e., α_L can be assumed to have a value one tenth of the plume length.

Gelhar et al. (1985) conducted a comprehensive compilation of dispersivity values published in many sources based on field-scale tracer tests and contaminant transport modeling efforts. The literature sources and pertinent data in each study are summarized in Gelhar et al. (1992). These data and information are also included in Batu (2006, pp. 585–598, Table 7-1) with their entirety. Gelhar et al. (1992, pp. 1971–1972) state that it does not seem reasonable that a single universal line (or curve-fitting) can be meaningfully identified by applying standard linear regression to all of the data.

Despite the above-given statements of Gelhar et al. (1992) regarding curve-fitting, the data for the longitudinal dispersivity in Gelhar et al. (1985) were interpreted by Neuman (1990) who presented formulas for longitudinal dispersivity (α_L). Later on, Xu and Eckstein (1995) also analyzed the longitudinal dispersivity data of Gelhar et al. (1992) based on the weighted least-squares method and presented some empirical formulas for the longitudinal dispersivity (α_L).

During the past decade, the controversial issue regarding the longitudinal dispersivity (α_L) was raised once again by some eminent scientists (Zech et al., 2015) by revisiting mainly the well-known α_L data in Gelhar et al. (1992) and concluded that transport, and particularly the macrodispersivity, is formation specific, and that modeling of transport cannot be relegated to a unique scaling law. Instead, transport requires characterization of aquifer properties, e.g., spatial distribution of hydraulic conductivity, and the use of adequate models. The approach of Zech et al. (2015) was commented on by Neuman (2016) who wrote that their theory is equally unsuitable for the analysis of nonstationary transport. In their response, Zech et al. (2016) stated their work reached the conclusion that there is no trend of increase α_L with distance L and that values are scattered and are aquifer-specific.

Further to their work in Zech et al. (2015), in a recent paper, Zech et al. (2023) presented a method for estimation of α_L with some guidelines based on the log hydraulic conductivity variance (σ_f^2) and longitudinal correlation scale λ_f, which is equivalent to I_h that of Zech et al. (2015), i.e., $\lambda_f \equiv I_h$.

1.3.5 Chapter 5

One of the most widely used remediation techniques for a contaminant plume in an aquifer is to take control of existing contaminant plumes in aquifers. In such cases, a situation often can occur such that control of contaminant migration in an aquifer by the ground water flow toward the region of higher concentration is required. In other words, the goal is to generate inward flow gradients toward the center of the contaminant plume in order to reduce the concentrations below the imposed concentrations by regulatory requirements. The question is: Are the inward hydraulic gradients, generated by extraction wells or other means, sufficient to take control of the plume? The purpose of this chapter is to provide answers to this question.

1.3.6 Chapter 6

During the past half-century, some methods were developed to estimate the degradation rate constant for various contaminant constituents under laboratory and field conditions using measured dissolved solute concentrations (Buscheck and Alcantar, 1995;

Chapelle et al., 1995; Zhang and Heathcote, 2003; Batu, 2010; Batu et al., 2013). These methods are based on the steady-state one-dimensional solutions of the advective–dispersive solute transport differential equation with degradation and sorption effects using the first-type (Dirichlet-type) source condition with the exception of Batu (2010) for which the third-type (Cauchy-type) source condition is used. The first-type source solutions do not satisfy the mass balance at the inlet, whereas the third-type source solution exactly satisfies the mass balance at the inlet. As a result, methods based on the first-type source solutions are associated with some errors in estimating degradation rates depending on the longitudinal dispersivity and retardation factor values. In Chapter 6, these methods are described along with their theoretical foundations with examples. And, it has been proven that the first-type source solutions can generate more than several times higher degradation rates depending on the values of dispersion coefficient (D_x) and retardation factor (R_d).

1.3.7 Chapter 7

Besides aquifer contamination, VOC contamination of unsaturated zones is a serious widespread problem as well. The chemicals entering the ground due to leaking storage tanks, pipelines, and other means generate contaminant sources to the underlying aquifers. The conventional method for remediation of VOC-contaminated soils is excavation and disposal at approved off-site locations. This method is not only expensive but also problematic in transferring contaminant sources to other locations. Beyond this, at many contaminated sites, soil excavation may not be feasible because of the presence of ground structures, buried pipelines, utilities, and similar facilities.

The method of *soil venting* is to remove volatile and semivolatile contaminants from soil. A positive or negative air pressure is applied either passively or actively to soil to remove vapors that are appropriately treated. It is well known that petroleum products and organic solvents are serious contaminant constituents, especially in industrialized countries. SVE and BV systems are being designed to remediate especially unsaturated zones.

The SVE method is one of the most effective and cost-efficient methods of removing VOCs from unsaturated soils. The BV method is a remediation technology that uses indigenous microorganisms to biodegrade organic constituents adsorbed to soils in the unsaturated zone. Soils in the capillary fringe and the saturated zone are not affected. When extraction wells are used for BV, the process is similar to SVE. However, while SVE removes constituents primarily through volatilization, BV systems promote biodegradation of constituents and minimize volatilization generally by using lower air flow rates than for SVE.

The dynamics of soil venting is the foundation of both SVE and BV. In Chapter 7, the mechanics of soil venting is presented in detail based on the scientific advancements during the past few decades.

1.3.8 Chapter 8

Containment and cleanup of contaminated aquifers are among the primary objectives of the environmental programs of many countries in the world. For example, in the United States, CERCLA (Comprehensive Environmental Response, Compensation, and Liability Act; also known as Superfund) and RCRA (Resource Conservation and Recovery Act) remediation programs are among them. A common, frequently used practice is to extract contaminated ground water from an aquifer and then clean it at the ground surface to discharge it back to the same aquifer. This is called conventional P&T remediation.

In Chapter 8, general description, data requirements, conceptual design, performance evaluation, and limitations of a P&T system are described.

2

Physical Characteristics of Fluids

2.1 Introduction

The term *fluids* covers both liquids and gases. A solid body can resist an applied shear stress by deforming its shape, whereas a fluid deforms continuously under the influence of shear stresses. Therefore, the distinction between a solid and a fluid is made on the basis of the ability of a substance to resist an applied shear stress that changes its shape.

Liquids and gases have some common characteristics, but they have some distinctive characteristics of their own. For example, it is easy to compress a gas, whereas the compressibility of liquids is negligible in most cases. A given amount of liquid occupies a fixed volume, irrespective of the size and shape of the container. On the contrary, a gas body has no fixed volume and expands continuously unless restrained by the containing vessel. For liquids, a free surface is formed in its container, whereas a gas will completely occupy any vessel in which it is placed, and therefore will not have a surface.

In this chapter, the physical characteristics of fluids along with their units are presented.

2.2 Dimensions and Units

A dimension is the measure by which a physical variable is expressed quantitatively, and the unit is a particular way of attaching a number to the quantities of dimension. The majority of countries in the world have been using the metric *Système International d'Unités* (*SI*). However, in the United States, the English FSS (foot-slug-second) system is still being used. Therefore, in this book, both systems are used. Below, the units of various quantities are described in both the *SI* and *English FSS* systems.

2.2.1 SI System

For the *SI* system, there are four basic dimensions, such as length (L, m), mass (M, kg), time (t, s), and temperature (T, Kelvin), with which fluid properties are expressed.

There are several conventions used in the *SI* system. The names of multiples and submultiples of the basic and derived *SI* units are related to these units. For example, the prefixes include mega (M)$=10^6$, kilo (k)$=10^3$, milli (m)$=10^{-3}$, and micro (μ)$=10^{-6}$. Therefore, 1 megawatt$=$1 MW$=10^3$ kW$=10^6$ W and square millimeter$=$1 mm$^2=10^{-6}$m$^2=10^{-6}$m^2.

2.2.2 English FSS System

Likewise, in the English FSS system, there are four basic dimensions, such as length (L, ft), mass (M, slug), time (t, s), and temperature (T, °R), with which fluid properties are expressed.

DOI: 10.1201/9781003739128-2

2.3 Density, Specific Weight, Specific Volume, and Specific Gravity

2.3.1 Density and Specific Weight of Fluids

The *density* ρ, or more strictly mass density, of a fluid is the mass per unit volume. The *specific weight* γ is its weight per unit volume and is the force exerted by gravity.

The density and specific weight of a fluid are related by

$$\rho = \frac{\gamma}{g} \quad \text{or} \quad \gamma = \rho g \tag{2-1}$$

in which g is the gravity of acceleration. The units in the SI system are

$$[\rho] = \frac{[\gamma]}{[g]} = \frac{\frac{\text{N}}{\text{m}^3}}{\frac{\text{m}}{\text{s}^2}} = \frac{\text{N.s}^2}{\text{m}^4} = \frac{M}{V} = \frac{\text{kg}}{\text{m}^3} \tag{2-2}$$

In the English FSS system,

$$[\rho] = \frac{[\gamma]}{[g]} = \frac{\frac{\text{lb}}{\text{ft}^3}}{\frac{\text{ft}}{\text{s}^2}} = \frac{\text{lb.s}^2}{\text{ft}^4} = \frac{M}{V} = \frac{\text{slugs}}{\text{ft}^3} \tag{2-3}$$

in which

$$\text{Mass } M \text{ in slugs} = \frac{\text{Weight } W \text{ in pounds (lb)}}{g\left(32.2\,\frac{\text{ft}}{\text{s}^2}\right)} \tag{2-4}$$

It is important to note that ρ is an absolute quantity because it depends on mass, which is independent of location. Specific weight γ is not an absolute quantity since it depends on gravitational acceleration g, which varies with location on the Earth, primarily latitude and elevation from the mean sea level.

Densities and specific weights of fluids vary with temperature. Table 2-1 presents the density of some fluids with temperature in the SI system. Table 2-2 presents them in the English FSS system. Tables 2-3 and 2-4 present the density and specific weight values with temperature for water in the SI and English FSS systems, respectively. For example, in the SI system at 5°C temperature, the density of water is $\rho = 1{,}000\ \text{kg/m}^3$ and specific weight $\gamma = 9.807\ \text{kN/m}^3$ (Table 2-3). In the English FSS system at 32°F temperature, the density of water is $\rho = 1.940\ \text{slug/ft}^3$ and specific weight $\gamma = 62.42\ \text{lbf/ft}^3$ (Table 2-4).

The unit of Newton (N) is defined based on Newton's second law and is given as

$$F = ma \tag{2-5}$$

in which F is the force acting on mass m and a is the acceleration. In the SI system, if the mass $m = 1$ kg and $a = 1\ \text{m/s}^2$, the corresponding force is defined as

$$1\,\text{N} = (1\,\text{kg})\left(1\,\frac{\text{m}}{\text{s}^2}\right) \tag{2-6}$$

TABLE 2-1

Density, Specific Gravity (s.g.), Modulus of Elasticity, and Viscosity of Some Fluids as a Function of Temperature in the SI Units

Fluid	Temperature T (°C)	Density ρ (kg/m³)	Specific Gravity (s.g.)	Modulus of Elasticity E (kPa)	Viscosity μ (Pa.s)
Benzene	20	876.2	0.88	1,034,250	6.56×10^{-4}
Carbon tetrachloride	20	1,587.4	1.59	1,103,200	9.74×10^{-4}
Crude oil	20	855.6	0.86	–	71.8×10^{-4}
Ethyl alcohol	20	788.6	0.79	1,206,625	12.0×10^{-4}
Gasoline	20	680.3	0.68	–	2.9×10^{-4}
Hydrogen	–257.2	73.7	–	–	0.21×10^{-4}
Mercury	15.6	13,555	13.56	26,201,000	0.51×10^{-4}
Oxygen	–195.6	1,206.0	–	–	2.78×10^{-4}
Sodium	315.6	876.2	–	–	3.30×10^{-4}
Water	20	998.2	1.00	2,170,500	10.0×10^{-4}

Source: Adapted from Vennard and Street (1982).

TABLE 2-2

Density, Specific Gravity (s.g.), Modulus of Elasticity, and Viscosity of Some Fluids as a Function of Temperature in the English FSS Units

Fluid	Temperature T (°F)	Density ρ (slug/ft³)	Specific Gravity (s.g.)	Modulus of Elasticity E (psi)	Viscosity μ (lbf.s/ft²)
Benzene	68	1.70	0.88	150,000	1.37×10^{-5}
Carbon tetrachloride	68	3.08	1.59	160,000	2.035×10^{-5}
Crude oil	68	1.66	0.86	–	15.0×10^{-5}
Ethyl alcohol	68	1.53	0.79	175,000	2.51×10^{-5}
Gasoline	68	1.32	0.68	–	0.61×10^{-5}
Hydrogen	–431	0.143	–	–	0.0435×10^{-5}
Mercury	60	26.3	13.56	3,800,000	3.26×10^{-5}
Oxygen	–320	2.34	–	–	0.58×10^{-5}
Sodium	600	1.70	–	–	0.69×10^{-5}
Water	68	1.936	1.00	318,000	2.10×10^{-5}

Source: Adapted from Vennard and Street (1982).

In the English FSS system, if the mass $m = 1$ slug and $a = 1$ ft/s^2, the corresponding force is defined as

$$1 \text{ lbf} = (1 \text{ slug})\left(1 \frac{\text{ft}}{\text{s}^2}\right) \tag{2-7}$$

2.3.2 Specific Volume

Specific volume v is the volume occupied by the unit mass of fluid. Thus

$$v = \frac{1}{\rho} \tag{2-8}$$

TABLE 2-3

Specific Weight, Density, Modulus of Elasticity, Dynamic Viscosity, and Kinematic Viscosity of Water as a Function of Temperature in the SI Units

Temperature T (°C)	Specific Weight γ (kN/m³)	Density ρ (kg/m³)	Modulus of Elasticity E (kPa)	Dynamic Viscosity μ (Pa.s)	Kinematic Viscosity ν (m²/s)
0	9.805	999.8	1.98×10^{6}	1.781×10^{-3}	1.785×10^{-6}
5	9.807	1,000.0	2.05×10^{6}	1.518×10^{-3}	1.518×10^{-6}
10	9.804	999.7	2.10×10^{6}	1.307×10^{-3}	1.306×10^{-6}
15	9.798	999.1	2.15×10^{6}	1.139×10^{-3}	1.139×10^{-6}
20	9.789	998.2	2.17×10^{6}	1.002×10^{-3}	1.003×10^{-6}
25	9.777	997.0	2.22×10^{6}	0.890×10^{-3}	0.893×10^{-6}
30	9.764	995.7	2.25×10^{6}	0.798×10^{-3}	0.800×10^{-6}
40	9.730	992.2	2.28×10^{6}	0.653×10^{-3}	0.658×10^{-6}
50	9.689	988.0	2.29×10^{6}	0.547×10^{-3}	0.553×10^{-6}
60	9.642	983.2	2.28×10^{6}	0.466×10^{-3}	0.474×10^{-6}
70	9.589	977.8	2.25×10^{6}	0.404×10^{-3}	0.413×10^{-6}
80	9.530	971.8	2.20×10^{6}	0.354×10^{-3}	0.364×10^{-6}
90	9.466	965.3	2.14×10^{6}	0.315×10^{-3}	0.326×10^{-6}
100	9.399	958.4	2.07×10^{6}	0.282×10^{-3}	0.294×10^{-6}

Source: Adapted from Vennard and Street (1982).

TABLE 2-4

Specific Weight, Density, Modulus of Elasticity, Dynamic Viscosity, and Kinematic Viscosity of Water as a Function of Temperature in the English FSS Units

Temperature T (°F)	Specific Weight γ (lbf/ft³)	Density ρ (slug/ft³)	Modulus of Elasticity E (psi)	Dynamic Viscosity μ (lbf.s/ft²)	Kinematic Viscosity ν (ft²/s)
32	62.42	1.940	287×10^{3}	3.746×10^{-5}	1.931×10^{-5}
40	62.43	1.940	296×10^{3}	3.229×10^{-5}	1.664×10^{-5}
50	62.41	1.940	305×10^{3}	2.735×10^{-5}	1.410×10^{-5}
60	62.37	1.938	313×10^{3}	2.359×10^{-5}	1.217×10^{-5}
70	62.30	1.936	319×10^{3}	2.050×10^{-5}	1.059×10^{-5}
80	62.22	1.934	324×10^{3}	1.799×10^{-5}	0.930×10^{-5}
90	62.11	1.931	328×10^{3}	1.595×10^{-5}	0.826×10^{-5}
100	62.00	1.927	331×10^{3}	1.424×10^{-5}	0.739×10^{-5}
110	61.86	1.923	332×10^{3}	1.264×10^{-5}	0.667×10^{-5}
120	61.71	1.918	332×10^{3}	1.168×10^{-5}	0.609×10^{-5}
130	61.55	1.913	331×10^{3}	1.069×10^{-5}	0.558×10^{-5}
140	61.38	1.908	330×10^{3}	0.981×10^{-5}	0.514×10^{-5}
150	61.20	1.902	328×10^{3}	0.905×10^{-5}	0.476×10^{-5}
160	61.00	1.896	326×10^{3}	0.838×10^{-5}	0.442×10^{-5}
170	60.80	1.890	322×10^{3}	0.780×10^{-5}	0.413×10^{-5}
180	60.58	1.893	318×10^{3}	0.726×10^{-5}	0.385×10^{-5}
190	60.36	1.876	313×10^{3}	0.678×10^{-5}	0.362×10^{-5}
200	60.12	1.868	308×10^{3}	0.637×10^{-5}	0.341×10^{-5}
212	59.83	1.860	300×10^{3}	0.593×10^{-5}	0.319×10^{-5}

Source: Adapted from Vennard and Street (1982).

which is the reciprocal of density ρ. Therefore, its units in the SI and English FSS systems, respectively, are

$$[v] = \frac{1}{[\rho]} = \frac{\text{m}^3}{\text{kg}} \tag{2-9}$$

$$[v] = \frac{1}{[\rho]} = \frac{\text{ft}^3}{\text{slug}} \tag{2-10}$$

2.3.3 Specific Gravity (*s.g.*) of Fluids

Specific gravity (s.g.) is the ratio of the density of a substance to the density of water at a specified temperature and pressure:

$$\text{s.g.} = \frac{\gamma}{\gamma_w} = \frac{\rho}{\rho_w} \tag{2-11}$$

Because these quantities vary with temperature, temperatures must be specified when s.g. is used. Specific gravities of some liquids are given in Tables 2-1 and 2-2.

2.3.4 Specific Weight and Density of a Perfect Gas

The specific weight of a perfect gas is

$$\gamma = \frac{gp}{RT} \tag{2-12}$$

in which p is the absolute pressure, T is the temperature, and R is the gas constant. This equation can also be written in terms of density ρ as

$$\rho = \frac{\gamma}{g} = \frac{p}{RT} \tag{2-13}$$

The R gas constant is a constant value if the gas is a perfect gas. Common gases encountered in many practices may be considered perfect for many engineering calculations.

Air density ρ, like air pressure, decreases with increasing altitude. It also varies with the atmospheric pressure, temperature, and humidity. According to the International Standard Atmosphere values, in the SI units, at 101.325 kPa absolute pressure and 15°C temperature, air has a density of approximately 1.225 kg/m^3. The corresponding value in the English FSS system is 0.0765 lbf/ft^3 or 0.00238 slug/ft^3.

2.3.5 Units of the *R* Specific Gas Constant and Its Values for Dry Air

2.3.5.1 SI Units

Since

$$[p] = \frac{\text{N}}{\text{m}^2} \quad [\rho] = \frac{[\gamma]}{[g]} = \frac{\frac{\text{N}}{\text{m}^3}}{\frac{\text{m}}{\text{s}^2}} = \frac{\text{N.s}^2}{\text{m}^4} \quad [T] = {}^\circ\text{K} \tag{2-14}$$

From Eq. (2-13), one can write

$$[R] = \frac{[p]}{[\rho T]} = \frac{\dfrac{\mathrm{N}}{\mathrm{m}^2}}{\left(\dfrac{\mathrm{N.s}^2}{\mathrm{m}^4}\right)(^\circ\mathrm{K})} = \frac{\mathrm{m}^2}{\mathrm{s}^2.^\circ\mathrm{K}} \tag{2-15}$$

Using

$$1\,\mathrm{N} = (1\,\mathrm{kg})\left(1\,\frac{\mathrm{m}}{\mathrm{s}^2}\right) \quad 1\,\frac{\mathrm{m}}{\mathrm{s}^2} = 1\,\frac{\mathrm{N}}{\mathrm{kg}} \tag{2-16}$$

Equation (2-14) becomes

$$[R] = \frac{\mathrm{m}^2}{\mathrm{s}^2.^\circ\mathrm{K}} = \left(\frac{\mathrm{m}}{\mathrm{s}^2}\right)\left(\frac{\mathrm{m}}{^\circ\mathrm{K}}\right) = \left(\frac{\mathrm{N}}{\mathrm{kg}}\right)\left(\frac{\mathrm{m}}{^\circ\mathrm{K}}\right) = \frac{\mathrm{N.m}}{\mathrm{kg}.^\circ\mathrm{K}} \tag{2-17}$$

Since $1\,\mathrm{N.m} = 1\,\mathrm{Joule}\ (\mathrm{J})$,

$$[R] = \frac{\mathrm{J}}{\mathrm{kg.K}} \tag{2-18}$$

For dry air

$$R = 287.058\,\frac{\mathrm{J}}{\mathrm{kg}.^\circ\mathrm{K}} \tag{2-19}$$

The temperature T in Kelvin (K) is equal to the temperature T in degrees Celsius (°C) plus 273.15:

$$T\ (^\circ\mathrm{K}) = T\,(^\circ\mathrm{C}) + 273.15 \tag{2-20}$$

2.3.5.2 English FSS Units

Since

$$[p] = \frac{\mathrm{lbf}}{\mathrm{ft}^2} \quad [\rho] = \frac{[\gamma]}{[g]} = \frac{\dfrac{\mathrm{lbf}}{\mathrm{ft}^3}}{\dfrac{\mathrm{ft}}{\mathrm{s}^2}} = \frac{\mathrm{lbf.s}^2}{\mathrm{ft}^4} \quad [T] = {}^\circ\mathrm{R} \tag{2-21}$$

in which °R stands for *Rankine degree*. From Eq. (2-13), one can write

$$[R] = \frac{[p]}{[\rho T]} = \frac{\dfrac{\mathrm{lbf}}{\mathrm{ft}^2}}{\left(\dfrac{\mathrm{lbf.s}^2}{\mathrm{ft}^4}\right)(^\circ\mathrm{R})} = \frac{\mathrm{ft}^2}{\mathrm{s}^2.^\circ\mathrm{R}} = \left(\frac{\mathrm{ft}}{\mathrm{s}^2}\right)\left(\frac{\mathrm{ft}}{^\circ\mathrm{R}}\right) \tag{2-22}$$

in which °R stands for Rankine degrees. Using

$$1\ \text{lbf} = (1\ \text{slug})\left(1\ \frac{\text{ft}}{\text{s}^2}\right) = 32.2\ \frac{\text{lbm.ft}}{\text{s}^2} \qquad 1\ \frac{\text{ft}}{\text{s}^2} = 1\ \frac{\text{lbf}}{\text{slug}} \tag{2-23}$$

From Eq. (2-13),

$$[R] = \frac{\text{lbf.ft}}{\text{slug.°R}} \tag{2-24}$$

For dry air

$$R = 1{,}716.49\ \frac{\text{lbf.ft}}{\text{slug.°R}} \tag{2-25}$$

Equation (2-25) can also be expressed as

$$\begin{aligned} R &= 1{,}716.49\ \frac{(\text{lbf.ft})}{(\text{slug.°R})}\frac{\left(\frac{\text{ft}}{\text{s}^2}\right)}{\left(\frac{\text{ft}}{\text{s}^2}\right)} = 1{,}716.49\left(\frac{\text{lbf}}{\text{slug.}\frac{\text{ft}}{\text{s}^2}}\right)\left(\frac{\text{ft}}{\text{°R}}\right)\left(\frac{\text{ft}}{\text{s}^2}\right) \\ &= 1{,}716.49\ \frac{(\text{lbf.ft})}{(\text{slug.°R})} = 1{,}716.49\ \frac{(\text{lbf.ft})}{\left(\text{lbf.}\frac{\text{s}^2}{\text{ft}}\right).(\text{°R})} \\ &= 1{,}716.49\ \frac{(\text{lbf.ft})}{\left(32.2\ \frac{\text{lbm.ft}}{\text{s}^2}.\frac{\text{s}^2}{\text{ft}}\right).(\text{°R})} = 53.307\ \frac{(\text{lbf.ft})}{\text{lbm.°R}} \end{aligned} \tag{2-26}$$

The temperature T in degrees Rankine (°R) is equal to the temperature T in degrees Fahrenheit (°F) plus 459.67:

$$T°R = T°\text{F} + 459.67 \tag{2-27}$$

Example 2-1

The density of water in the SI system is $\rho = 1{,}000\ \text{kg/m}^3$. What is its corresponding value in the FSS system?

Solution

Since 1 lb = 0.4536 kg and 1 ft = 0.3048 m,

$$\rho = 1{,}000\ \frac{\text{kg}}{\text{m}^3} = 1{,}000\frac{\frac{1\ \text{lb}}{0.4536}}{\left(\frac{1\ \text{ft}}{0.3048}\right)^3} = 62.427\ \frac{\text{lb}}{\text{ft}^3}$$

From Eq. (2-7),

$$1\ \text{slug} = \frac{1\ \text{lbf.s}^2}{1\ \text{ft}}$$

With this,

$$\rho = \frac{62.427\ \dfrac{\text{lbf}}{\text{ft}^3}}{32.2\ \dfrac{\text{ft}}{\text{s}^2}} = 1.94\ \frac{\text{lbf.s}^2}{\text{ft}^4} = 1.94\left(\frac{\text{lbf.s}^2}{\text{ft}}\right)\left(\frac{1}{\text{ft}^3}\right) = 1.94\ \frac{\text{slug}}{\text{ft}^3}$$

2.4 Compressibility of Fluids

2.4.1 Compressible and Incompressible Fluids

Fluids are elastic media. For compressible fluids such as gases, the density changes with pressure, whereas for incompressible fluids such as water, the compressibility effects can be neglected for most cases.

2.4.2 Compressibility and Elasticity of Fluids

2.4.2.1 The Stress–Strain Diagram for Liquids and Gases

The mechanics of elastic compression of a fluid is shown in Figure 2-1, in which a perfectly rigid cylinder under a perfectly rigid piston includes an initial V_{in} volume of elastic fluid. Application of a force F to the piston having cross-sectional area A increases the piston pressure $p = F/A$ in the fluid and causes the initial volume V_{in} to decrease. By plotting p against V/V_{in} generates the stress–strain diagram shown in Figure 2-2, in which the modulus of elasticity of the fluid at any point on the curve is defined as the geometric slope of the curve at that point as

$$E = -\frac{dp}{\dfrac{dV}{V_{in}}} \tag{2-28}$$

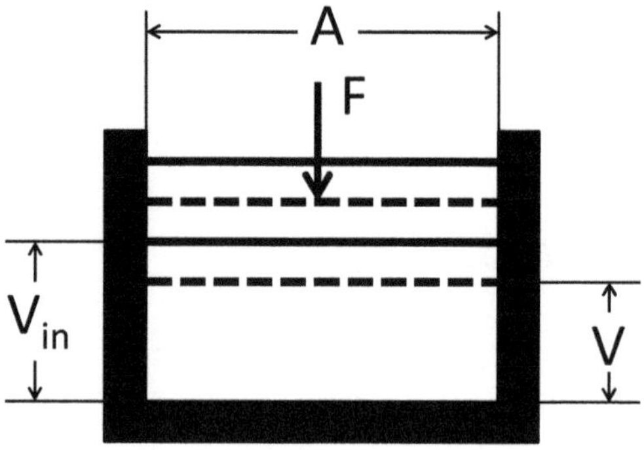

FIGURE 2-1
Schematic for the compression of a liquid in a cylinder.

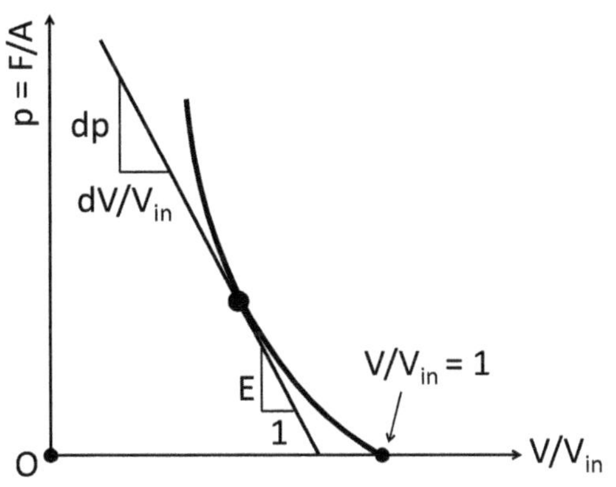

FIGURE 2-2
Stress-strain diagram for a compressible liquid.

The steepness of the curve with the increasing pressure shows that as the fluid body is compressed, it becomes increasingly difficult to compress further. One should notice that the geometric slope of the curve in Figure 2-2 is negative.

Although the schematic curve in Figure 2-2 is equally applicable to both liquids and gases, the most important part of the curve is near $V/V_{in}=1$ for liquids. The slope of the curve in this region is taken as the *modulus of elasticity* E in practice and for common liquids, their values are given in Table 2-1 in the SI units and Table 2-2 in the English FSS units.

As the relative increase of ρ or γ is exactly equal to the relative decrease of volume, one can write

$$E=\frac{dp}{\frac{d\rho}{\rho}}=\frac{dp}{\frac{d\gamma}{\gamma}} \tag{2-29}$$

2.4.2.2 Pressure–Density Relationship for Adiabatic Gases

An *adiabatic process* is generally used for a perfect gas with a constant-specific heat ratio k. An adiabatic process is one in which no heat is transferred to or from the system. The well-known pressure–density expression is

$$p=c\rho^k \tag{2-30}$$

in which k is the *specific heat ratio* and is also called the *adiabatic exponent,* and c is a constant. The value of k for Earth's atmosphere is about 1.4.

Example 2-2

Find the change in volume of 6.45 ft^3 of water at 60°F when subjected to a pressure increase in 235 psi.

Solution

From Table 2-4, E at 60°F is 313×10^3 psi. From Eq. (2-28),

$$dV=-\frac{dpV_{in}}{E}=-\frac{(235\text{ psi})(6.45\text{ ft}^3)}{313{,}000\text{ psi}}=-0.004843\text{ ft}^3$$

Example 2-3

Find the change in volume of 0.82 m^3 of water at 20°C when subjected to a pressure increase in 1,425 kPa.

Solution

From Table 2-3, E at 20°C is 2.17×10^6 kPa. From Eq. (2-28),

$$dV = -\frac{dpV_{in}}{E} = -\frac{(1{,}425\ \text{kPa})(0.82\ \text{m}^3)}{2.17 \times 10^6\ \text{kPa}} = -0.000538\ \text{m}^3$$

2.5 Viscosity of Fluids

2.5.1 Definition of Viscosity

The viscosity of a fluid is a measure of its resistance to flow, and it describes the internal friction of a moving fluid. For example, honey has a higher viscosity than water because its resistance to flow is higher than water. Viscosity can be quantified as the internal frictional force that arises between adjacent layers of fluid that are in relative motion.

Isaac Newton (1643–1727) was the first to quantify viscosity and is known as *Newton's viscosity law.* And it states that the shear stress between adjacent fluid layers in a *laminar flow* is proportional to the velocity gradients between the two layers. The ratio of shear stress to shear rate is a constant, for a given temperature and pressure, and is defined as the *viscosity* or *coefficient of viscosity.*

2.5.2 Newton's Viscosity Law for Laminar Flows

2.5.2.1 Newton's Viscosity Law Equation

As shown in Figure 2-3, consider two parallel plates separated by a d distance with a fluid between them. To keep the upper plate moving at a constant velocity U, a constant force F must be applied. This force causes a viscous interaction between the plate and the fluid.

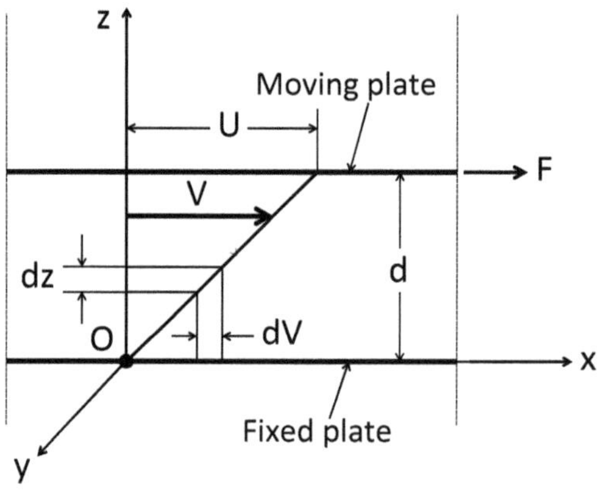

FIGURE 2-3
Schematic for Newton's viscosity law for laminar fluid flows.

The fluid in contact with the fixed plate will have zero velocity, namely, $V = 0$. Under laminar flow conditions, the $V = V(z)$ velocity profile will be a straight line, as shown in Figure 2-3. Based on experiments, the shearing force F varies directly with the area of the plate A, with velocity U, and inversely with distance d; mathematically, it can be expressed as

$$F \propto \frac{AU}{d} \tag{2-31}$$

From similar triangles, one can write

$$\frac{U}{d} = \frac{dV}{dz} \tag{2-32}$$

With this, Eq. (2-31) becomes

$$F \propto \frac{AU}{d} = A\frac{dV}{dz} \tag{2-33}$$

Since

$$\frac{F}{A} = \tau \tag{2-34}$$

is the shear stress, Eq. (2-33) takes the form

$$\tau \propto \frac{dV}{dz} \tag{2-35}$$

If a proportionality constant μ is introduced, it becomes

$$\tau = \mu\frac{dV}{dz} \tag{2-36}$$

in which μ is the *dynamic viscosity* and this equation is known as *Newton's viscosity law.*

Another viscosity, called *kinematic viscosity*, is defined as

$$\nu = \frac{\mu}{\rho} \tag{2-37}$$

in which ρ is the fluid density and defined by Eq. (2-1) and with that it takes the form

$$\nu = \frac{\mu}{\frac{\gamma}{g}} = \frac{\mu g}{\gamma} \tag{2-38}$$

2.5.2.2 Dimensions of Viscosity in the SI System

From Eq. (2-36), the dimensions of dynamic viscosity,

$$[\mu] = \frac{[\tau]}{\left[\frac{dV}{dz}\right]} = \frac{\frac{\text{N}}{\text{m}^2}}{\frac{\frac{\text{m}}{\text{s}}}{\text{m}}} = \frac{\text{N.s}}{\text{m}^2} = \text{Pa.s} \tag{2-39}$$

Using Eq. (2-6), it can also be expressed as

$$[\mu]=\frac{\text{N.s}}{\text{m}^2}=\frac{(1\ \text{kg})\left(1\ \frac{\text{m}}{\text{s}^2}\right).\text{s}}{\text{m}^2}=\frac{\text{kg}}{\text{m.s}} \tag{2-40}$$

From Eq. (2-38), the dimensions of the kinematic viscosity,

$$[\nu]=\frac{[\mu][g]}{[\gamma]}=\frac{\left(\frac{\text{N.s}}{\text{m}^2}\right)\left(\frac{\text{m}}{\text{s}^2}\right)}{\frac{\text{N}}{\text{m}^3}}=\frac{\text{m}^2}{\text{s}} \tag{2-41}$$

The dynamic viscosity values for some fluids in the SI system are given in Table 2-1 depending on temperature. In Table 2-3, both the dynamic viscosity and kinematic viscosity values of water as a function of temperature are given in the SI system as well.

The "poise" symbol P is the unit of dynamic viscosity in the cm–g–s system of units. It is named after Jean Léonard Marie Poiseuille:

$$1\ \text{P}=0.1\frac{\text{kg}}{\text{m.s}}=1\ \frac{\text{g}}{\text{cm.s}}=1\ \frac{\text{dyn.s}}{\text{cm}^2} \tag{2-42}$$

The analogous unit in the SI system is

$$1\ \text{Pa.s}=1\frac{\text{N.s}}{\text{m}^2}=1\ \frac{\text{kg}}{\text{m.s}}=10\ \text{P} \tag{2-43}$$

2.5.2.3 Dimensions of Viscosity in the English FSS System

From Eq. (2-36), the dimensions of dynamic viscosity,

$$[\mu]=\frac{[\tau]}{\left[\frac{dV}{dz}\right]}=\frac{\frac{\text{lbf}}{\text{ft}^2}}{\frac{\text{ft}}{\frac{\text{s}}{\text{ft}}}}=\frac{\text{lbf.s}}{\text{ft}^2} \tag{2-44}$$

Using Eq. (2-7), it can also be expressed as

$$[\mu]=\frac{\text{lbf.s}}{\text{ft}^2}=\frac{(1\ \text{slug})\left(1\ \frac{\text{ft}}{\text{s}^2}\right).\text{s}}{\text{ft}^2}=\frac{\text{slug}}{\text{ft.s}} \tag{2-45}$$

From Eq. (2-38), the dimensions of the kinematic viscosity,

$$[\nu]=\frac{[\mu][g]}{[\gamma]}=\frac{\left(\frac{\text{lbf.s}}{\text{ft}^2}\right)\left(\frac{\text{ft}}{\text{s}^2}\right)}{\frac{\text{lbf}}{\text{ft}^3}}=\frac{\text{ft}^2}{\text{s}} \tag{2-46}$$

The dynamic viscosity values for some fluids in the English FSS system are given in Table 2-2 depending on temperature. In Table 2-4, both the dynamic viscosity and kinematic viscosity values of water as a function of temperature are given in the English FSS system as well. In terms of poise (P), from Eq. (2-45)

$$1\,\frac{\text{lbf.s}}{\text{ft}^2} = 1\,\frac{\text{slug}}{\text{ft.s}} = 478.80\ \text{P} \tag{2-47}$$

Derivation of Eq. (2-47) is derived in Example 2-4 below.

Example 2-4

Derive Eq. (2-47).

Solution

Since 1 lbf = 444,822 dyn and 1 ft = 30.48 cm,

$$1\,\frac{\text{lbf.s}}{\text{ft}^2} = \frac{(444{,}822\ \text{dyn}).\text{s}}{(30.48\ \text{cm})^2} = 478.80\,\frac{\text{dyn.s}}{\text{cm}^2}$$

From Eq. (2-42),

$$1\ \text{P} = 1\,\frac{\text{dyn.s}}{\text{cm}^2}$$

Then,

$$1\,\frac{\text{lbf.s}}{\text{ft}^2} = 478.80\ \text{P}$$

which is exactly the same as Eq. (2-47).

2.5.2.4 Newtonian and Non-Newtonian Fluids

Equation (2-36) is the base of the Newtonian fluids, which is diagrammatically shown in Figure 2-4 in which μ is the geometric slope of a straight line passing through the origin. Here, dV is the displacement per unit time and dV/dz is the *time rate of strain*. According to

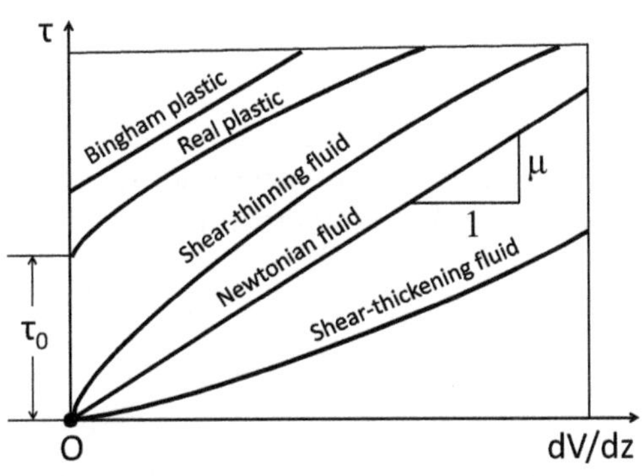

FIGURE 2-4
The $\tau - dV/dz$ diagrams of Newtonian and non-Newtonian fluids.

Newton's law of viscosity as given by Eq. (2-36), fluids that follow this law are commonly known as *Newtonian fluids*. Throughout this book, only the mechanics of Newtonian fluids will be presented.

Other fluids, which do not obey Eq. (2-36), are called *non-Newtonian fluids*. More specifically, a non-Newtonian fluid is a fluid that does not follow Newton's law of viscosity, and viscosity can change without obeying Eq. (2-36). In Figure 2-4, the diagrams of some typical non-Newtonian fluids are shown.

2.6 Surface Tension and Capillarity

2.6.1 Surface Tension

2.6.1.1 Definition

A molecule inside a liquid is under attractive forces in all directions, and they are in equilibrium. In other words, their vector sum is zero. But a molecule at the surface of a liquid is in contact with another fluid or solid having different cohesive forces. Because of the cohesive force differences, surface forces occur on the surfaces of liquids. These forces are known as *surface tension*. Although these forces are negligible in many cases, in some cases, such as the capillary rise of liquids in narrow spaces, they must be taken into account.

2.6.1.2 Mathematical Formulation

Surface tension is defined as the ratio of the surface force F to the length L over which the force acts. Mathematically, surface tension can be expressed as

$$\sigma = \frac{F}{L} \tag{2-48}$$

where F is the force per unit length, L is the length over which force acts, and σ is the surface tension of the liquid. From Eq. (2-48), the unit of σ in the SI system is N/m and in the English FSS system is lbf/ft. For example, at 0°C the surface tension of water in the SI system is $\sigma = 0.0756$ N/m; and in the FSS system, it is $\sigma = 0.00518$ lbf/ft.

2.6.2 Capillarity

2.6.2.1 Definition

The driving mechanism of capillarity is the surface tension. *Capillarity* is defined as the tendency of a liquid in a capillary tube or absorbent material to rise or fall as a result of surface tension.

2.6.2.2 Mathematical Formulation for a Double Curvature

The conceptual model of capillarity is shown in Figure 2-5, in which in the $x-y$ plane a $dxdy$ differential curvature area is perpendicular to the z-axis having R_1 and R_2 radii (or radiuses). The pressures above and below the curvature are p_i and p_0, respectively.

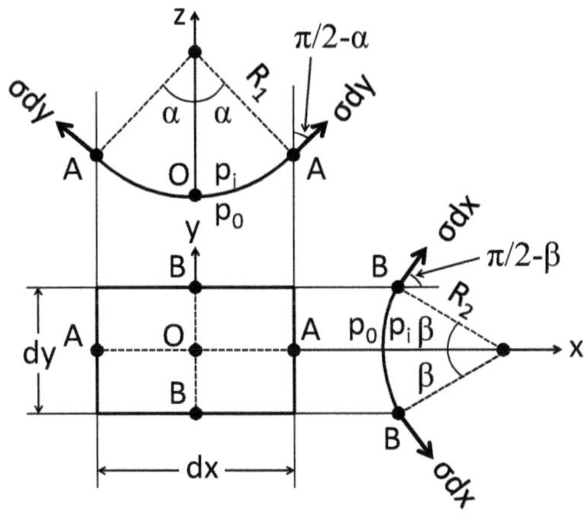

FIGURE 2-5
The conceptual model for capillarity.

In general, p_i is the atmospheric pressure. The forces acting on the curvature area perpendicularly in the vertical direction must be in equilibrium, i.e.,

$$\sum F_z = 0 \tag{2-49}$$

With the forces in Figure 2-5,

$$\sum F_z = (p_i - p_0)dxdy - 2\sigma dy \cos\left(\frac{\pi}{2} - \alpha\right) - 2\sigma dx \cos\left(\frac{\pi}{2} - \beta\right) = 0 \tag{2-50}$$

or

$$\begin{aligned} &(p_i - p_0)dxdy - 2\sigma dy\left[\cos\left(\frac{\pi}{2}\right)\cos(\alpha) + \sin\left(\frac{\pi}{2}\right)\sin(\alpha)\right] \\ &-2\sigma dx\left[\cos\left(\frac{\pi}{2}\right)\cos(\beta) - \sin\left(\frac{\pi}{2}\right)\sin(\beta)\right] = 0 \end{aligned} \tag{2-51}$$

After simplifications,

$$(p_i - p_0)dxdy = 2\sigma dy \sin(\alpha) + 2\sigma dx \sin(\beta) \tag{2-52}$$

in which α and β are small angles. From Figure 2-5,

$$\sin(\alpha) = \frac{dx}{2R_1} \quad \sin(\beta) = \frac{dy}{2R_2} \tag{2-53}$$

Using these expressions in Eq. (2-52), one gets

$$p_i - p_0 = \sigma\left(\frac{1}{R_1} + \frac{1}{R_2}\right) \tag{2-54}$$

2.6.2.3 Capillary Rise or Drop Equation

Figure 2-6 illustrates capillary rise in a vertical glass tube for water. Since the water molecules have a lower affinity for each other (cohesion) than for the glass molecules (adhesion), water rises in the glass tube.

With the assumption of spherical curvature, i.e., $R_1 = R_2 = R$; and $p_0 = -\gamma h$ and $p_i = 0$, Eq. (2-54) takes the form

$$\gamma h = \frac{2\sigma}{R} \tag{2-55}$$

Since from Figure 2-6, $r = R\cos(\theta)$ it becomes

$$h = \frac{2\sigma\cos(\theta)}{\gamma r} \tag{2-56}$$

There are some potential questions about this equation. The first one is the meaning of the θ angle, which is called the angle of contact. The assumption of a spherical liquid surface limits the equation to very small tubes, and in large tubes, the liquid surface is far from the spherical form.

Figure 2-7 corresponds to the case when mercury drops in a vertical glass tube. Since the mercury molecules have a greater affinity for each other (cohesion) than for the glass molecules (adhesion), mercury drop occurs in the glass tube.

Good experimental confirmation of Eq. (2-56) is found for small tubes having $r < 2.5$ mm with the conditions that the liquid and tube surfaces are very clean. However, such conditions are difficult to encounter and h will be found considerably smaller than predicted by Eq. (2-56). If the tube is clean, θ is 0° for water and about 140° for mercury.

Example 2-5

For water at 20°C, the following values are given: $\gamma = 9.789$ kN/m^3, $\sigma = 0.072$ N/m, and $\theta = 0°$. Determine the capillary rise in a vertical glass tube of 3 mm diameter.

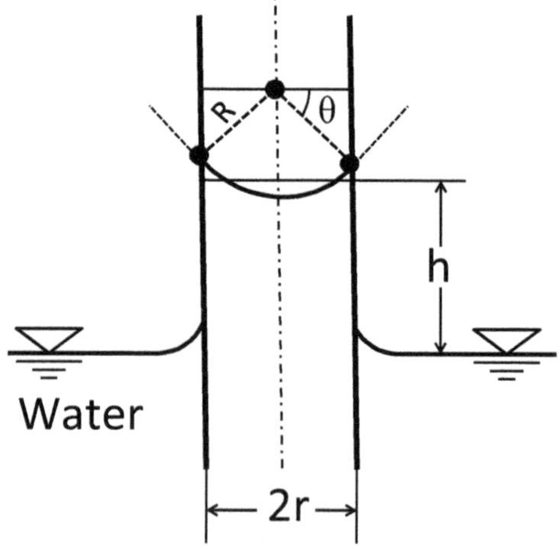

FIGURE 2-6
Capillary rise for water.

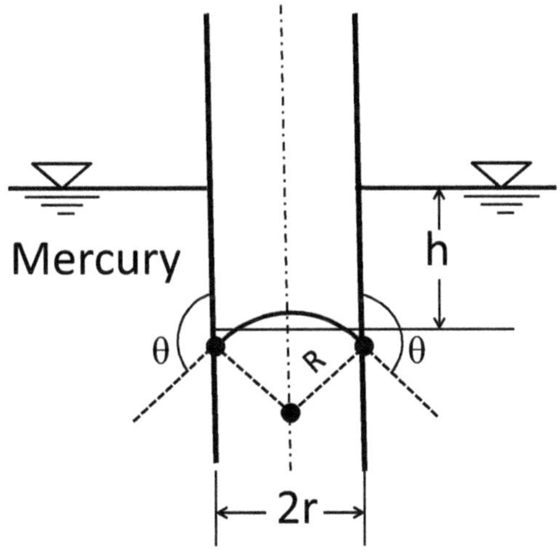

FIGURE 2-7
Capillary drop for mercury.

Solution

From Eq. (2-56),

$$h = \frac{2\sigma\cos(\theta)}{\gamma r} = \frac{2\left(0.072\,\dfrac{\text{N}}{\text{m}}\right)\cos(0°)}{\left(9{,}789\,\dfrac{\text{N}}{\text{m}^3}\right)(0.0015)} = 0.0098\text{ m} = 9.81\text{ mm}$$

Example 2-6

For water at 32°F, the following values are given: $\gamma = 62.42$ lbf/ft^3, $\sigma = 0.00518$ lbf/ft, and $\theta = 0°$. Determine the capillary rise in a vertical glass tube of 0.06 diameter.

Solution

From Eq. (2-56),

$$h = \frac{2\sigma\cos(\theta)}{\gamma r} = \frac{2\left(0.00518\,\dfrac{\text{lbf}}{\text{ft}}\right)\cos(0°)}{\left(62.42\,\dfrac{\text{lbf}}{\text{ft}^3}\right)\left(\dfrac{0.03}{12}\,\text{ft}\right)} = 0.066\text{ ft} = 0.80\text{ in}$$

Problems

2.1 The volume of a fluid body is 8 m^3 and its mass is 6,000 kg. Determine its density, specific volume, and specific gravity in the SI and FSS systems.

2.2 A material weight on the Earth is $W_e = 10{,}500$ N. Determine the mass of the material on the Earth and on the Moon and its weight on the Moon as well. The gravity acceleration of the Moon g_m is $g_m \approx g_e / 6$, in which g_e is the acceleration of the Earth.

2.3 A 15 m^3 volume V of oil weighs $W = 120$ kN. Determine its specific weight γ, density ρ, and specific gravity s.g. Use $\gamma_w = 9.807\ \frac{kN}{m^3}$.

2.4 A compressibility test was conducted for water and the following data were obtained: At 720 psi, the volume was 0.950 ft^3, and at 44,000 psi the volume was 0.805 ft^3. Determine the bulk modulus of elasticity.

2.5 Initially $V_{in} = 2.5\ m^3$ is placed under an absolute pressure 11,200 kPa at 20°C. Determine the volume of water at this pressure.

2.6 The volume of water is reduced 0.028% by applying 920 kPa pressure. Determine the modulus of elasticity.

2.7 Determine the kinematic viscosity of carbon tetrachloride in both the SI and English FSS systems at 20°C(68°F) by obtaining the given data from the appropriate tables.

2.8 Determine the kinematic viscosity of ethyl alcohol in both the SI and English FSS systems at 20°C(68°F) by obtaining the given data from the appropriate tables.

2.9 The dynamic viscosity of crude oil is $\mu = 15.0 \times 10^{-5}$ lbf.s/ft^2 (Table 2-2). Convert it to the units of poise (P).

2.10 For water at 60°F, the following values are given: $\gamma = 62.37$ lbf/ft^3, $\sigma = 0.00504$ lbf/ft, and $\theta = 0°$. Determine the capillary rise in a vertical glass tube of 0.10 diameter.

3

Fundamental Principles of Water Flow in Porous Media

3.1 Introduction

3.1.1 Porous Materials

A porous medium or a porous material is a material containing pores. The solid portion of the material is often called the *solid matrix*. The pores are typically filled with a fluid, which can be liquid or gas. Natural porous materials are the ones below the ground surface, and they are soils and rocks. Bricks, cement, and ceramics are typical examples of man-made porous materials.

3.1.2 The Purpose of This Chapter

The main purpose of this chapter is to present the water flow mechanics through saturated natural porous materials along with the related physical laws, principles, and governing differential equations. After presenting the concepts of discharge velocity and ground water velocity, all details of Darcy's law are presented.

3.2 Natural Porous Media and Their Characteristics

3.2.1 Definitions for Natural Porous Media

Naturally, water that exists below the ground surface is called the *subsurface water*. The subsurface water may be divided into the *vadose zone* and the *saturated zone*. In the vadose zone, the pore space of the porous medium is filled partly with water and partly with air and is also called the *unsaturated zone*. In the saturation zone, all the interconnected pores are completely filled with water. In general, the vadose zone is underlain by the zone of saturation and extends upward to the ground surface. Generally, the saturated zone is underlain by impervious or semipervious geologic strata. The saturated zone is generally overlain at the top by either impervious or semipervious geologic strata. If the strata do not exist on the top saturated zone, the top of the saturated zone is called the *water table*.

The water body below the water table is called an *aquifer*. Etymologically, aquifer means water-bearing formation. The word "aquifer" is generated from two Latin words "aqua" (water) and "ferre" (to bear). In subsurface hydrology, an aquifer is defined as a single geologic formation or a group of geologic formations transmitting and yielding a significant amount of water.

DOI: 10.1201/9781003739128-3

3.2.2 Types of Natural Water-Bearing Formations

In this section, definitions of some natural water-bearing formations will be given (e.g., Hantush, 1964, pp. 282–284; Bear, 1979, pp. 24–26; Freeze and Cherry, 1979, pp. 47–49; Batu, 1998, pp. 22–25; Batu, 2024, pp. 990–993).

If a geologic formation does not have the capability to transmit a significant amount of water, it is called an *aquiclude.*

An *aquitard,* or a *confining layer,* is defined as a geologic formation that can transmit water at a relatively low rate compared with aquifers.

If a geologic formation neither absorbs nor transmits water, it is called an *aquifuge.*

If an aquifer loses or gains water through adjacent semipervious layers, it is called a *leaky aquifer.*

An *unconfined aquifer,* or *water-table aquifer,* is an aquifer in which the water table forms the upper boundary of the aquifer. Unconfined aquifers occur near the ground surface.

A *confined aquifer,* which is also called an *artesian aquifer* or a *pressure aquifer,* is bounded by impervious or semipervious layers.

Figure 3-1 shows the schematic picture of a subsurface flow system, which includes unconfined and confined aquifers, and a confining layer between them. The zone between the ground surface and the water table is the *unsaturated zone.* The confined aquifer rests on an impermeable base.

When the hydraulic conductivities of the aquifers are more than two orders of magnitude greater than that of the aquitard, the errors introduced by this assumption are less than 5%. Under these conditions, the flow directions in the aquifer are horizontal, and the flow direction through the confining layer is vertical (Neuman and Witherspoon, 1969).

In Figure 3-1, the water table elevation of the unconfined aquifer is higher than the potentiometric head surface elevation of the confined aquifer, and it loses water to the confined aquifer. Under these conditions, the unconfined aquifer is a *leaky aquifer.* If the potentiometric head surface of the confined aquifer is higher than the water table of the unconfined aquifer, the flow direction through the confining layer will be upward, and under these conditions, the confined aquifer will also be a *leaky aquifer.*

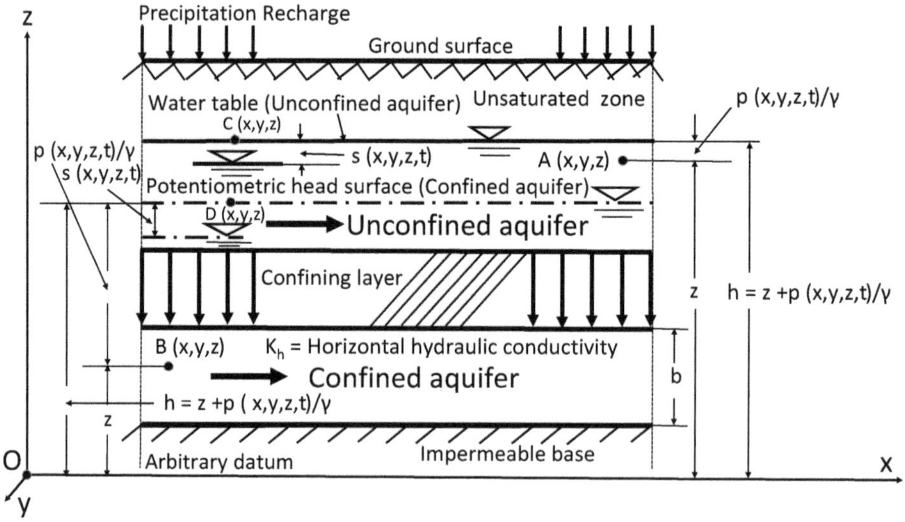

FIGURE 3-1
Schematic picture of a subsurface flow system.

3.3 Hydraulic Head and Drawdown

Hydraulic head; drawdown; porosity; discharge velocity and ground water velocity; hydraulic conductivity, permeability, and transmissivity; capillarity and capillary fringe are the key hydraulic and hydrogeological quantities in characterizing aquifers, and they are defined below.

3.3.1 Hydraulic Head

Hydraulic head, denoted by h in Figure 3-1 for both unconfined and confined aquifers, is defined as

$$h(x,y,z,t) = \frac{p(x,y,z,t)}{\gamma} + z \tag{3-1}$$

in which $p(x,y,z,t)/\gamma$ is the *pressure head*; x, y, z are the Cartesian coordinates; γ is the unit weight of water; z is the geometric elevation with respect to an arbitrary datum; and t is the time of observation since an arbitrarily chosen reference time. As shown in Figure 3-1, the hydraulic head $h(x,y,z,t)$ at point $A(x,y,z)$ in an unconfined aquifer consists of the pressure head $p\ (x,y,z,t)/\gamma$ and its z elevation with respect to the arbitrary datum. Likewise, at point $B(x,y,z)$ in the confined aquifer, the same quantities are shown.

For the term "hydraulic head," the terms *head* and *piezometric head* are also widely used.

The dimension of hydraulic head is "length" or (L). In practice, the hydraulic head is generally expressed as "meters of water" or "feet of water."

3.3.2 Drawdown

Under extraction or injection of water conditions from or to a well, the initial hydraulic head at a point in an aquifer changes. Under extraction conditions from a well, the vertical distance between the initial hydraulic head at a point in an aquifer and the lowered position of the hydraulic head for the same point is called *drawdown*. In Figure 3-1, the water level at point $C(x,y,z)$ on the water table is dropped by a vertical distance $s(x,y,z,t)$. This vertical distance is the drawdown at the same point. For the confined aquifer case, the drawdown is defined with respect to the potentiometric head surface at point $D(x,y,z)$.

3.4 Porosity and Effective Porosity

Porosity or *total porosity* of an aquifer n, or porous medium, is defined as the percentage of void spaces as

$$n = \frac{V_v}{V} \tag{3-2}$$

in which V_v is the volume of voids of a given sample and V is the total volume of the sample. The *effective porosity* n_e is the portion of pore space in a saturated porous material

TABLE 3-1

Porosity of Various Porous Materials (Morris and Johnson, 1967; Mercer et al., 1982)

Group	Porous Material	Number of Analysis	Range of Porosity (n)	Arithmetic Mean Average Porosity (n_e)
Igneous rocks	Weathered granite	8	0.34–0.57	0.45
	Weathered gabbro	4	0.42–0.45	0.43
	Basalt	94	0.03–0.35	0.17
Sedimentary materials	Sandstone	65	0.14–0.49	0.34
	Siltstone	7	0.21–041	0.35
	Sand (fine)	243	0.26–0.53	0.43
	Sand (coarse)	26	0.31–0.46	0.39
	Gravel (fine)	38	0.25–038	0.34
	Gravel (coarse)	15	0.24–036	0.28
	Silt	281	0.34–0.61	0.46
	Clay	74	0.35–0.57	0.42
	Limestone	74	0.07–0.56	0.30
Metamorphic rocks	Schist	18	0.04–0.49	0.39

in which water flows. Namely, it is the volume of the interconnected voids of a sample. Likewise, its definition is

$$n_e = \frac{V_{vi}}{V} \tag{3-3}$$

in which V_{vi} is the volume of the interconnected voids of the sample. It is clear that n_e is smaller than n.

Table 3-1 presents representative porosity (n) ranges as well as their arithmetic mean (n_{av}) values for various aquifer materials (Morris and Johnson, 1967; Mercer et al., 1982). In general, clayey materials have smaller pores than sandy materials, but they may have much greater effective porosity values. Most aquifers generally have values of effective porosity in the range of 0.10–0.30, but higher and lower values can also be found.

3.5 Discharge Velocity and Ground water Velocity

Flow can occur only through interconnected pores of saturated porous media shown in Figure 3-2 having the area A and Q is the discharge rate passing through it. The total area and the area of its pores are A and A_p, respectively. The *effective porosity* of the A area is

$$n_e = \frac{A_p}{A} \tag{3-3}$$

The flow rate passing through the area A_p is

$$Q = A_p v = n_e A v \tag{3-4}$$

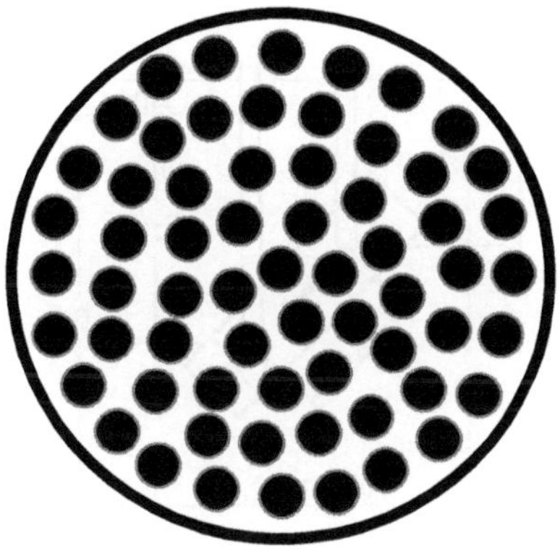

FIGURE 3-2
Schematic picture for the derivation of the relationship between the discharge velocity and ground water velocity.

in which v is the *ground water velocity* or *seepage velocity*. Here $n_e v = q$ is the *discharge velocity* and $A_p = An_e$. Then, Eq. (3-4) becomes

$$Q = An_e v = qA \tag{3-5}$$

And from here,

$$v = \frac{q}{n_e} \tag{3-6}$$

Equation (3-6) is the relationship between the *discharge velocity* and *ground water velocity*. The term *seepage velocity* is also used for ground water velocity. Equation (3-6) is also called the *Dupuit-Forchheimer's seepage equation*.

3.6 Darcy's Law

3.6.1 Description of Darcy Experiment

Unlike in porous media hydraulics, Darcy's law is presented with pressure. The basic law expressing the relationship between the fluid flux and hydraulic head gradient is Darcy's law. Darcy's law was established in 1856 by Henry Darcy based on an experimental study for flow through sands (Darcy, 1856). Here, the one-dimensional form of Darcy's law will be presented with pressure along with the fundamental principles of hydromechanics.

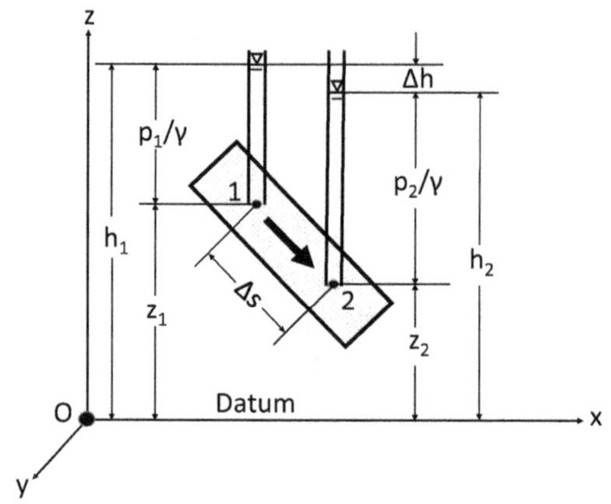

FIGURE 3-3
Schematic of Darcy's experiment.

The schematic picture of the experimental system shown in Figure 3-3 is equivalent to that used by Darcy in his original experiments, except that Darcy's system was vertical. The pipe shown in Figure 3-3 is filled with a porous material, and water flows in the arrow direction. Two piezometers are located at points 1 and 2 in the pipe, which are separated by the Δs distance. The geometric elevations of these two points are z_1 and z_2 with respect to a datum. During water flow through the porous medium, head losses occur between points 1 and 2. The water levels in the piezometers are h_1 and h_2 from the datum.

3.6.2 Bernoulli's Equation for Porous Media Flow

In porous media flow, due to the resistance against the flow within pores, the Bernoulli equation is

$$\frac{V_1^2}{2g} + \frac{p_1}{\gamma} + z_1 = \frac{V_2^2}{2g} + \frac{p_2}{\gamma} + z_2 + \Delta h \tag{3-7}$$

in which Δh is the head loss over the distance Δs.

The *hydraulic gradient I*, or the *gradient of piezometric head*, is defined as

$$I = -\lim_{\Delta s \to 0} \frac{\Delta h}{\Delta s} = -\frac{dh}{ds} \tag{3-8}$$

In porous media flow, velocities are much lower than surface water flow velocities. As a result, the velocity heads in Eq. (3-7) may be neglected without appreciable error. Therefore, from Eq. (3-7), one can write

$$\Delta h \cong \left(\frac{p_1}{\gamma} + z_1\right) - \left(\frac{p_2}{\gamma} + z_2\right) \tag{3-9}$$

According to Figure 3-3,

$$h_1 = \frac{p_1}{\gamma} + z_1 \quad h_2 = \frac{p_2}{\gamma} + z_2 \tag{3-10}$$

which can be generalized as

$$h = \frac{p}{\gamma} + z \tag{3-11}$$

Therefore, Eq. (3.9) becomes

$$\Delta h = h_1 - h_2 \tag{3-12}$$

Equation (3-11) is one of the key expressions in porous media hydraulics, whereas Eq. (3-7) is the hydraulic head expression for surface water flows such as open channels, pipes, and rivers. In a typical river or stream, the average flow velocity is around 1 m/s. However, in most aquifers even 1 m/d is a high velocity. In some tight formations, the velocity of water is less than 1 m/yr. Therefore, the velocity head $v^2/(2g)$ of Eq. (3-7) in porous media flow is extremely low. As a result, the velocity head terms in Eq. (3-7) can safely be neglected for flow in aquifers as well as all other porous media.

Equation (3-12) can also be written as

$$\Delta p = \Upsilon \Delta h \tag{3-13}$$

in which Υ is the specific weight.

3.6.3 Darcy Experiment

In 1856, Henry Darcy experimentally found that the discharge rate Q is proportional to the difference of water levels Δh and inversely proportional to Δs (Figure 3-3),

$$Q \propto \Delta h \quad Q \propto \frac{1}{\Delta s} \tag{3-14}$$

Introducing a proportionality constant K, they lead to the equation

$$Q = -KA\frac{h_2 - h_1}{\Delta s} = -KA\frac{\Delta h}{\Delta s} \tag{3-15}$$

In the limiting case,

$$Q = -KA\frac{dh}{ds} \tag{3-16}$$

in which A is the cross-sectional area of the column in Figure 3-3.

From Eq. (3-16), the discharge velocity is

$$q = \frac{Q}{A} = -K\frac{dh}{ds} = -KI \tag{3-17}$$

where q is the *discharge velocity* or *Darcy velocity* described by Eq. (3-17) and $I = dh/ds$ is the *hydraulic gradient*. For q, the terms *Darcy flux* and *specific discharge* are also used.

Equation (3-17) is known as *Darcy's law,* which means that there is a linear relationship between the hydraulic gradient I and discharge velocity q. The negative sign in Eq. (3-17) indicates that the flow of water is in the direction of decreasing hydraulic head. It must be emphasized that Darcy's law does not describe the state within individual pores. Darcy's law represents the statistical macroscopic equivalent of the Navier–Stokes equations of motion for viscous flow of water in porous media.

3.6.4 Validity of Darcy's Law

Darcy's law given by Eq. (3-17) states that the discharge velocity q is proportional to the first power of the hydraulic gradient. Since the velocity in laminar flow is also proportional to the first power of the hydraulic gradient, it can be inferred that the flow in a porous medium must be laminar for Darcy's law to be valid. In hydrodynamics, the criterion for determining whether the flow is laminar or turbulent is the Reynolds number. For homogeneous porous materials, Darcy's law is valid only under the condition of the inequality for the Reynolds number given as

$$\mathrm{Re} = \frac{qD\rho}{\mu} = \frac{qD}{\nu} \leq \mathrm{Re}_{pm} \tag{3-18}$$

in which q is the discharge velocity, D is the average diameter of porous medium particles, μ is the dynamic viscosity, ν is the kinematic viscosity, and Re_{pm} (*pm* stands for porous media) is a number for which various authors give different values ranging from 3 to 10 (Polubarinova-Kochina, 1962, p. 17; Bear, 1979, p. 34). For example, at 15°C, $\nu \times 10^6 = 1.139 \text{ m}^2/\text{s}$. From here, $\nu = 0.0139 \text{ cm}^2/\text{s}$. With these values, Eq. (3-18) becomes

$$qD \leq 0.0417 - 0.1390 \tag{3-19}$$

in which q is expressed in centimeters per second and D is expressed in centimeters.

3.6.5 Hydraulic Conductivity, Permeability, and Transmissivity

Hydraulic conductivity, permeability, and transmissivity are some of the important quantities used in aquifer hydraulics. Their definitions with the units used in practice are presented below.

3.6.5.1 Hydraulic Conductivity

Definition of *hydraulic conductivity* is based on Darcy's law. From Eq. (3-16),

$$K = -\frac{Q}{A\left(\dfrac{dh}{ds}\right)} = -\frac{q}{I} \tag{3-20}$$

which indicates that K has the dimensions of "length/time" (L/T) or velocity. Based on this expression, hydraulic conductivity can be defined as the rate of discharge of water under laminar flow conditions through a unit cross-sectional area of a porous medium under a unit hydraulic gradient and standard temperature conditions (usually 20°C).

The quantity of K is called *hydraulic conductivity* or *saturated hydraulic conductivity,* and it is a function of properties of both the porous medium and the fluid flowing through it.

TABLE 3-2

Hydraulic Conductivity of Various Porous Materials (Morris and Johnson, 1967; Mercer et al., 1982)

Group	Porous Material	Number of Analysis	Range of Hydraulic Conductivity K (cm/s)	Arithmetic Mean Average Hydraulic Conductivity K_{av} (cm/s)
Igneous rocks	Weathered granite	7	$(3.3–52) \times 10^{-4}$	1.65×10^{-3}
	Weathered gabbro	4	$(0.5–3.8) \times 10^{-4}$	1.98×10^{-4}
	Basalt	93	$(0.2–4{,}250) \times 10^{-6}$	9.45×10^{-6}
Sedimentary materials	Sandstone	20	$(0.5–2{,}270) \times 10^{-6}$	3.31×10^{-4}
	Siltstone	8	$(0.1–142) \times 10^{-8}$	1.9×10^{-7}
	Sand (fine)	159	$(0.2–189) \times 10^{-4}$	2.28×10^{-3}
	Sand (medium)	255	$(0.9–567) \times 10^{-4}$	1.65×10^{-2}
	Sand (coarse)	158	$(0.9–6{,}610) \times 10^{-4}$	5.20×10^{-2}
	Gravel	40	$(0.3–31.2) \times 10^{-1}$	4.03×10^{-1}
	Silt	39	$(0.09–7{,}090) \times 10^{-7}$	2.83×10^{-5}
	Clay	19	$(0.1–47) \times 10^{-8}$	9.00×10^{-8}
Metamorphic rocks	Schist	17	$(0.002–1{,}130) \times 10^{-6}$	1.90×10^{-4}

The saturated hydraulic conductivity of a given porous material can be determined in the laboratory by a device whose schematic is shown in Figure 3-3. This device can also be vertical as well as horizontal. Tables and nomographs for saturated hydraulic conductivity values for different formations have been presented in a number of publications, especially during the past century. It is important to emphasize that these tables mostly use the term "hydraulic conductivity" rather than making it specific for horizontal and vertical hydraulic conductivity values. It is a well-known fact that vertical hydraulic conductivities may be one to three orders of magnitude less than horizontal hydraulic conductivities, depending on the type of formation.

Table 3-2 presents saturated hydraulic conductivity (K) ranges as well as their arithmetic mean (K_{av}) values for some typically encountered porous materials (Morris and Johnson, 1967; Mercer et al., 1982).

In the geotechnical engineering literature and practice, the term *coefficient of permeability* (e.g., Lambe and Whitman, 1969, p. 251) is mostly used instead of *hydraulic conductivity*.

3.6.5.2 Permeability

3.6.5.2.1 Permeability Versus Coefficient of Permeability

As mentioned above, the term coefficient of permeability is used instead of hydraulic conductivity in geotechnical engineering, and this causes confusion because hydraulic conductivity K depends not only on the permeability of the porous medium but also on the properties of the fluid flowing through the porous medium.

3.6.5.2.2 The Relationship between Hydraulic Conductivity and Permeability

The fluid properties affecting the flow are its dynamic viscosity (μ) and its specific weight (γ). The dynamic viscosity μ of the fluid is a measure of the resistive force within the pores of the porous medium. The specific weight γ may be considered the driving force exerted by gravity. Also, the size of a typical pore is proportional to the grain diameter d. Using the method of dimensional analysis, the relationship between K, μ, γ, and d is determined as [Jacob, 1950, p. 324, Eq. (2)]

$$K = \frac{Cd^2\gamma}{\mu} = k\frac{\gamma}{\mu} = k\frac{\rho g}{\mu} = k\frac{g}{\nu} \tag{3-21}$$

in which

$$k = Cd^2 \tag{3-22}$$

and

$$\nu = \frac{\mu}{\rho} \tag{3-23}$$

is the *kinematic viscosity* having $[\nu] = L^2/T$. k depends only on the character of the porous medium. Since $[d] = L, [k] = L^2$. Since k depends only on the character of the porous medium, it is called the coefficient of permeability or just permeability. The term intrinsic permeability is also used by some authors. In Eq. (3-21), both γ and μ change with temperature. Therefore, the ratio γ/μ depends on the temperature as well. However, for most cases in ground water hydraulics, the temperature does not vary significantly, and the hydraulic conductivity K may then be considered constant in the related equations.

3.6.5.2.3 Units of Permeability

Units of permeability k can be in square centimeters (cm^2) or square feet (ft^2). In petroleum engineering, the unit of permeability is *darcy* according to the relationship written from Eq. (3-20) and Eq. (3-21) as

$$k = \frac{\mu}{\gamma} \frac{Q}{A\left(\frac{\Delta h}{\Delta s}\right)} \tag{3-24}$$

By replacing $h = p/\gamma$ where p is the pressure, it takes the form

$$k = \frac{\frac{\mu Q}{A}}{\frac{\Delta p}{\Delta s}} \tag{3-25}$$

Based on this equation, 1 darcy is defined by taking $\mu = 1$ centipoise $= 0.01$ dyne.s/cm^2, $Q = 1\ cm^3/s, A = 1\ cm^2, \Delta p = 1$ atmosphere $= 1.0132 \times 10^6$ dyne/cm^2, and $\Delta s = 1$ cm. Therefore, substitution of the above values into Eq. (3-25) gives the definition of *Darcy* as

$$k = \frac{\frac{(1\ \text{centipoise})\left(1\ \frac{cm^3}{S}\right)}{(1\ cm^2)}}{\frac{\left(1.0132\times 10^6\ \frac{dyne}{cm^2}\right)}{(1\ cm)}} = \frac{\frac{\left(0.01\ \frac{dynes.s}{cm^2}\right)\left(1\ \frac{cm^3}{s}\right)}{(1\ cm^2)}}{\frac{\left(1.0132\times 10^6\ \frac{dyne}{cm^2}\right)}{(1\ cm)}} = 9.86972\times 10^{-9}\ cm^2 \tag{3-26}$$

$$\cong 1.0\times 10^{-8}\ cm^2 = 1\ \text{darcy}$$

3.6.5.2.4 Transmissivity

For a confined aquifer of thickness b, the *transmissivity* (or *transmissibility*) T is defined as

$$T = K_h b \tag{3-27}$$

in which K_h is the horizontal hydraulic conductivity of the confined aquifer. The transmissivity concept can also be used for unconfined aquifers. In Eq. (3-27), b is now the saturated thickness of the unconfined aquifer or the height of the water table above the underlying aquitard or impermeable boundary.

Since $[K_h] = L/T$ and $[b] = L$, $[T] = L^2/T$. In practice, for transmissivity T, the units m^2/d and ft^2/d are widely used.

3.6.6 Darcy and Flow Rate Equations in Terms of Pressure

3.6.6.1 One-Dimensional Form of Darcy Equation

Using Eq. (3-21) in Eq. (3-17)

$$q = -k\frac{\gamma}{\mu}\frac{dp}{ds} \tag{3-28}$$

Since $\gamma dh = dp$, it becomes

$$q = -\frac{k}{\mu}\frac{dp}{ds} \tag{3-29}$$

which is a one-dimensional form of Darcy's law in terms of pressure.

3.6.6.2 Flow Rate Equation

The flow rate through the sand box in Figure 3-3 is

$$Q = \frac{k}{\mu}A\frac{\Delta p}{\Delta s} \tag{3-30}$$

in which A is the cross-sectional area perpendicular to the flow direction in Figure 3-3.

3.6.6.3 Three-Dimensional Form of Darcy Equation in Terms of Pressure

The three-dimensional vector form of Darcy's equation is

$$\boldsymbol{q} = q_x\boldsymbol{i} + q_y\boldsymbol{j} + q_z\boldsymbol{k} \tag{3-31}$$

in which $\boldsymbol{q}$ is the Darcy velocity vector and q_x, q_y, and q_z are the Darcy velocity components along the x, y, and z coordinate directions, respectively. $\boldsymbol{i}$, $\boldsymbol{j}$, and $\boldsymbol{k}$ are the unit vectors in respective coordinate directions as well. Therefore, the three-dimensional form of Eq. (3-29) becomes

$$\boldsymbol{q} = -\frac{k}{\mu}\nabla p \tag{3-32}$$

in which ∇ is the vector differential *operator del*, defined by

$$\nabla = \frac{\partial}{\partial x}\boldsymbol{i} + \frac{\partial}{\partial y}\boldsymbol{j} + \frac{\partial}{\partial z}\boldsymbol{k} \equiv \boldsymbol{i}\frac{\partial}{\partial x} + \boldsymbol{j}\frac{\partial}{\partial y} + \boldsymbol{k}\frac{\partial}{\partial z} \tag{3-33}$$

which is also known as *nabla*. This operator has analogous properties to those of ordinary vectors. The nabla operator was first introduced by an Irish mathematician and physicist Sir William Rowan Hamilton (1805–1865). With Eq. (3-31), Eq. (3-32) becomes

$$\boldsymbol{q} = -\frac{k}{\mu}\left(\frac{\partial p}{\partial x}\boldsymbol{i} + \frac{\partial p}{\partial y}\boldsymbol{j} + \frac{\partial p}{\partial z}\boldsymbol{k}\right) \tag{3-34}$$

From Eqs. (3-31) and (3-34), one can write

$$q_x = -\frac{k}{\mu}\frac{\partial p}{\partial x} \tag{3-35}$$

$$q_y = -\frac{k}{\mu}\frac{\partial p}{\partial y} \tag{3-36}$$

$$q_z = -\frac{k}{\mu}\frac{\partial p}{\partial x} \tag{3-37}$$

Example 3-1

In a Darcy experiment shown in Figure 3.2, the following values are given: $D = 20$ cm, $\Delta s = 50$ cm, $k = 0.2$ darcy, $\Delta p = 3.5$ atm, and $\mu = 5$ centipoise. Determine the flow rate Q $\left(\text{cm}^3/\text{min}\right)$ and Darcy velocity q (cm/s).

Solution

$$1 \text{ darcy} = 9.87 \times 10^{-9} \ \text{cm}^2 = 9.87 \times 10^{-13} \ \text{m}^2$$

$$1 \text{ atmosphere} = 1.0132 \times 10^6 \ \text{dyne/cm}^2$$

$$1 \text{ centipoise} = 0.01 \ \text{dyne.s/cm}^2$$

Therefore,

$$k = 0.2 \text{ darcy} = (0.2)\left(9.87 \times 10^{-9} \ \text{cm}^2\right) = 1.974 \times 10^{-9} \ \text{cm}^2$$

$$\Delta p = 3.5 \text{ atm} = 3.5\left(1.0132 \times 10^6 \ \frac{\text{dyne}}{\text{cm}^2}\right) = 3.3462 \times 10^6 \ \frac{\text{dyne}}{\text{cm}^2}$$

$$\mu = 5 \text{ centipoise} = 5\left(0.01 \ \frac{\text{dyne.s}}{\text{cm}^2}\right) = 0.05 \ \frac{\text{dyne.s}}{\text{cm}^2}$$

$$A = \frac{\pi D^2}{4} = \frac{\pi (20 \text{ cm})^2}{4} = 314.159 \text{ cm}^2$$

$$\Delta s = 50 \text{ cm}$$

From Eq. (3-30),

$$Q = \frac{k}{\mu} A \frac{\Delta p}{\Delta s} = \frac{(1.974 \times 10^{-9} \text{ cm}^2)}{\left(0.05 \frac{\text{dyne.s}}{\text{cm}^2}\right)} (314.159 \text{ cm}^2) \frac{\left(3.3462 \times 10^6 \frac{\text{dyne}}{\text{cm}^2}\right)}{(50 \text{ cm})} = 0.830058 \frac{\text{cm}^3}{\text{s}}$$

In terms of cm^3/min,

$$Q = 0.830058 \frac{\text{cm}^3}{\text{s}} = 0.830058 \frac{\text{cm}^3}{\frac{1}{60} \text{min}} = 49.803 \frac{\text{cm}^3}{\text{min}}$$

Darcy velocity,

$$q = \frac{Q}{A} = \frac{0.830058 \frac{\text{cm}^3}{\text{s}}}{314.159 \text{ cm}^2} = 0.002642 \frac{\text{cm}}{\text{s}}$$

3.6.6.4 Three-Dimensional Form of Darcy Equation in Terms of Hydraulic Head

The general form of Eq. (3-34) is

$$\boldsymbol{q} = -\frac{1}{\mu}\left(k_x \frac{\partial p}{\partial x} \boldsymbol{i} + k_y \frac{\partial p}{\partial y} \boldsymbol{j} + k_z \frac{\partial p}{\partial z} \boldsymbol{k} \right) \tag{3-38}$$

in which k_x, k_y, and k_z are the principal permeabilities in the x, y, and z directions, respectively. Using Eqs. (3-13) and (3-21), Eq. (3-38) can be written as

$$\boldsymbol{q} = -\left(K_x \frac{\partial h}{\partial x} \boldsymbol{i} + K_y \frac{\partial h}{\partial y} \boldsymbol{j} + K_z \frac{\partial h}{\partial z} \boldsymbol{k} \right) \tag{3-39}$$

From Eqs. (3-31) and (3-39), one can write

$$q_x = -K_x \frac{\partial h}{\partial x} \tag{3-40}$$

$$q_y = -K_y \frac{\partial h}{\partial y} \tag{3-41}$$

$$q_z = -K_z \frac{\partial h}{\partial x} \tag{3-42}$$

in which K_x, K_y, and K_z are the principal hydraulic in the x, y, and z directions, respectively, and their definitions are given in some well-known books (e.g., Verruijt, 1970; Bear, 1972; Batu, 1998, 2006, 2024).

Problem

3.1 In a Darcy experiment shown in Figure 3-3, the following values are given: $D = 15$ cm, $\Delta s = 40$ cm, $k = 0.4$ darcy, $\Delta p = 3$ atm, and $\mu = 4$ centipoise. Determine the flow rate Q $\left(\text{cm}^3/\text{min}\right)$ and Darcy velocity q (cm/s).

4

Fundamental Principles of Solute Transport in Porous Media

4.1 Introduction

Dissolved substances in liquids such as water are called *solutes*. Quantification of solute transport in porous media is based on the advective–dispersive solute transport theory that is based on *Fick's first law*, which was established by the mid-19th century. The original Fick's first law was established for molecular diffusion in surface water, and later on by the mid-20th century, it was extended to solute transport in porous media by including the dispersion effects.

The term "porous media" represents both manmade and natural porous materials. The term "manmade porous materials" includes cement, ceramic, and others. The term "natural materials" includes mainly the unsaturated zone below the ground surface and the saturated zone below the water table of an unconfined aquifer.

The transport mechanism of solutes in porous media flow is more complex than the one in surface water flow. Different solute constituents have different migration rates, depending on their physical and chemical characteristics as well as the characteristics of the porous materials. Also, some solute constituents may be subjected to decay or degradation under transport conditions, especially in natural porous media.

The purpose of this chapter is to present the fundamental principles of solute transport in porous media along with its mathematical modeling foundations with details of the governing differential equations, which are crucial to be understood in order to use various numerical and analytical mathematical models with their initial and boundary conditions for effective usage of the remediation methods. In this chapter, only some well-known one-dimensional analytical solute transport solutions are presented. Some of these solutions will be used in Chapter 6 to determine the degradation rates of contaminant constituents.

4.2 Concentration Units

4.2.1 Solute and Solvent

Consider two substances called substance 1 and substance 2. In a solution of substance 1 in substance 2, substance 1 is the dissolved substance and is called the *solute*. Substance 2, in which substance 1 is dissolved, is called the *solvent*. When the relative amount of one substance in a solution is much greater than the amount of the other dissolved substance, the substance present in greater amounts is generally regarded as the solvent. If the amounts

DOI: 10.1201/9781003739128-4

of substance 1 and substance 2 are of the same order, it becomes difficult to specify which one is the solvent (e.g., Rosenberg, 1980).

In the case of ground water, the solvent is always water; the dissolved inorganic and organic constituents are the solutes. To analyze ground water quality problems efficiently, the relative amounts of solute (the dissolved parts) and solvent (water) need to be specified. This can be achieved by the use of the concentration concepts and their units. Units of concentration are given in the following sections (e.g., Rosenberg, 1980).

4.2.2 Relative Concentration Units

Concentrations are often expressed in terms of relative units, e.g., percentages. Generally, there are three different types of percentage concentrations commonly used, and they are described below.

4.2.2.1 Mass/Mass Concentration

The mass/mass concentration is used to express the concentration of a solution when the mass of a solute and the mass of a solution are given. The mass/mass concentration is

$$\text{Mass/mass} = \frac{\text{Mass of solute}}{\text{Mass of solution}} \tag{4-1}$$

Example 4-1

In a solution, there is $70.47\ \text{mL}\,(69.76\ \text{g})$ solvent and $13.44\ \text{mL}\,(14.38\ \text{g})$ solute. Determine the mass/mass concentration.

Solution

From Eq. (4-1),

$$\begin{aligned}\text{Mass/mass} &= \frac{\text{Mass of solute}}{\text{Mass of solution}}\\ &= \frac{14.38\ \text{g}}{69.76\ \text{g} + 14.38\ \text{g}}\\ &= 0.1709\\ &= 17.09\%\end{aligned}$$

4.2.2.2 Volume/Volume Concentration

The volume/volume concentration is used to express the concentration of a solution when the volume of a solute and the volume of a solution are given. The volume/volume concentration can be determined as

$$\text{Volume/volume} = \frac{\text{Volume of solute}}{\text{Volume of solution}} \tag{4-2}$$

Example 4-2

For Example 4-1, determine the volume/volume concentration.

Solution

From Eq. (4-2),

$$\text{Volume/volume} = \frac{\text{Volume of solute}}{\text{Volume of solution}}$$

$$= \frac{13.44 \text{ mL}}{70.47 \text{ mL} + 13.44 \text{ mL}}$$

$$= 0.1602$$

$$= 16.02\%$$

4.2.2.3 Mass/Volume Concentration Units

The *mass concentration* is the mass of solute dissolved in a specified unit volume of solution. Kilograms per cubic meter with the symbol kg/m^3, grams per liter with the symbol g/L, milligrams per liter with the symbol mg/L, and micrograms per liter with the symbol µg/L are commonly used SI units in ground water studies. Therefore, the definition of mass concentration is

$$\text{Mass concentration} = \frac{\text{Mass of solute (g)}}{\text{Volume of solution (mL)}} \tag{4-3}$$

In the SI units,

$$10^{-3} \text{ kg} = 1 \text{ g} = 1{,}000 \text{ mg} = 1{,}000{,}000 \text{ µg} = 1{,}000{,}000{,}000 \text{ ng} \tag{4-4}$$

Since $1 \text{ m}^3 = 1{,}000 \text{ L}$ in the *SI* units system, mass concentration conversions are

$$\frac{1 \text{ g}}{1 \text{ L}} = \frac{1{,}000 \text{ mg}}{1 \text{ L}} = \frac{1{,}000 \text{ g}}{1{,}000 \text{ L}} = \frac{1{,}000 \text{ g}}{1 \text{ m}^3} = \frac{1{,}000{,}000 \text{ µg}}{1 \text{ L}}$$
$$= \frac{1{,}000{,}000{,}000 \text{ ng}}{1{,}000 \text{ L}} = \frac{1{,}000{,}000{,}000 \text{ ng}}{1 \text{ m}^3} \tag{4-5}$$

in which m, μ, and n stand for *milli, micro,* and *nano,* respectively.

In practice, grams per liter (g/L), milligrams per liter (mg/L), micrograms (µg/L), and grams per milliliter (g/mL) are widely used.

4.2.3 Dilute Mass Concentration Units

When solutions are too dilute, their percentage concentrations become too low. Therefore, instead of using low percentage concentrations such as 0.000001% or 0.000000001%, another way is used to express the concentrations in a more convenient manner. Their units are ppm, ppb, and ppt, which are frequently used. Details for ppm, ppb, and ppt are given below.

4.2.3.1 Parts Per Million (ppm) Concentration

A concentration of a solution containing 1 g of solute and 1,000,000 mL solution, which is the same as 1 mg solute and 1 Lsolution, will generate a very small percentage concentration.

Because a solution like this will be so dilute, the density of the solution is well approximated by the density of the solvent for water that is 1 g/mL, but will be different for different solvents. By assuming water is the solvent and converting the milliliters (mL) of a solution into grams (g) of solution, one can write

$$\frac{1\text{ g solute}}{1{,}000{,}000\text{ mL solution}} \times \frac{1\text{ mL}}{1\text{ g}} = \frac{1\text{ g solute}}{1{,}000{,}000\text{ g solution}} \tag{4-6}$$

Because both the solute and the solution are expressed in terms of grams (g), it can be said that the solute concentration is 1 part per million (ppm). Therefore, parts per million (ppm) is the number of grams of a solute constituent per million grams of solution, and Eq. (4-6) can be written as

$$1\text{ ppm} = \frac{1\text{ g solute constituent}}{10^6\text{ g solution}} \tag{4-7}$$

For nonsaline waters,

$$1\text{ ppm} = \frac{1\text{ g solute constituent}}{1\text{ m}^3\text{ solution}} = \frac{1\text{ mg solute constituent}}{1\text{ L solution}} \tag{4-8}$$

For the soil case, parts per million (ppm) is the number of grams of soil constituent per million grams of soil,

$$1\text{ ppm} = \frac{1\text{ g of soil constituent}}{10^6\text{ g soil}} \tag{4-9}$$

Also,

$$1\text{ ppm} = 1{,}000\text{ ppb} = 1{,}000{,}000\text{ ppt} \tag{4-10}$$

in which ppb and ppt are described in the following sections.

4.2.3.2 Parts Per Billion (PPB) Concentration

For the solute case, parts per billion (ppb) is the number of grams of a solute constituent per billion grams of solution,

$$1\text{ ppb} = \frac{1\text{ g solute constituent}}{10^9\text{ g solution}} = \frac{1\,\mu\text{g solute constituent}}{10^3\text{ g solution}} \tag{4-11}$$

For nonsaline waters,

$$1\text{ ppb} = \frac{1\,\mu\text{g solute constituent}}{1\text{ L solution}} = \frac{1\text{ mg solute constituent}}{1\text{ m}^3\text{ solution}} \tag{4-12}$$

For the soil case, parts per billion (ppb) is the number of grams of soil constituent per billion grams of soil,

$$1\text{ ppb} = \frac{1\text{ g soil constituent}}{10^9\text{ g soil}} \tag{4-13}$$

4.2.3.3 Parts Per Trillion (PPT) Concentration

For the solute case, parts per trillion (ppt) is the number of grams of a solute constituent per trillion grams of solution,

$$1\text{ ppt} = \frac{1\text{ g of solute constituent}}{10^{12}\text{ g solution}} = \frac{10^{-6}\text{ g solute constituent}}{10^{6}\text{ g solution}} \tag{4-14}$$

For nonsaline waters,

$$1\text{ ppt} = \frac{1\,\mu\text{g solute constituent}}{1\text{ m}^3\text{ solution}} = \frac{10^{-3}\,\mu\text{g solute constituent}}{1\text{ L solution}} \tag{4-15}$$

For the soil case, parts per billion (ppt) is the number of grams of soil constituent per trillion grams of soil,

$$1\text{ ppt} = \frac{1\text{ g of soil constituent}}{10^{12}\text{ g of soil}} \tag{4-16}$$

4.2.4 Concentration Units Based on Moles

4.2.4.1 Mole Fraction

The mole fraction of a substance is the fraction of all of its molecules (or atoms) out of the total number of molecules (or atoms). For a substance A, the mole fraction is

$$X_A = \frac{\text{Number of moles of substance } A}{\text{Number of total moles in solution}} \tag{4-17}$$

The sum of each of the substances $A, B, C, \ldots$ in the solution is equal 1:

$$X_A + X_B + X_c + \cdots = 1 \tag{4-18}$$

4.2.4.2 Molar Concentration of a Solution (Molarity)

The *molar concentration* or *molarity* (M) of a solution is used to represent the amount of moles of solute per liter of solution, and its expression is (e.g., Rosenberg, 1980, p. 159; CLO, Internet, 2023)

$$M = \text{Molar concentration of a solution} = \frac{\text{Number of moles of solute}}{\text{Number of liters of solution}} \tag{4-19}$$

In other words, molarity is the number of moles of solute in 1 m^3 of solution. The SI unit for molarity is designated as mol/m^3, which is equal mmol/L. Moles per liter, with the symbol mol/L, is the unit for molarity in the SI system and is commonly used in ground water studies.

Note in Eq. (4-19) that the volume is in liters of solution and not liters of solvent. When a molarity is reported, the unit is the symbol M and is read as "molar." For example, a solution is labeled 1.8 M NH_3 and is read as "1.8 molar ammonia solution."

Example 4-3

What is the molarity of a solution containing 64.25 g NH_4Cl in 2 Lof solution.

Solution

From the standard atomic weight table,

$$N = 14.0067\ g$$

$$H = 1.0079\ g$$

$$Cl = 35.453\ g$$

Molecular weight of NH_4Cl

$$= N + 4H + Cl\ = 14.0067\ g + 4(1.0079\ g) + 35.453\ g = 53.4913\ g$$

From Eq. (4-19),

$$\text{Molar concentration} = \frac{\text{Number of moles of solute}}{\text{Number of liters of solution}} = \frac{\dfrac{64.25\ g}{53.4913\ \frac{g}{mol}}}{2\ L} = 0.601\ M$$

Example 4-4

What is the molarity of a solution containing 22.0 g CH_3OH in 600 cm^3 of solution.

Solution

From the standard atomic weight tables,

$$C = 12.011\ g$$

$$H = 1.0079\ g$$

$$O = 15.9995\ g$$

Molecular weight of CH_3OH

$$= C + 3H + O + H = 12.011\ g + 3(1.0079\ g) + 15.9995\ g + 1.0079\ g = 32.0421\ g$$

From Eq. (4-19),

$$\text{Molar concentration} = \frac{\text{Number of moles of solute}}{\text{Number of liters of solution}} = \frac{\dfrac{22.0\ g}{32.0421\ \frac{g}{mol}}}{0.600\ L} = 1.1443\ M$$

Example 4-5

What is the molarity of a solution containing 36.0 g $AgNO_3$ in 2 Lof solution.

Solution

From the standard atomic weight tables,

$$Ag = 107.868 \text{ g}$$

$$N = 14.0067 \text{ g}$$

$$O = 15.9995 \text{ g}$$

Molecular weight of $AgNO_3$

$$= Ag + N + 3O = 107.868 \text{ g} + 14.0067 \text{ g} + 3(15.9995 \text{ g}) = 168.8732 \text{ g}$$

From Eq. (4-19),

$$\text{Molar concentration} = \frac{\text{Number of moles of solute}}{\text{Number of liters of solution}} = \frac{\dfrac{36.0 \text{ g}}{168.8732 \dfrac{\text{g}}{\text{mol}}}}{2 \text{ L}} = 0.1066 \text{ M}$$

4.2.4.3 Molality

The *molality* is defined as the number of moles of solute dissolved in a 1 kilogram (kg) mass of solvent or solution, and its expression in SI unit is (e.g., Rosenberg, 1980, p. 159; CLO, Internet, 2023)

$$\text{Molality of a solution} = m = \frac{\text{Number of moles of solute}}{\text{Number of kilograms of solvent}} \quad (4\text{-}20)$$

Its symbol is mol/kg. As can be seen from Eqs. (4-19) and (4-20), the molarity and molality differ only in their denominators.

Example 4-6

Determine the molality of a solution of 48.56g of glucose ($C_6H_{12}O_6$) by dissolving into 700 g of water.

Solution

From the standard atomic weight table,

$$C = 12.011 \text{ g}$$

$$H = 1.0079 \text{ g}$$

$$O = 15.9994 \text{ g}$$

Molecular weight of $C_6H_{12}O_6$

$$= 6C + 12H + 6O = 6\ (12.011 \text{ g}) + 12\ (1.0079 \text{ g}) + 6(15.9994 \text{ g}) = 180.1572 \text{ g}$$

$$48.56\text{ g}\quad C_6H_{12}O_6 \times \frac{1\text{ mol } C_6H_{12}O_6}{180.1572\text{ g } C_6H_{12}O_6} = 0.2695\text{ mol } C_6H_{12}O_6$$

From Eq. (4-20),

$$\text{Molality of a solution} = m$$

$$= \frac{\text{Number of moles of solute}}{\text{Number of kilograms of solvent}}$$

$$= \frac{0.2695\text{ mol } C_6H_{12}O_6}{0.700\text{ kg } H_2O} = 0.3850\text{ m } C_6H_{12}O_6$$

Example 4-7

What is the molality of a solution that contains 45.0 g of cane sugar ($C_{12}H_{22}O_{11}$) dissolved in 350 g of water?

Solution

From the standard atomic weight table,

$$C = 12.011\text{ g}$$

$$H = 1.0079\text{ g}$$

$$O = 15.9994\text{ g}$$

Molecular weight of $C_{12}H_{22}O_{11}$

$$= 12C + 22H + 11O = 12\,(12.011\text{ g}) + 22\,(1.0079\text{ g}) + 11(15.9994\text{ g}) = 342.2992\text{ g}$$

$$45.0\text{ g}\quad C_{12}H_{22}O_{11} \times \frac{1\text{ mol } C_{12}H_{22}O_{11}}{342.2992\text{ g } C_{12}H_{22}O_{11}} = 0.1315\text{ mol } C_{12}H_{22}O_{11}$$

From Eq. (4-20),

$$\text{Molality of a solution} = m$$

$$= \frac{\text{Number of moles of solute}}{\text{Number of kilograms of solvent}}$$

$$= \frac{0.1315\text{ mol } C_{12}H_{22}O_{11}}{0.350\text{ kg } H_2O} = 0.3757\text{ m } C_{12}H_{22}O_{11}$$

4.3 Type of Concentrations

4.3.1 Introduction

Distinguishing between volume-averaged or resident concentrations (C_r) and flux-averaged or flowing concentrations (C_f) is widely accepted and has been used for decades. These are two types of concentrations that are described in this section.

Distinctions between C_r and C_f have been made by scientists in the field of petroleum and chemical engineering (Brigham, 1974; Kreft and Zuber, 1978, 1979). Parker and van Genuchten (1984) showed that it is of fundamental importance to make such a distinction in order to stipulate boundary conditions appropriate for specific experimental solute detection models.

4.3.2 Volume-Averaged Concentration

The mass of solute dissolved in a specified unit volume of solution is the *mass concentration*. This concentration is also called *volume-averaged concentration* or *resident concentration*, which is denoted by C_r. The resident concentration must be defined in the context of representative elementary volume. If in a unidirectional ground water flow, Δm_A is the mass of solute for a constituent A in a volume ΔV, the volume-averaged or resident concentration is defined as (Parker and van Genuchten, 1986; Batu, 2006)

$$C_r = \frac{\Delta m_A}{\Delta V} \tag{4-21}$$

in which

$$\Delta V = \varphi_e S \Delta x \tag{4-22}$$

Here, φ_e is the effective porosity, S is the cross-sectional area of the tube (Figure 4-1), and Δx is the length of the fluid volume.

The *mass concentration* defined by Eq. (4-3) in Section 4.2.2.3 is nothing but the *volume-averaged concentration* defined by Eq. (4-22).

4.3.3 Flux-Averaged Concentration

The definition of flux-averaged concentration is shown in Figure 4-2 and is denoted by C_f. Let F_x be the mass flux, which is mass per unit area per unit time. Then the definition of flux-averaged concentration C_f is (Kreft and Zuber, 1986)

$$C_f = \frac{F_x \Delta t S}{Q \Delta t} = \frac{F_x S}{Q} \tag{4-23}$$

in which Q is the volumetric flow rate in the tube, S is the cross-sectional area of the tube, and Δ is the time interval. Since $Q = qS$ in Eq. (4-23), one can write (Parker, 1984)

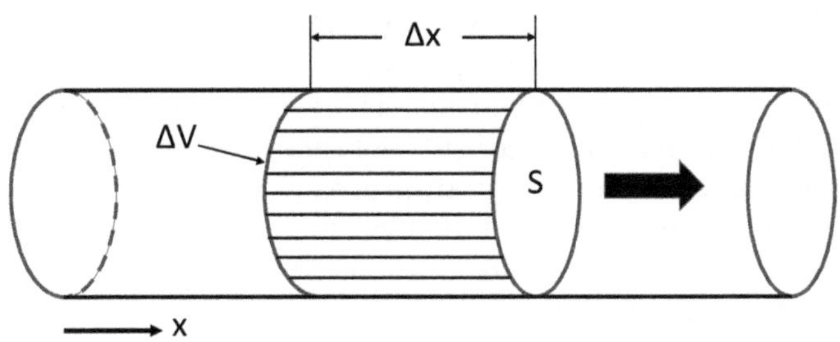

FIGURE 4-1
Definition of volume-averaged concentration for unidirectional flow. Adapted from Kreft and Zuber (1986).

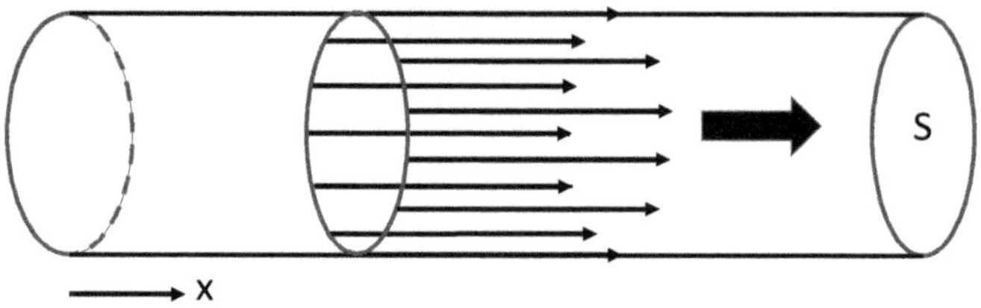

FIGURE 4-2
Definition of flux-averaged concentration for unidirectional flow. Adapted from Kreft and Zuber (1986).

$$C_f = \frac{F_x S}{Q} = \frac{F_x}{q} = \frac{F_x}{\varphi_e U} \tag{4-24}$$

in which q is the Darcy velocity, φ_e is the effective porosity, and U is the ground water velocity (or seepage velocity). In Eq. (4-24), since the mass flux F_x is divided by the Darcy velocity q, C_f is called the *flux-averaged concentration.*

4.4 Physical Mechanism of Solute Transport in Saturated Porous Media without Sorption and Decay: One-Dimensional Case

4.4.1 Solute Flux Equation Due to Diffusive Solute Transport

The effects of diffusion can easily be observed by a simple experiment such as taking a glass of water and dropping a small amount of dye into it. During the early stages, a sharp interface can be observed between the colored water and the rest of the water body in the glass. As time goes on, gradually, the sharp interface disappears, and the whole glass of water eventually takes on the same color. This phenomenon is called *molecular diffusion,* and the transport mechanism is called *diffusive transport.* Diffusive transport can occur in fluids at rest as well as in motion.

4.4.2 Quantification of Molecular Diffusion: Fick's First Law

4.4.2.1 Fick's First Law for Pure Water

The first quantitative study of diffusion was made by a German physician and physiologist Adolph E. Fick (1829–1901). Fick's law is analogous to the relationships discovered at the same epoch by other eminent scientists: Darcy's law for water flow through porous media, Ohm's law for electrical currents, and Fourier's law for heat conduction.

To describe the diffusion of a solute constituent quantitatively, a mathematical relationship between the solute mass flux per unit time area (F_s) and concentration gradient along the distance ($\Delta C/\Delta s$) must be established. In 1855, Fick's work was inspired by the earlier experiments of Scottish chemist Thomas Graham (1805–1869) known for his pioneering work in dialysis and the diffusion of gases.

The schematic representation of Fick's first law with a solute source in a stagnant liquid is shown in Figure 4-3. Figure 4-3 shows that Fick's first law relates the diffusive solute

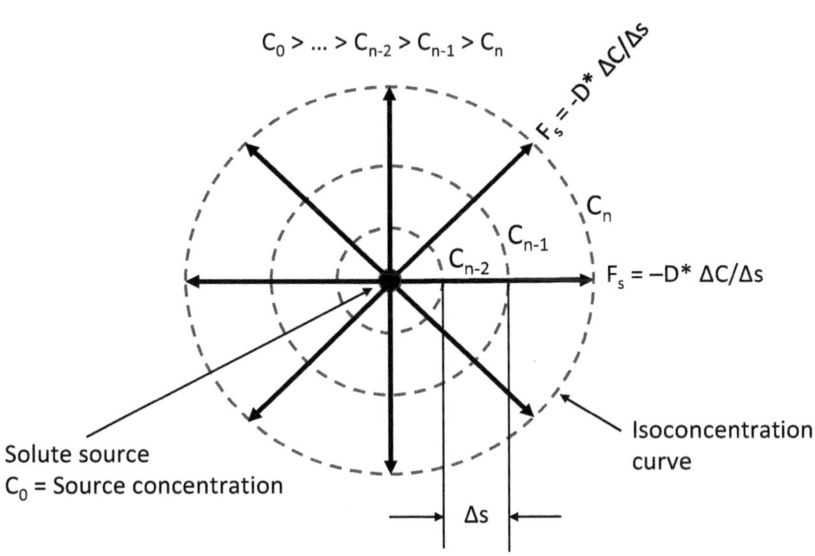

FIGURE 4-3
Schematic representation of Fick's first law with a solute source in a stagnant liquid.

flux to the concentration gradient, and it postulates that the solute flux occurs from regions of high concentration to regions of low concentration. And the law states that the rate of transfer of diffusing substance ($F_{s_{dif}}$) through the unit area of a section is proportional to the concentration gradient ($\Delta C/\Delta s$) normal to the section as

$$F_{s_{dif}} \propto \frac{\Delta C}{\Delta s} \tag{4-25}$$

A proportionality constant shown by D^* leads to the equation

$$F_{s_{dif}} = -D^* \frac{\Delta C}{\Delta s} = -D^* \frac{C_{n-1} - C_{n-2}}{\Delta s} \tag{4-26}$$

The negative sign indicates that solute transport is in the direction of decreasing concentration. In Eq. (4-26), D^* is the ionic or effective molecular diffusion coefficient of the porous medium and s is the distance. From dimensional analysis, it can be shown that the dimension of D^* is L^2/T.

In the limiting case, $\Delta C/\Delta s$ becomes dC/ds, then Eq. (4-26) becomes

$$F_{s_{dif}} = -D^* \frac{dC}{ds} \tag{4-27}$$

which is called *Fick's first law*. The term *Fickian diffusion* is also widely used.

4.4.2.2 Fick's Modified First Law for Porous Media

By modifying Fick's first law expressed by Eq. (4-27), in the presence of a solid phase, the diffusion flux in a porous medium having θ volumetric water content becomes

$$F_{s_{dif}} = -\theta D^* \frac{dC}{ds} \tag{4-28}$$

which is valid for saturated and unsaturated porous media flow cases. Equation (4-28) is exactly the same as in the pertinent literature [e.g., van Genuchten and Wierenga, 1986, p. 1026, Eq. (2)].

Since $\theta = \varphi_e$ for saturated porous media, Eq. (4-28) becomes [e.g., Batu, 2006, p. 13, Eq. (2.21)]

$$F_{sdif} = -\varphi_e D^* \frac{dC}{ds} \tag{4-29}$$

in which φ_e is the effective porosity.

Due to a tortuous diffusion pathway and the presence of the solution and interface, D^* in granular materials is comparatively lower than the molecular diffusion coefficient in pure water D_0 or diffusion coefficient in aqueous solution. Therefore,

$$D^* = D_0 \tau \quad 0 < \tau < 1 \tag{4-30}$$

in which τ is a dimensionless *tortuosity factor*, ranging roughly from 0.3 to 0.7 for most soils (van Genuchten and Wierenga, 1986).

4.4.2.3 Fick's Second Law

Fick's second law predicts the spatial and temporal variation of concentration of a solute, and it is a one-dimensional differential equation given as

$$\frac{\partial C}{\partial t} = D^* \frac{\partial^2 C}{\partial x^2} \tag{4-31}$$

for which derivation details are given in Section 4.7.3.3.

4.4.2.4 Effective Molecular Diffusion Coefficient D*

4.4.2.4.1 Effective Molecular Diffusion Coefficients in Pure Water

According to Eq. (4-30), for liquids, $\tau = 1$ and $D^* = D_0$. The major ions in ground water (Na^+, K^+, Mg^{2+}, Cl^-, HCO_3^-, and SO_4^{2-}) have diffusion coefficients in the range 1.0×10^{-9} to 2.0×10^{-9} m^2/s at 25°C (Robinson and Stokes, 1965). At 5°C, for example, the coefficients are about 50% smaller.

4.4.2.4.2 Effective Molecular Diffusion Coefficients in Porous Media

Because of the tortuosity factor, the effective molecular diffusion coefficient D^* is the dominant parameter for diffusive transport in porous media. In other words, D^* is smaller than in water because the ions follow longer paths of diffusion through the granular structure of porous media. According to Freeze and Cherry (1979, p. 393), $D^* = 1.0 \times 10^{-11}$ to 1.0×10^{-10} m^2/s, which are representative of a range typical of nonreactive chemical species in clayey geological deposits. Values of effective molecular diffusion coefficient for coarse-grained unconsolidated materials can be higher than 1.0×10^{-10} m^2/s, but are less than D_0 values in water, i.e., $< 2.0 \times 10^{-9}$ m^2/s (Freeze and Cherry, 1979, p. 393).

4.4.3 Solute Flux Equation Due to Advective Transport

In the solute transport mechanism, *advective transport* means that solutes are transported only by the motion of the flowing water in the pores of a porous medium. Therefore, under

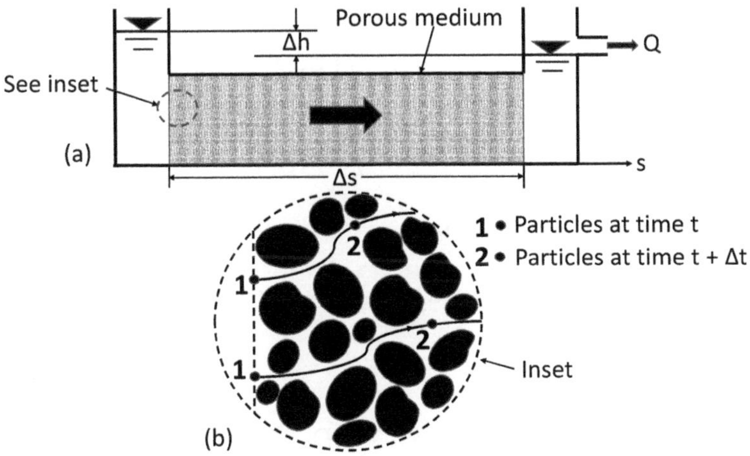

FIGURE 4-4
Physical mechanism of mechanical dispersion in a porous medium flow: (a) Porous media flow between two reservoirs and (b) fluctuating streamlines around the grains of the porous medium on a pore scale.

pure advection conditions, ground water and dissolved solute move at the same average velocity, and the solute flux for advection $F_{s_{adv}}$ is

$$F_{s_{adv}} = v_s \varphi_e C = q_s C \tag{4-32}$$

in which $F_{s_{adv}}$ represents the mass of solute per unit cross-sectional area transported in the *s* direction (Figure 4-4) per unit time, v_s is the ground water velocity, C is the concentration, φ_e is the effective porosity, and q_s is the Darcy velocity in the same direction.

4.4.4 Solute Flux Equation Due to Dispersive Transport

Consider an average one-dimensional ground water flow shown in Figure 4-4. In Figure 4-4a, the mean ground water flow direction is shown in the sense of Darcy velocity. In Figure 4-4b, the path lines of individual fluid particles around the grains of a porous medium, which are filled with water, are shown. Due to the irregular shapes of grains and pores, local fluid velocities inside individual pores deviate from the average pore velocity. Such velocity variations can cause the solute to be transported at different rates, thus leading to a mixing process that is macroscopically similar to mixing caused by diffusive transport. Because of the passive nature of the dispersion process, the term *mechanical dispersion* is often used to describe mixing caused by local velocity variations.

The *mechanical dispersive flux* or *dispersive flux* is described by an equation similar to the equation for diffusive flux given by Eq. (4-29),

$$F_{s_{disp}} = -\varphi_e D_m \frac{dC}{ds} \tag{4-33}$$

in which D_m is the *mechanical dispersion coefficient* (e.g., Bear, 1972, 1979). It is clear that Eq. (4-33) is based on the assumption that the mechanical dispersion process is a *Fickian process*. The mechanical dispersion is the most complex and controversial component for the transport of solutes in porous media. Like in Eq. (4-29), the negative sign in Eq. (4-33) indicates that the solute moves toward the zone of lower concentration.

4.4.5 Solute Flux Equation under Hydrodynamic Dispersion Conditions

Based on the details in Sections 4.4.2. 4.4.3, and 4.4.4, the solute flux F_s is the sum of the fluxes in Eqs. (4-29), (4-32), and (4-33):

$$F_s = F_{s_{dif}} + F_{s_{adv}} + F_{s_{disp}} \tag{4-34}$$

Substitution of the fluxes in this equation,

$$F_s = -\varphi_e D^* \frac{dC}{ds} + v_s \varphi_e C - \varphi_e D_m \frac{dC}{ds} \tag{4-35}$$

After the arrangement,

$$F_s = v_s \varphi_e C - \varphi_e D_s \frac{dC}{ds} \tag{4-36}$$

in which

$$D_s = D^* + D_m \tag{4-37}$$

Here D_s is the *longitudinal dispersion coefficient* along the *s* coordinate.

4.4.6 Solute Flux Equation in Unsaturated Porous Media for Constant Water Content under Hydrodynamic Dispersion Conditions

If the water content θ in an unsaturated porous medium is constant, with $\varphi_e \equiv \theta$, Eq. (4-36) becomes

$$F_s = v_s \theta C - \theta D_s \frac{dC}{ds} \tag{4-38}$$

which is the solute flux equation in unsaturated porous media having a constant water content under hydrodynamic dispersion conditions.

4.5 Three-Dimensional Advective–Dispersive Solute Flux Equations for Saturated Porous Media—General Case

The one-dimensional solute flux equation given by Eq. (4-36) has been generalized for three-dimensional cases for saturated porous media based on experiments conducted since the 1960s. Based on the studies of Bear (1961) and Scheidegger (1961), it is assumed that the hydrodynamic dispersion coefficient is approximated by a tensor having nine components. Using the standard notation for second-order tensors, the three-dimensional tensorial form of Eq. (4-36) is (Bear, 1961; Scheidegger, 1961; Ogata, 1970)

$$F_i = \varphi_e v_i C - \varphi_e D_{ij} \frac{\partial C}{\partial x_j} \quad i, j = 1, 2, 3 \tag{4-39}$$

With $i = x, y, z$ and $j = x, y, z$, the advective–dispersive solute flux components in the Cartesian coordinates can be written as

$$F_x = \varphi_e v_x C - \varphi_e \left(D_{xx} \frac{\partial C}{\partial x} + D_{xy} \frac{\partial C}{\partial y} + D_{xz} \frac{\partial C}{\partial z} \right) \tag{4-40a}$$

$$F_y = \varphi_e v_y C - \varphi_e \left(D_{yx} \frac{\partial C}{\partial x} + D_{yy} \frac{\partial C}{\partial y} + D_{yz} \frac{\partial C}{\partial z} \right) \tag{4-40b}$$

$$F_z = \varphi_e v_z C - \varphi_e \left(D_{zx} \frac{\partial C}{\partial x} + D_{zy} \frac{\partial C}{\partial y} + D_{zz} \frac{\partial C}{\partial z} \right) \tag{4-40c}$$

and the hydrodynamic dispersion tensor D_{ij} is

$$[D_{ij}] = \begin{bmatrix} D_{xx} & D_{xy} & D_{xz} \\ D_{yx} & D_{yy} & D_{yz} \\ D_{zx} & D_{zy} & D_{zz} \end{bmatrix} \tag{4-41}$$

Equations (4-40) describe the state-of-the-art forms of the advective–dispersive solute flux components, which constitute almost all analytical and numerical solute transport models existing in the literature since the 1950s.

4.6 Three-Dimensional Advective–Dispersive Solute Flux Equations for Saturated Porous Media—Special Cases

There are some important special cases of the advective–dispersive solute flux components that have importance in practice. These cases are given below.

4.6.1 Solute Flux Equations for Homogeneous and Anisotropic Porous Media

Consider points 1 and 2 in a homogeneous and anisotropic aquifer schematically shown in Figure 4-5. If the hydraulic conductivities in the x-, y-, and z-coordinate directions satisfy

$$K_x\left(x_1, y_1, z_1\right) = K_x\left(x_2, y_2, z_2\right) = K_x \tag{4-42a}$$

$$K_y\left(x_1, y_1, z_1\right) = K_y\left(x_2, y_2, z_2\right) = K_y \tag{4-42b}$$

$$K_z\left(x_1, y_1, z_1\right) = K_z\left(x_2, y_2, z_2\right) = K_z \tag{4-42c}$$

the aquifer is said to be a *homogeneous and anisotropic aquifer*. The corresponding form of the hydrodynamic dispersion tensor given by Eq. (4-41) will be ($D_{xx} \equiv D_x$, $D_{yy} \equiv D_y$, and $D_{zz} \equiv D_z$, $D_{xy} = D_{xz} = D_{yx} = D_{yz} = D_{zx} = D_{zy} = 0$)

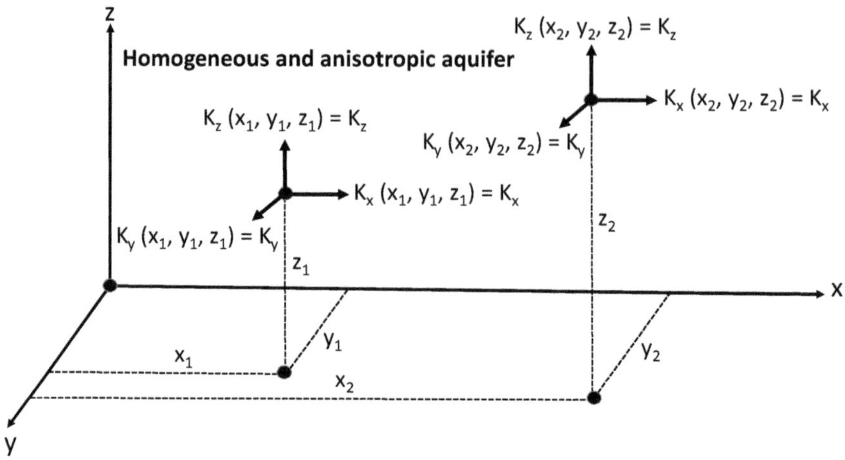

FIGURE 4-5
Homogeneous and anisotropic aquifer.

$$[D_{ij}] = \begin{bmatrix} D_x & 0 & 0 \\ 0 & D_y & 0 \\ 0 & 0 & D_z \end{bmatrix} \tag{4-43}$$

As a result, the advective–dispersive solute flux components given by Eqs. (4-40) take, respectively, the following forms:

$$F_x = \varphi_e v_x C - \varphi_e D_x \frac{\partial C}{\partial x} \tag{4-44a}$$

$$F_y = \varphi_e v_y C - \varphi_e D_y \frac{\partial C}{\partial x} \tag{4-44b}$$

$$F_z = \varphi_e v_z C - \varphi_e D_{zx} \frac{\partial C}{\partial x} \tag{4-44c}$$

in which v_x, v_y, and v_z are the ground water velocity components in the x-, y-, and z-coordinate directions, respectively, and are expressed in relation to the Darcy velocity components as

$$v_x = \frac{q_x}{\varphi_e} \quad v_y = \frac{q_y}{\varphi_e} \quad v_z = \frac{q_z}{\varphi_e} \tag{4-45}$$

q_x, q_y, and q_z are the Darcy velocity components in the x-, y-, and z-coordinate directions, respectively. The Darcy velocity components are

$$q_x = K_x I_x \quad q_y = K_y I_y \quad q_z = K_z I_z \tag{4-46}$$

in which I_x, I_y, and I_z are the hydraulic gradients in the x-, y-, and z-coordinate directions, respectively.

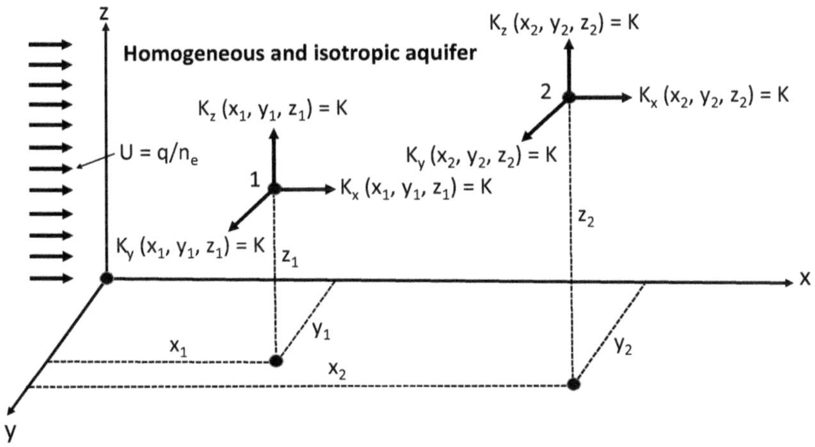

FIGURE 4-6
Homogeneous and isotropic aquifer having $U = q/n_e$ uniform ground water velocity.

4.6.2 Solute Flux Equations for Homogeneous and Isotropic Porous Media

Consider points 1 and 2 in a homogeneous and isotropic aquifer schematically shown in Figure 4-6. If the hydraulic conductivities in the *x*-, *y*-, and *z*-coordinate directions satisfy

$$K_x\left(x_1,y_1,z_1\right)=K_y\left(x_1,y_1,z_1\right)=K_z\left(x_1,y_1,z_1\right)=K \qquad \text{(4-47a)}$$

$$K_x\left(x_2,y_2,z_2\right)=K_y\left(x_2,y_2,z_2\right)=K_z\left(x_2,y_2,z_2\right)=K \qquad \text{(4-47b)}$$

for points 1 and 2, respectively, the aquifer is said to be a *homogeneous and isotropic* aquifer.

For a unidirectional flow field, for example, in the *x*-coordinate direction,

$$I_x = I \quad I_y = I_z = 0 \qquad \text{(4-48)}$$

and

$$K_x = K_y = K_z = K \qquad \text{(4-49)}$$

Equations (4-46) take the forms

$$q_x = q = KI \quad q_y = q_z = 0 \qquad \text{(4-50)}$$

Then, Eqs. (4-45) take the forms

$$v_x = U = \frac{q}{\varphi_e} \quad v_y = v_z = 0 \qquad \text{(4-51)}$$

Substitution of Eqs. (4-51) into Eqs. (4-44),

$$F_x = \varphi_e UC - \varphi_e D_x \frac{\partial C}{\partial x} \qquad \text{(4-52a)}$$

$$F_y = -\varphi_e D_y \frac{\partial C}{\partial x} \tag{4-52b}$$

$$F_z = -\varphi_e D_z \frac{\partial C}{\partial x} \tag{4-52c}$$

4.7 Solute Transport in Saturated Porous Media without Sorption and Decay

In this section, the differential equations for solute transport are presented with the following assumptions: (1) mass transfer for a solute constituent does not exist between the liquid and solid portions of the porous medium system (sorption) and (2) the solute constituent does not degrade (decay) during its transport.

Historically, the solute transport differential equations were derived without taking into account the sorption and decay effects. Later on, based on extensive research efforts, the effects of sorption and decay were added to the differential equations. Therefore, in order to derive the differential equations for solute transport in a more understandable manner, in this section, the sorption and decay effects are not taken into account. In this section, by taking into account the effects of sorption and decay, the solute transport differential equations will be generalized.

The differential equations for solute transport in saturated porous media in Cartesian coordinates are presented in the following sections. With the advective–dispersive solute flux components presented in the previous sections, the differential equations for solute transport in porous media are derived for the general case. Differential equations for the other cases are presented as special cases.

4.7.1 General Case: Homogeneous and Anisotropic Dispersion

The solute transport differential equation for the most general case in porous media without sorption and decay is

$$\begin{aligned}\varphi_e \frac{\partial C}{\partial t} &= \frac{\partial}{\partial x}\left(\varphi_e D_{xx}\frac{\partial C}{\partial x} + \varphi_e D_{xy}\frac{\partial C}{\partial y} + \varphi_e D_{xz}\frac{\partial C}{\partial z}\right)\\ &+ \frac{\partial}{\partial y}\left(\varphi_e D_{yx}\frac{\partial C}{\partial x} + \varphi_e D_{yy}\frac{\partial C}{\partial y} + \varphi_e D_{yz}\frac{\partial C}{\partial z}\right)\\ &+ \frac{\partial}{\partial z}\left(\varphi_e D_{zx}\frac{\partial C}{\partial x} + \varphi_e D_{zy}\frac{\partial C}{\partial y} + \varphi_e D_{zz}\frac{\partial C}{\partial z}\right)\\ &- \frac{\partial}{\partial x}(\varphi_e v_x C) - \frac{\partial}{\partial y}(\varphi_e v_y C) - \frac{\partial}{\partial z}(\varphi_e v_z C)\end{aligned} \tag{4-53}$$

which is derived in Example 4-8.

Example 4-8

Derive the solute transport differential equation for the most general case in porous media without sorption and decay given by Eq. (4-53).

Solution

Consider the differential parallelepiped porous medium element having (dx, dy, dz) dimensions in Figure 4-7 showing in and out solute fluxes. The concentration of a solute is defined as the mass of solute per unit volume of solution. The effective porosity of the medium is φ_e.

The total mass of solute entering the element per unit time is

$$m_{in} \text{ per unit time} = F_x dydz + F_y dxdz + F_z dxdy \tag{E4-8-1}$$

The total mass of solute leaving the element per unit time is

$$m_{out} \text{ per unit time} = \left(F_x + \frac{\partial F_x}{\partial x} dx \right) dydz + \left(F_y + \frac{\partial F_y}{\partial y} dy \right) dxdz + \left(F_z + \frac{\partial F_z}{\partial z} dz \right) dxdy \tag{E4-8-2}$$

The difference in the mass rate entering and leaving the element is

$$(m_{out} - m_{in}) \text{ per unit time} = \left(\frac{\partial F_x}{\partial x} + \frac{\partial F_y}{\partial y} + \frac{\partial F_z}{\partial z} \right) dxdydz \tag{E4-8-3}$$

Since the dissolved substance is assumed to be nonreactive, the difference between the solute flux into the element and the flux out of the element equals the amount of

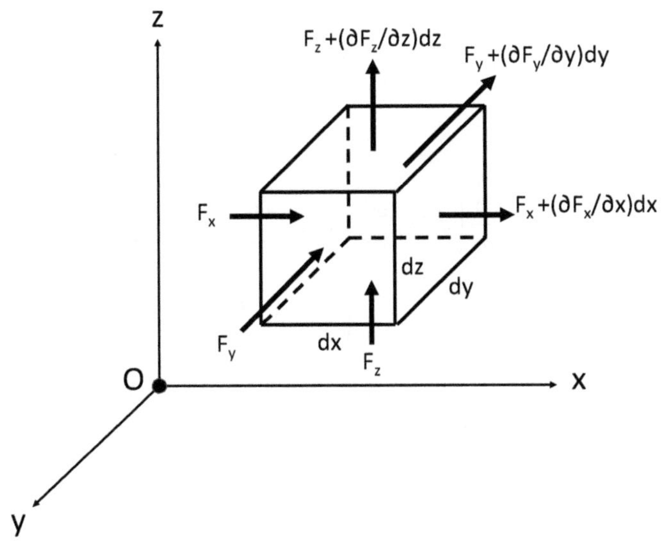

FIGURE 4-7
A differential parallelepiped porous medium element showing in and out solute fluxes.

dissolved substance accumulated in the element. The rate of mass (m) change in the element is

$$m \text{ change per unit time} = -\frac{\partial C}{\partial t}\varphi_e dxdydz \tag{E4-8-4}$$

Continuity or conservation of mass requires that the per unit volume of the porous medium per unit time be equal to the rate of solute concentration per unit volume of the porous medium. In other words, the difference of mass per unit time, as given by Eq. (E4-8-3), must be equal to the mass change per unit time, as given by Eq. (E4-8-4):

$$\left(\frac{\partial F_x}{\partial x}+\frac{\partial F_y}{\partial y}+\frac{\partial F_z}{\partial z}\right)dxdydz = -\frac{\partial C}{\partial t}\varphi_e dxdydz \tag{E4-8-5}$$

The infinitesimal volume term $dxdydz$ exists on both sides; therefore, these terms can be dropped out, which means that Eq. (E4-8-5) is valid for any porous medium volume. Therefore,

$$\frac{\partial F_x}{\partial x}+\frac{\partial F_y}{\partial y}+\frac{\partial F_z}{\partial z} = -\frac{\partial C}{\partial t}\varphi_e \tag{E4-8-6}$$

Substitution of the advective–dispersive solute flux components, as given by Eqs. (4-40), into Eq. (E4-8-6) results

$$\begin{aligned}
-\varphi_e\frac{\partial C}{\partial t} &= \frac{\partial}{\partial x}\left[\varphi_e v_x C - \varphi_e\left(D_{xx}\frac{\partial C}{\partial x}+D_{xy}\frac{\partial C}{\partial y}+D_{xz}\frac{\partial C}{\partial z}\right)\right] \\
&+\frac{\partial}{\partial y}\left[\varphi_e v_y C - \varphi_e\left(D_{yx}\frac{\partial C}{\partial x}+D_{yy}\frac{\partial C}{\partial y}+D_{yz}\frac{\partial C}{\partial z}\right)\right] \\
&+\frac{\partial F_z}{\partial z}\left[\varphi_e v_z C - \varphi_e\left(D_{zx}\frac{\partial C}{\partial x}+D_{zy}\frac{\partial C}{\partial y}+D_{zz}\frac{\partial C}{\partial z}\right)\right]
\end{aligned} \tag{E4-8-7}$$

After some manipulations,

$$\begin{aligned}
\varphi_e\frac{\partial C}{\partial t} &= \frac{\partial}{\partial x}\left(\varphi_e D_{xx}\frac{\partial C}{\partial x}+\varphi_e D_{xy}\frac{\partial C}{\partial y}+\varphi_e D_{xz}\frac{\partial C}{\partial z}\right) \\
&+\frac{\partial}{\partial y}\left(\varphi_e D_{yx}\frac{\partial C}{\partial x}+\varphi_e D_{yy}\frac{\partial C}{\partial y}+\varphi_e D_{yz}\frac{\partial C}{\partial z}\right) \\
&+\frac{\partial}{\partial z}\left(\varphi_e D_{zx}\frac{\partial C}{\partial x}+\varphi_e D_{zy}\frac{\partial C}{\partial y}+\varphi_e D_{zz}\frac{\partial C}{\partial z}\right) \\
&-\frac{\partial}{\partial x}(\varphi_e v_x C)-\frac{\partial}{\partial y}(\varphi_e v_y C)-\frac{\partial}{\partial z}(\varphi_e v_z C)
\end{aligned} \tag{E4-8-8}$$

which is exactly the same as Eq. (4-53).

4.7.2 Homogeneous and Isotropic Dispersion with Three Constant Velocity Components

Assuming that the effective porosity φ_e and the velocity components (v_x, v_y, v_z) are not spatially variable quantities, Eq. (4-53) takes the form

$$\frac{\partial C}{\partial t} = D_x \frac{\partial^2 C}{\partial x^2} + D_y \frac{\partial^2 C}{\partial y^2} + D_Z \frac{\partial^2 C}{\partial z^2} - v_x \frac{\partial C}{\partial x} - v_y \frac{\partial C}{\partial y} - v_z \frac{\partial C}{\partial z} \tag{4-54}$$

which is derived in Example 4-9.

Example 4-9

Derive the solute transport differential equation for the homogeneous and isotropic dispersion case for constant (v_x, v_y, v_z) porous medium flow velocity components given by Eq. (4-54).

Solution

For a homogeneous and isotropic porous medium, the advective–dispersive flux components are given by Eqs. (4-44). Substitution of these expressions into Eq. (E4-8-6) of Example E4-8 gives

$$-\frac{\partial C}{\partial t}\varphi_e = \frac{\partial}{\partial x}\left(\varphi_e v_x C - \varphi_e D_x \frac{\partial C}{\partial x}\right) + \frac{\partial}{\partial y}\left(\varphi_e v_y C - \varphi_e D_y \frac{\partial C}{\partial x}\right)$$
$$+ \frac{\partial}{\partial y}\left(\varphi_e v_z C - \varphi_e D_z \frac{\partial C}{\partial x}\right) \tag{E4-9-1}$$

Assuming that the effective porosity φ_e and (v_x, v_y, v_z) are not spatially variable quantities, Eq. (E4-9-1) takes the form

$$\frac{\partial C}{\partial t} = D_x \frac{\partial^2 C}{\partial x^2} + D_y \frac{\partial^2 C}{\partial y^2} + D_Z \frac{\partial^2 C}{\partial z^2} - v_x \frac{\partial C}{\partial x} - v_y \frac{\partial C}{\partial y} - v_z \frac{\partial C}{\partial z} \tag{E4-9-2}$$

which is exactly the same as Eq. (4-54).

4.7.3 Homogeneous and Isotropic Dispersion with Uniform Flow

4.7.3.1 Three-Dimensional Uniform Flow Case

For the three-dimensional uniform porous media flow case based on Eqs. (4-51), Eq. (4-54) becomes

$$\frac{\partial C}{\partial t} = D_x \frac{\partial^2 C}{\partial x^2} + D_y \frac{\partial^2 C}{\partial y^2} + D_Z \frac{\partial^2 C}{\partial z^2} - U \frac{\partial C}{\partial x} \tag{4-55}$$

which is derived in Example 4-10.

Example 4-10

Derive the three-dimensional solute transport differential equation for the uniform porous media flow case given by Eq. (4-55).

Solution

For a uniform porous media flow case, the advective–dispersive flux components are given by Eqs. (4-51). Substitution of these expressions into Eq. (E4-8-6) of Example E4-8 gives

$$-\frac{\partial C}{\partial t}\varphi_e = \frac{\partial}{\partial x}\left(\varphi_e UC - \varphi_e D_x \frac{\partial C}{\partial x}\right) + \frac{\partial}{\partial y}\left(-\varphi_e D_y \frac{\partial C}{\partial x}\right) + \frac{\partial}{\partial z}\left(-\varphi_e D_z \frac{\partial C}{\partial x}\right) \qquad \text{(E4-10-1)}$$

After simplifications,

$$\frac{\partial C}{\partial t} = D_x \frac{\partial^2 C}{\partial x^2} + D_y \frac{\partial^2 C}{\partial y^2} + D_Z \frac{\partial^2 C}{\partial z^2} - U\frac{\partial C}{\partial x} \qquad \text{(E4-10-2)}$$

which is exactly the same as Eq. (4-55).

4.7.3.2 Two-Dimensional Uniform Flow Case

For the case of two-dimensional solute transport, say in the x–y plane, Eq. (4-55) takes the form

$$\frac{\partial C}{\partial t} = D_x \frac{\partial^2 C}{\partial x^2} + D_y \frac{\partial^2 C}{\partial y^2} - U\frac{\partial C}{\partial x} \qquad \text{(4-56)}$$

4.7.3.3 One-Dimensional Uniform Flow Case

For the case of one-dimensional solute transport, say in the x-coordinate direction, Eq. (4-56) takes the form

$$\frac{\partial C}{\partial t} = D_x \frac{\partial^2 C}{\partial x^2} - U\frac{\partial C}{\partial x} \qquad \text{(4-57)}$$

Under only molecular diffusion effects $U = 0$ and $D_m = 0$. With these conditions, from Eq. (4-37), with $D_s \equiv D_x$, Eq. (4-57) turns out to be the same as Eq. (4-31), which is the second Fick's law.

4.8 Solute Transport in Saturated Porous Media with Sorption and Decay Conditions

In this section, the differential equations for solute transport will be presented with the following assumptions: (1) mass transfer for a solute constituent exists between the liquid and solid portions of the porous medium system and (2) the solute constituent degrades during its transport.

4.8.1 Solute Transport Equations under Sorption Conditions

The mass transfer between the liquid and solid portions of a porous medium is called *sorption*. In order to account for sorption effects, modifications are necessary in the solute transport differential equations presented in Section 4.7. As will be shown below, under sorption conditions, only the temporal term $\partial C/\partial t$ in the differential equations in Section 4.7 will be affected. In other words, its effects are the same in one-, two-, and three-dimensional solute transport cases. Therefore, inclusion of the sorption effects will be made in the one-dimensional solute transport differential equation given by Eq. (4-57).

4.8.1.1 One-Dimensional Solute Transport with Sorption and Desorption

The one-dimensional advective–dispersive solute transport differential equation for homogeneous media, as given by Eq. (4-57), modified to account for the effect of chemical reaction, can be written as (e.g., Gillham and Cherry, 1982b)

$$\frac{\partial C}{\partial t} = D_x \frac{\partial^2 C}{\partial x^2} - U \frac{\partial C}{\partial x} - G \tag{4-58}$$

in which $G(M/L^3/T)$ is a source/sink term representing the rate at which the dissolved species is removed from solution. Namely, G represents the mass of solute per unit volume of solution per unit time. The G term is expressed as (e.g., van Genuchten, 1974; Gillham and Cherry, 1982b)

$$G = \frac{\rho_b}{\varphi} \frac{\partial S}{\partial t} \tag{4-59}$$

in which $S(M/M)$ is the concentration of the solute in the solid phase, $\rho_b(M/L^3)$ is the dry bulk density of the porous medium, and φ is the total porosity. Assuming that (1) the chemical processes occur rapidly relative to the flow rate and chemical equilibrium is achieved, and (2) under isothermal conditions, the concentration of the solute in the solution is a function only of the concentration in the solid phase, it follows that

$$\frac{\partial S}{\partial t} = \frac{\partial S}{\partial C} \frac{\partial C}{\partial t} \tag{4-60}$$

Substitution of Eq. (4-60) into Eq. (4-59),

$$G = \frac{\rho_b}{\varphi} \frac{\partial S}{\partial C} \frac{\partial C}{\partial t} \tag{4-61}$$

Substitute this into Eq. (4-58),

$$\frac{\partial C}{\partial t} = D_x \frac{\partial^2 C}{\partial x^2} - U \frac{\partial C}{\partial x} - \frac{\rho_b}{\varphi} \frac{\partial S}{\partial C} \frac{\partial C}{\partial t} \tag{4-62}$$

After the arrangement, it takes the form

$$\left(1 + \frac{\rho_b}{\varphi} \frac{\partial S}{\partial C}\right) \frac{\partial C}{\partial t} = D_x \frac{\partial^2 C}{\partial x^2} - U \frac{\partial C}{\partial x} \tag{4-63}$$

For solute constituents undergoing transfer between the liquid and solid phases of the porous medium system, the equilibrium partitioning of the species between the two phases is commonly measured by batch tests. In batch tests, a specified volume of solution is brought in contact with a known mass of soil for enough time to allow equilibrium partitioning of the solute species to occur between the liquid and solid phases.

From these test results, graphs of the concentration in solution C against the concentration of the solids S can be determined. These graphs are named as *isotherms*. In Eq. (4-63), $\partial S/\partial C$ is the geometric slope of the isotherm.

The reversal of the sorption process is called *desorption*. Namely, solute molecules sorbed to a solid phase may be released back into solution because of changes in conditions.

4.8.1.2 Freundlich Nonlinear Sorption Isotherm

In the pertinent literature, there are various isotherm equations, but most of the models used in practice are based on the *Freundlich isotherm* (Freundlich, 1909) given as

$$S = KC^N \tag{4-64}$$

in which K and N are constants. In Eq. (4-64), S represents the total amount of solute retained by the soil (mg/kg), C is the solute concentration in solution (mg/L), K is the partitioning coefficient (L/kg), and N is the dimensionless reaction order commonly less than one.

With Eq. (4-64), Eq. (4-63) becomes

$$R_d \frac{\partial C}{\partial t} = D_x \frac{\partial^2 C}{\partial x^2} - U \frac{\partial C}{\partial x} \tag{4-65}$$

in which

$$R_d = 1 + \frac{\rho_b}{\varphi} KNC^{N-1} \tag{4-66}$$

is called the *retardation factor*. With R_d given by Eq. (4-66), Eq. (4-65) is a nonlinear differential equation.

4.8.1.3 Linear Sorption Isotherm: The K_d Approach

For most solute constituent cases, the isotherm given by Eq. (4-64), $N \approx 1$, which means that Eq. (4-64) becomes a linear equation. With $K \equiv K_d$ and $N = 1$, Eq. (4-64) becomes

$$S = K_d C \tag{4-67}$$

Likewise, for $N = 1$, Eq. (4-66) becomes

$$R_d = 1 + \frac{\rho_b}{\varphi} K_d \tag{4-68}$$

According to Eq. (4-67), the S vs. C function will be a straight line on an arithmetic plot (Figure 4-8). From Eq. (4-67), one can write

$$K_d = \frac{\Delta S}{\Delta C} \tag{4-69}$$

which means that with the S vs. C data are plotted as shown in Figure 4-8, the value of K_d can be determined graphically for a given solute constituent. Fortunately, many contaminant constituents in ground water studies satisfy these requirements. This approach is

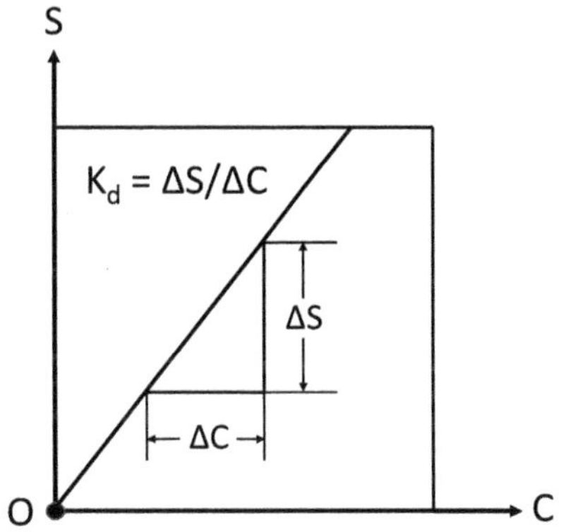

FIGURE 4-8
Linear sorption isotherm with *C* vs. *S*.

called the K_d *approach*. If $K_d = 0$ in Eq. (4-68), which corresponds to no sorption conditions, R_d becomes equal to unity. Thus, Eq. (4-65) becomes the same as Eq. (4-57).

The proportionality parameter K_d is the *distribution coefficient*, and it can be expressed as

$$K_d = \frac{\text{Mass of solute on the solid phase per unit mass of solute phase}}{\text{Concentration of solute in solution}} \tag{4-70}$$

In dimensional form,

$$[K_d] = \frac{\dfrac{M}{M}}{\dfrac{M}{L^3}} = \frac{L^3}{M} \tag{4-71}$$

With units, Eqs. (4-70) and (4-71) can be written as

$$[K_d] = \frac{\dfrac{\mu\text{g}\,(\text{Sorbed solute mass})}{\text{g}\,(\text{Mass of solid})}}{\dfrac{\mu\text{g}}{\text{mL}}\,(\text{Aqueous concentration})} = \frac{\text{mL}}{\text{g}} \tag{4-72}$$

Alternatively, Eqs. (4-70) and (4-71) can also be written as

$$[K_d] = \frac{\dfrac{\text{mg}\,(\text{Sorbed solute mass})}{\text{kg}\,(\text{Mass of solid})}}{\dfrac{\text{mg}}{\text{L}}\,(\text{Aqueous concentration})} = \frac{\text{L}}{\text{kg}} \tag{4-73}$$

The units of K_d given by Eqs. (4-72) and (4-73) are widely used in practice.

4.8.1.4 Retardation Factor (R_d)

The term R_d in Eqs. (4-66) and (4-68) is the *retardation factor*. The term *retardation coefficient* is also used occasionally. The retardation factor equation in the form of Eq. (4-68) was introduced in the chemical literature by Mayer and Tompkins (1947) and was first incorporated into the advection–dispersion equation for use in chromatography studies in chemical engineering by Vermeulen and Hiester (1952).

If there are no interactions between the chemical and the soil, K_d becomes zero and R_d becomes one. In some cases, R_d may become less than one, which indicates that only a fraction of the liquid phase participates in the solute transport process. This may be the case when the chemical is subject to anion exclusion or when relatively immobile liquid regions are present, for example, inside dense aggregates, that do not contribute to advective solute transport. In case of anion exchange, $(1-R_d)$ may be viewed as the relative anion exclusion volume (van Genuchten and Wierenga, 1986, pp. 1028–1029).

4.8.1.5 Representative Parameter Values in the Retardation Factor (R_d) Equation

The dry bulk density of the porous medium ρ_b can be expressed in terms of the ρ_r of porous medium grain as

$$\rho_b = (1-\varphi)\rho_r \tag{4-74}$$

With this, Eq. (4-68) becomes

$$R_d = 1 + \frac{(1-\varphi)\rho_r}{\varphi} K_d \tag{4-75}$$

For unconsolidated granular deposits, the porosity φ generally ranges from 0.20 to 0.40. The average density ρ_r of minerals or porous aquifer grains is approximately 2.65 g/cm^3. For $\varphi = 0.40$, Eq. (4-75) gives the minimum retardation factor expression as a function of only K_d as

$$R_{d_{min}} = 1 + \frac{(1-0.40)\left(2.65\,\frac{g}{cm^3}\right)}{0.40} K_d = 1 + \left(3.975\,\frac{g}{cm^3}\right) K_d \tag{4-76}$$

Likewise, for $\varphi = 0.20$, Eq. (4-75) gives the maximum retardation factor expression as a function of only K_d as

$$R_{d_{max}} = 1 + \frac{(1-0.20)\left(2.65\,\frac{g}{cm^3}\right)}{0.20} K_d = 1 + \left(10.600\,\frac{g}{cm^3}\right) K_d \tag{4-77}$$

In Eqs. (4-76) and (4-77), the units of K_d must be cm^3/g. Eqs. (4-76) and (4-77) are approximately the same as Freeze and Cherry [1979, p. 405, Eq. (9.15)].

Example 4-11

If the measured K_d value of tetrachloroethylene is 0.28 cm^3/g and $\varphi = 0.38$, determine the value of R_d.

Solution

With $\rho_r = 2.65\ g/cm^3$, Eq. (4-75) gives

$$R_d = 1 + \frac{(1-\varphi)\rho_r}{\varphi} K_d = 1 + \frac{(1-0.38)\left(2.65\,\frac{\text{g}}{\text{cm}^3}\right)}{0.38}\left(0.28\,\frac{\text{cm}^3}{\text{g}}\right) = 2.21$$

4.8.2 Solute Transport Equations under Degradation Conditions

The term "degradation" is a general term and refers to both *radioactive decay* and *biodegradation*. In this section, the general principles of radioactive decay and biodegradation along with their incorporation into the solute transport differential equations are presented (e.g., Bear, 1972).

4.8.2.1 Radioactive Decay

4.8.2.1.1 Types of Decay and Half-Life

Radioactive decay is the random process by which a nucleus loses energy by emitting radiation. The emission is usually in the form of alpha particles (helium nuclei), beta particles (electrons or positrons), or gamma rays (high-energy photons). As the energy of the nucleus reduces, it becomes more stable. In all decay processes, mass, charge, and lepton number are conserved.

The activity A or number of decays per second is given by

$$A = \nu N \tag{4-78}$$

in which ν (T^{-1}) is the decay constant and N is the number of undecayed nuclei of that type after time t. The number of remaining nuclei is (e.g., Mercer et al., 1982

$$N = N_0 e^{-\nu t} \tag{4-79}$$

in which t (T) is time and N_0 is the number of these nuclei at $t = 0$.

Information for a radioactive constituent is usually given in terms of its half-life instead of ν. The half-life $t_{1/2}$ of a sample is defined as the time at which the number of nuclei remaining is just one half of the original number, i.e., $N = N_0/2$. Therefore, from Eq. (4-79),

$$\frac{N}{N_0} = \frac{1}{2} = e^{-\nu t} \tag{4-80}$$

Taking the natural logarithms (ln) of both sides gives

$$\nu = \frac{0.693}{t_{1/2}} \tag{4-81}$$

With the known value of half-life $t_{1/2}$ for a radioactive substance, the decay constant can be determined from Eq. (4-81).

4.8.2.1.2 Radioactive Decay Equation

All pure radioactive substances decay in accordance with the same law. For a radioactive substance, experiments have shown that

$$\Delta C = -\nu C \Delta t \tag{4-82}$$

in which C is the mass concentration (M/L^3) of material still undecayed, ΔC is the mass concentration (M/L^3) that decays in a time interval Δt (T), and ν (T^{-1}) is the decay constant. The decay constant is specific to each radioactive solute constituent. In the limiting case, i.e., $\Delta t \to 0$, Eq. (4-82) takes the form

$$\frac{\partial C}{\partial t} = -\nu C \tag{4-83}$$

which is exactly the same as Bear [1972, p. 618, Eq. (10-5-3)] with the exception of notation.

4.8.2.2 Biodegradation and Natural Attenuation

The natural process of *biodegradation* is the breaking down of organic pollutants in porous materials such as soils and rocks by microorganisms that eat contaminants and change them to other materials. Depending on the type of material, this eating process can take days, weeks, or even centuries. Potentially, the biodegradation process helps minimize the environmental impact of harmful products by converting them into less harmful products.

Reducing the concentrations of a contaminant constituent by biodegradation or similar means is called *natural attenuation.* Natural attenuation occurs at most contaminated sites, and these conditions are monitored to make sure natural attenuation is working. Monitoring activities typically involve collecting soil and ground water samples to analyze them for the presence of contaminants and other site characteristics. The entire process is called *monitored natural attenuation* or *MNA*.

4.8.3 Generalized Solute Transport Differential Equations in Saturated Porous Media

In Section 4.7, multidimensional solute transport differential equations are presented without including sorption and decay effects. In this section, the generalized forms of the same equations are presented by including the sorption and decay effects.

4.8.3.1 General Case: Homogeneous and Anisotropic Dispersion

The multidimensional solute transport differential equation is

$$\begin{aligned}
\varphi_e R_d \frac{\partial C}{\partial t} = {} & \frac{\partial}{\partial x}\left(\varphi_e D_{xx}\frac{\partial C}{\partial x} + \varphi_e D_{xy}\frac{\partial C}{\partial y} + \varphi_e D_{xz}\frac{\partial C}{\partial z}\right) \\
& + \frac{\partial}{\partial y}\left(\varphi_e D_{yx}\frac{\partial C}{\partial x} + \varphi_e D_{yy}\frac{\partial C}{\partial y} + \varphi_e D_{yz}\frac{\partial C}{\partial z}\right) \\
& + \frac{\partial}{\partial z}\left(\varphi_e D_{zx}\frac{\partial C}{\partial x} + \varphi_e D_{zy}\frac{\partial C}{\partial y} + \varphi_e D_{zz}\frac{\partial C}{\partial z}\right) \\
& - \frac{\partial}{\partial x}(\varphi_e v_x C) - \frac{\partial}{\partial y}(\varphi_e v_y C) - \frac{\partial}{\partial z}(\varphi_e v_z C) - \varphi_e \nu R_d C
\end{aligned} \tag{4-84}$$

which is derived in Example 4-12.

Equation (4-84) is the generalized three-dimensional solute transport differential equation under homogeneous and anisotropic dispersion, and sorption and decay conditions, which is the main differential equation for many three-dimensional numerical solute transport models currently used in practice.

Example 4-12

Derive Eq. (4-84).

Solution

As mentioned in Section 4.8.1, under sorption conditions, only the term $\partial C/\partial t$ in Eq. (E4-8-6) of Example 4-8 changes to $R_d\,\partial C/\partial t$. The effects of decay are represented by Eq. (4-83). Incorporating these into Eq. (E4-8-6) of Example 4-8 gives

$$\frac{\partial F_x}{\partial x}+\frac{\partial F_y}{\partial y}+\frac{\partial F_z}{\partial z}=-\varphi_e R_d\frac{\partial C}{\partial t}-\varphi_e \nu R_d C \qquad \text{(E4-12-1)}$$

Substitution of the advective–dispersive solute flux components, as given by Eqs. (4-40), into Eq. (E4-12-1) results

$$\begin{aligned}\varphi_e R_d\frac{\partial C}{\partial t}=&\frac{\partial}{\partial x}\left(\varphi_e D_{xx}\frac{\partial C}{\partial x}+\varphi_e D_{xy}\frac{\partial C}{\partial y}+\varphi_e D_{xz}\frac{\partial C}{\partial z}\right)\\&+\frac{\partial}{\partial y}\left(\varphi_e D_{yx}\frac{\partial C}{\partial x}+\varphi_e D_{yy}\frac{\partial C}{\partial y}+\varphi_e D_{yz}\frac{\partial C}{\partial z}\right)\\&+\frac{\partial}{\partial z}\left(\varphi_e D_{zx}\frac{\partial C}{\partial x}+\varphi_e D_{zy}\frac{\partial C}{\partial y}+\varphi_e D_{zz}\frac{\partial C}{\partial z}\right)\\&-\frac{\partial}{\partial x}(\varphi_e v_x C)-\frac{\partial}{\partial y}(\varphi_e v_y C)-\frac{\partial}{\partial z}(\varphi_e v_z C)-\varphi_e \nu R_d C\end{aligned} \qquad \text{(E4-12-2)}$$

which is exactly the same as Eq. (4-84).

4.8.3.2 Special Case: Homogeneous and Isotropic Dispersion

Like in Section 4.7.2, assuming that the effective porosity φ_e and the velocity components (v_x, v_y, v_z) are not spatially variable quantities, Eq. (4-84) takes the form

$$R_d\frac{\partial C}{\partial t}=D_x\frac{\partial^2 C}{\partial x^2}+D_y\frac{\partial^2 C}{\partial y^2}+D_Z\frac{\partial^2 C}{\partial z^2}-v_x\frac{\partial C}{\partial x}-v_y\frac{\partial C}{\partial y}-v_z\frac{\partial C}{\partial z}-\nu R_d C \qquad \text{(4-85)}$$

which is derived in Example 4-13.

Example 4-13

Derive the solute transport differential equation for the homogeneous and isotropic dispersion case for constant (v_x, v_y, v_z) porous media flow velocity components given by Eq. (4-85).

Solution

For a homogeneous and isotropic porous medium, the advective–dispersive flux components are given by Eqs. (4-44) based on Eq. (4-43). Under these conditions, Eq. (4-84) becomes

$$\varphi_e R_d \frac{\partial C}{\partial t} = \frac{\partial}{\partial x}\left(\varphi_e D_x \frac{\partial C}{\partial x}\right) + \frac{\partial}{\partial y}\left(\varphi_e D_{yy} \frac{\partial C}{\partial y}\right) + \frac{\partial}{\partial z}\left(\varphi_e D_z \frac{\partial C}{\partial z}\right)$$
$$- \frac{\partial}{\partial x}(\varphi_e v_x C) - \frac{\partial}{\partial y}(\varphi_e v_y C) - \frac{\partial}{\partial z}(\varphi_e v_z C) - \varphi_e \nu R_d C \quad \text{(E4-13-1)}$$

Assuming that the effective porosity φ_e and (v_x, v_y, v_z) are not spatially variable quantities, Eq. (E4-13-1) takes the form

$$R_d \frac{\partial C}{\partial t} = D_x \frac{\partial^2 C}{\partial x^2} + D_y \frac{\partial^2 C}{\partial y^2} + D_Z \frac{\partial^2 C}{\partial z^2} - v_x \frac{\partial C}{\partial x} - v_y \frac{\partial C}{\partial y} - v_z \frac{\partial C}{\partial z} - \nu R_d C \quad \text{(E4-13-2)}$$

which is exactly the same as Eq. (4-85).

4.8.3.3 Special Cases: Homogeneous and Isotropic Dispersion with Uniform Flow Field

4.8.3.3.1 Three-Dimensional Case

For this case, the velocity field is given by Eq. (4-51). With these, Eq. (4-85) becomes

$$R_d \frac{\partial C}{\partial t} = D_x \frac{\partial^2 C}{\partial x^2} + D_y \frac{\partial^2 C}{\partial y^2} + D_z \frac{\partial^2 C}{\partial z^2} - U \frac{\partial C}{\partial x} - \nu R_d C \quad \text{(4-86)}$$

4.8.3.3.2 Two-Dimensional Case

For the case of two-dimensional solute transport, i.e., in the x–y plane, Eq. (4-86) becomes

$$R_d \frac{\partial C}{\partial t} = D_x \frac{\partial^2 C}{\partial x^2} + D_y \frac{\partial^2 C}{\partial y^2} - U \frac{\partial C}{\partial x} - \nu R_d C \quad \text{(4-87)}$$

4.8.3.3.3 One-Dimensional Case

For the case of one-dimensional solute transport in the x-coordinate direction, Eq. (4-87) becomes

$$R_d \frac{\partial C}{\partial t} = D_x \frac{\partial^2 C}{\partial x^2} - U \frac{\partial C}{\partial x} - \nu R_d C \quad \text{(4-88)}$$

4.9 Mechanical Dispersion Coefficients for Saturated Porous Media

As mentioned in Section 4.4.4, the mechanical dispersion is the most complex and controversial component for the transport of solutes in porous media. The *mechanical dispersive flux* or *dispersive flux* is described by Eq. (4-33), which is similar to the equation for diffusive flux given by Eq. (4-29). Namely, it has been assumed that the mechanical dispersion process is a *Fickian process*. Based on this assumption, since the 1950s, research has been conducted to determine expressions for the components of the hydrodynamic dispersion tensor, which is given by Eq. (4-41).

In this section, the state-of-the-art knowledge for the mechanical dispersion coefficients are presented. Based on these, the components of the hydrodynamic dispersion tensor given by Eq. (4-41) are specified.

4.9.1 Mechanical Dispersion Equations Based on Laboratory Experiments

4.9.1.1 Mechanical Dispersion Coefficient Equations from One-Dimensional Experiments

For a one-dimensional solute transport case, Eq. (4-37) can be written as

$$D_x = D^* + D_{m_x} \quad (4\text{-}89)$$

where D_x is the longitudinal hydrodynamic dispersion coefficient in the x-coordinate direction, D^* is the effective molecular diffusion coefficient, and D_{m_x} is the mechanical dispersion coefficient.

Based on laboratory experiments, Rumer (1962) determined the values of D_x for one-dimensional solute transport in a uniform porous medium flow field within a certain range of Reynolds numbers. The media used were quartz gravel and glass beds. The quartz gravel and glass beads had mean sizes of 1.65 and 0.39 mm, respectively. The porosity of both materials was 0.39. The seepage velocity U for quartz gravel ranged from 0.04 to 0.40 cm/s, whereas for glass beads U ranged from 0.01 to 0.20 cm/s. In the experiments, NaCl was used as the tracer solution. For relatively coarse materials, D^* is negligible as compared with D_{m_x} in Eq. (4-89). In other words, $D_{m_x} \cong D_x$. Rumer (1962) used Eq. (4-163) in Section 4.13.2 to analyze the results. The general equation relating the mechanical dispersion coefficient to the seepage velocity is given by (Rumer, 1962)

$$D_{m_x} = \alpha_L U^n \quad (4\text{-}90)$$

in which α_L denotes a constant for a given porous medium (expressed in cm when $n = 1$) and n is an exponent which is constant for a given porous medium in place. The terms α_L and n vary for each porous medium and also vary with porosity for a given porous medium. The correlation of n and α_L porous media characteristics is not completely known.

For quartz gravel, Rumer (1962) determined that the values of α_L and n are 0.2 and 1.083, respectively. For glass beads, the values of α_L and n are 0.027 and 1.105, respectively. The factor α_L, which is measured in cm when $n = 1$, can be thought of as an intrinsic dispersivity. The *intrinsic dispersivity* is a parameter expressing the mixing ability or mass transport properties of the media. It is clear that α_L does not completely describe the dispersive properties of a medium when n is different from unity, but n has been found by other investigators to be always equal to unity. With Eq. (4-90), Eq. (4-89) becomes

$$D_x = D^* + \alpha_L U^n \quad (4\text{-}91)$$

4.9.1.2 Mechanical Dispersion Coefficient Equations from Two-Dimensional Experiments

Harleman and Rumer (1963), based on the previous investigations (Ebach and White, 1958; Raimondi et al., 1963; Rumer, 1962), generalized all related experiment-led longitudinal mechanical dispersion equations in the form

$$\frac{D_{m_x}}{\nu} = \psi_1 R^{n_1} \quad (4\text{-}92)$$

in which

$$R = \frac{Ud}{\nu} \quad (4\text{-}93)$$

is the Reynolds number based on the average grain size d as the characteristic length and the seepage velocity U. In Eq. (4-93), ν is the kinematic viscosity; and in Eq. (4-92), ψ_1 and n_1 are parameters characterizing the medium. Harleman and Rumer (1963) showed that ψ_1 is independent of particle size for nearly uniform media and is dependent on particle shape and possible size distribution.

Harleman and Rumer (1963) postulated that Eq. (4-92) can be written for the transverse mechanical dispersion coefficient as

$$\frac{D_{m_y}}{\nu} = \psi_2 R^{n_2} \tag{4-94}$$

From Eqs. (4-92) and (4-94), the ratio of the longitudinal and transverse dispersion coefficients should be of the form

$$\frac{D_{m_x}}{D_{m_y}} = \lambda R^n \tag{4-95}$$

in which

$$\lambda = \frac{\psi_1}{\psi_2} \qquad n = n_1 - n_2 \tag{4-96}$$

In accordance with the above points, λ and n should be functions only of the porous media characteristics.

Harleman and Rumer (1963) used a porous medium composed of Dow Chemical "Palespan 8" plastic spheres, which had a mean size of 0.96 mm and their average hydraulic conductivity was 0.8 cm/s. The parameters in Eqs. (4-92)–(4-96) are also given in Batu (2006, p. 42, Tables 2.7 and 2.8). Harleman and Rumer (1963) noted that the extension of Eq. (4-95) for R values greater than 10 is not justified.

Substitution of Eq. (4-93) into Eq. (4-92),

$$D_{m_x} = \lambda_1 U^{n_1} \tag{4-97}$$

in which

$$\lambda_1 = \psi_1 \nu \left(\frac{d}{\nu} \right)^{n_1} \tag{4-98}$$

Similarly, substitution of Eq. (4-93) into Eq. (4-94),

$$D_{m_y} = \lambda_2 U^{n_2} \tag{4-99}$$

in which

$$\lambda_2 = \psi_2 \nu \left(\frac{d}{\nu} \right)^{n_2} \tag{4-100}$$

In Eqs. (4-97) and (4-99), λ_1, λ_2, n_1, and n_2 are experimentally determined constants for the medium. Since the 1960s, many researchers have conducted similar laboratory experiments on various porous materials and the equations they obtained have similar forms (see, e.g., Guvanasen and Volker, 1983). The results of the experiments of Rumer (1962) and Harleman and Rumer (1963) showed that the exponents of n_1 and n_2 are approximately equal to 1.0 and λ_1 and λ_2 depend on several factors and are denoted by α_L and α_T, respectively. Therefore, Eqs. (4-97) and (4-99), respectively, take the forms

$$D_{m_x} \equiv D_{m_L} \cong \alpha_L U \tag{4-101}$$

$$D_{m_y} \equiv D_{m_T} \cong \alpha_T U \tag{4-102}$$

in which α_L and α_T are called the *longitudinal dispersivity* and *transverse dispersivity*, respectively.

4.9.2 Generalized Mechanical Dispersion Coefficient Equations Based on Theory and Postulation

In Section 4.8.1, equations for mechanical dispersion coefficients based on laboratory experiments are presented. Equations (4-101) and (4-102) are the final mechanical dispersion coefficient equations, which are being used in practice. After the early 1960s, these equations were generalized based on theory and postulation (Bear, 1961; Scheidegger, 1961).

The general form of the mechanical dispersion coefficients equation is [e.g., Bear, 1979, p. 234, Eq. (7-11)]

$$D_{m_{ij}} = \alpha_T v \delta_{ij} + (\alpha_L - \alpha_T)\frac{v_i v_j}{v} \tag{4-103}$$

in which δ_{ij} is the Kronecker delta, namely, $\delta_{ij} = 0$ if $i \neq j$, $\delta_{ij} = 1$ if $i = j$. Like Eq. (4-41), $D_{m_{ij}}$ can be written in matrix form as

$$\left[D_{m_{ij}}\right] = \begin{bmatrix} D_{m_{xx}} & D_{m_{xy}} & D_{m_{xz}} \\ D_{m_{yx}} & D_{m_{yy}} & D_{m_{yz}} \\ D_{m_{zx}} & D_{m_{zy}} & D_{m_{zz}} \end{bmatrix} \tag{4-104}$$

In Cartesian coordinates, the velocity vector is

$$\boldsymbol{v} = v_x \boldsymbol{i} + v_y \boldsymbol{j} + v_z \boldsymbol{k} \tag{4-105}$$

in which v_x, v_y, and v_z are the velocity components in the *x*-, *y*-, and *z*-coordinate directions, respectively; and ***i***, ***j***, ***k*** are the unit vectors in the same coordinate directions, respectively, as well. The magnitude of the $\boldsymbol{v}$ velocity vector is

$$|\boldsymbol{v}| = v = \left(v_x^2 + v_y^2 + v_z^2\right)^{\frac{1}{2}} \tag{4-106}$$

From Eqs. (4-103) and (4-106), the mechanical dispersion coefficient in the *x*-coordinate direction can be expressed as

$$\begin{aligned} D_{m_{xx}} &= \frac{\alpha_T v^2 + (\alpha_L - \alpha_T) v_x^2}{v} \\ &= \frac{\alpha_T \left(v_x^2 + v_y^2 + v_z^2\right) + (\alpha_L - \alpha_T) v_x^2}{v} \\ &= \alpha_L \frac{v_x^2}{v} + \alpha_T \frac{v_y^2}{v} + \alpha_T \frac{v_z^2}{v} \end{aligned} \tag{4-107}$$

Similarly, the other mechanical dispersion coefficient tensor components in Eq. (4-103) can be written as

$$D_{m_{yy}} = \alpha_L \frac{v_y^2}{v} + \alpha_T \frac{v_x^2}{v} + \alpha_T \frac{v_z^2}{v} \tag{4-108}$$

$$D_{m_{zz}} = \alpha_L \frac{v_z^2}{v} + \alpha_T \frac{v_x^2}{v} + \alpha_T \frac{v_y^2}{v} \tag{4-109}$$

$$D_{m_{xy}} = D_{m_{yx}} = (\alpha_L - \alpha_T)\frac{v_x v_y}{v} \tag{4-110}$$

$$D_{m_{xz}} = D_{m_{zx}} = (\alpha_L - \alpha_T)\frac{v_x v_z}{v} \tag{4-111}$$

$$D_{m_{yz}} = D_{m_{zy}} = (\alpha_L - \alpha_T)\frac{v_y v_z}{v} \tag{4-112}$$

Equations (4-107)–(4-112) are exactly the same as Bear [1979, p. 235, Eq. (7-11a)] with the exception of notation.

4.10 Generalized Hydrodynamic Dispersion Coefficient Equations with Three Different Dispersivities: α_L, α_{TH}, and α_{TV}

As will be mentioned later on, based on field experiments, the transverse dispersivities in the y- and z-coordinate directions are not the same. Therefore, Eqs. (4-108)–(4-112) can be generalized by distinguishing the transverse dispersivities in the y- and z-coordinate directions and showing them by α_{TH} and α_{TV}, respectively, as proposed by Burnett and Frind (1987). With these as well as with the effective molecular diffusion coefficient D^*, like Eq. (4-104), the hydrodynamic dispersion coefficient tensor $\left[D_{ij}\right]$ can be written as

$$\left[D_{ij}\right] = \begin{bmatrix} D_{m_{xx}} + D^* & D_{m_{xy}} + D^* & D_{m_{xz}} + D^* \\ D_{m_{yx}} + D^* & D_{m_{yy}} + D^* & D_{m_{yz}} + D^* \\ D_{m_{zx}} + D^* & D_{m_{zy}} + D^* & D_{m_{zz}} + D^* \end{bmatrix} \tag{4-113}$$

Based on this, the generalized forms of Eqs. (4-107)–(4-112) can be written as

$$D_{xx} = D_{m_{xx}} + D^* = \alpha_L \frac{v_x^2}{v} + \alpha_{TH} \frac{v_y^2}{v} + \alpha_{TV} \frac{v_z^2}{v} + D^* \tag{4-114}$$

$$D_{yy} = D_{m_{yy}} + D^* = \alpha_L \frac{v_y^2}{v} + \alpha_{TH} \frac{v_x^2}{v} + \alpha_{TV} \frac{v_z^2}{v} + D^* \tag{4-115}$$

$$D_{zz} = D_{m_{yy}} + D^* = \alpha_L \frac{v_z^2}{v} + \alpha_{TV} \frac{v_x^2}{v} + \alpha_{TV} \frac{v_y^2}{v} + D^* \tag{4-116}$$

$$D_{xy} = D_{yx} = (\alpha_L - \alpha_{TH})\frac{v_x v_y}{v} \tag{4-117}$$

$$D_{xz} = D_{zx} = (\alpha_L - \alpha_{TV})\frac{v_x v_z}{v} \tag{4-118}$$

$$D_{yz} = D_{zy} = (\alpha_L - \alpha_{TV})\frac{v_y v_z}{v} \tag{4-119}$$

Equations (4-114)–(4-119) are exactly the same as Zheng and Wang [1999, p. 9, Eqs. (10a)–(10f)]. Also, the mechanical dispersion coefficients of Eqs. (4-114)–(4-119) are exactly the same as Langevin et al. [2022, p. 11, Eq. (4-24)].

Equations (4-113)–(4-119) are the state-of-the-art hydrodynamic dispersion coefficients, which are being used in many numerical solute transport models. In the above equations, α_L is the *longitudinal dispersivity*, α_{TH} is the *transverse horizontal dispersivity*, and α_{TV} is the *transverse vertical dispersivity*.

4.11 Hydrodynamic Dispersion Coefficient Equations under Uniform Flow Conditions

In the $x - y - z$ Cartesian coordinate system shown in Figure 4-6, consider that the uniform U velocity in the aquifer coincides with the x-coordinate direction. Therefore, $v_y = v_z = 0$. Then, Eq. (4-106) becomes

$$|\mathbf{v}| = v_x \equiv U = \frac{q}{n_e} \tag{4-120}$$

in which q is the Darcy velocity and n_e is the effective porosity. Under these conditions, Eq. (4-114) becomes

$$D_{xx} \equiv D_x = \alpha_L \frac{U^2}{U} + \alpha_{TH}\frac{(0)^2}{U} + \alpha_{TV}\frac{(0)^2}{U} + D^* = \alpha_L U + D^* \tag{4-121}$$

Similarly, Eqs. (4-115) and (4-116) become

$$D_{yy} \equiv D_y = \alpha_L \frac{(0)^2}{U} + \alpha_{TH}\frac{U^2}{U} + \alpha_{TV}\frac{(0)^2}{U} + D^* = \alpha_{TH} U + D^* \tag{4-122}$$

$$D_{zz} \equiv D_z = \alpha_L \frac{(0)^2}{U} + \alpha_{TV}\frac{(0)^2}{U} + \alpha_{TV}\frac{U^2}{U} + D^* = \alpha_{TV} U + D^* \tag{4-123}$$

The rest of the terms all become zero:

$$D_{xy} = D_{yx} = (\alpha_L - \alpha_{TH})\frac{U(0)}{U} = 0 \tag{4-124}$$

$$D_{xz} = D_{zx} = (\alpha_L - \alpha_{TV})\frac{U(0)}{U} = 0 \tag{4-125}$$

$$D_{yz} = D_{zy} = (\alpha_L - \alpha_{TV})\frac{(0)(0)}{U} = 0 \tag{4-126}$$

Therefore, for the uniform velocity case, the hydrodynamic dispersion coefficient tensor given by Eq. (4-41) takes the form

$$[D_{ij}] = \begin{bmatrix} \alpha_L U + D^* & 0 & 0 \\ 0 & \alpha_{TH} U + D^* & 0 \\ 0 & 0 & \alpha_{TV} U + D^* \end{bmatrix} \tag{4-127}$$

Equations (4-121)–(4-123) are the hydrodynamic dispersion coefficients of the three-dimensional differential equation given by Eq. (4-86).

4.12 Estimation Methods of Dispersivities in Aquifers

4.12.1 Introduction

Since the 1960s, significant efforts have been made by various investigators to generate methods for estimating the values of longitudinal dispersivity (α_L), transverse horizontal dispersivity (α_{TH}), and transverse vertical dispersivity (α_{TV}) for the prediction of spatial and temporal variations of contaminant concentrations of solutes in natural porous media, such as aquifers, by analytical and numerical mathematical models. Initially, in the 1950s and 1960s, estimation of these quantities started under laboratory conditions by using analytical solute transport solutions based on the Fickian approach. However, especially after the 1970s, field experiments indicated that laboratory-scale determined dispersivity values are significantly less than the values determined under field conditions for the prediction of contaminant plume migration in aquifers. Among these studies, Pickens and Grisak (1981) recommended a simple "1/10" plume length formula for estimation of the longitudinal dispersivity (α_L), i.e., $\alpha_L$can be assumed to have a value one tenth of the plume length.

As a result of the aforementioned facts, further research has been conducted using mostly stochastic approaches to find out why higher dispersivity values were required to predict the spatial and temporal variations of contaminant solute concentration in natural porous formations, such as aquifers. With the guidelines of these approaches, field-measured dispersivity values were used to develop charts as well as empirical equations for estimating the values of the longitudinal dispersivity (α_L), transverse horizontal dispersivity (α_{TH}), and transverse vertical dispersivity (α_{TV}).

Gelhar et al. (1985) conducted a comprehensive compilation of dispersivity values published in many sources based on field-scale tracer tests and contaminant transport modeling efforts. The literature sources and pertinent data in each study are summarized in Gelhar et al. (1992). These data and information are also included in Batu (2006, pp. 585–598, Table 7-1) with their entirety. Gelhar et al. (1992, pp. 1971–1972) state that it does not seem reasonable that a single universal line (or curve-fitting) can be meaningfully identified by applying standard linear regression to all of the data.

Despite the above-given statements of Gelhar et al. (1992) regarding curve-fitting, the data for the longitudinal dispersivity in Gelhar et al. (1985) were interpreted by Neuman

(1990) who presented formulas for longitudinal dispersivity (α_L). Later on, Xu and Eckstein (1995) also analyzed the longitudinal dispersivity data of Gelhar et al. (1992) based on the weighted least-squares method and presented some empirical formulas for the longitudinal dispersivity (α_L).

During the past decade, the controversial issue regarding the longitudinal dispersivity (α_L) was raised once again by some eminent scientists (Zech et al., 2015) by revisiting mainly the well-known α_L data in Gelhar et al. (1992) and concluded that transport, and particularly the macrodispersivity, is formation specific, and that modeling of transport cannot be relegated to a unique scaling law. Instead, transport requires characterization of aquifer properties, e.g., spatial distribution of hydraulic conductivity, and the use of adequate models. The approach of Zech et al. (2015) was commented by Neuman (2016) who wrote that their theory is equally unsuitable for the analysis of nonstationary transport. In their response, Zech et al. (2016) stated their work reached the conclusion that there is no trend of increase α_L with distance L and that values are scattered and are aquifer-specific.

Further to their work in Zech et al. (2015), in a recent paper, Zech et al. (2023) presented a method for estimation of α_L with some guidelines based on the log hydraulic conductivity variance (σ_f^2) and longitudinal correlation scale λ_f, which is equivalent of I_h of Zech et al. (2015), i.e., $\lambda_f \equiv I_h$.

In the following sections, the aforementioned methods are described with examples.

4.12.2 Longitudinal Dispersivity Estimation Methods Based on Scale

4.12.2.1 Pickens and Grisak Method for Estimation of Longitudinal Dispersivity with Scale

Pickens and Grisak (1981) developed a theoretical basis for the quantification of scale-dependent dispersion in stratified granular media and showed from laboratory column and field tracer tests that longitudinal dispersivity α_L is dependent on the scale of ground water sampling.

The scale-dependent α_L expressions for the three stratification types are summarized in Pickens and Grisak (1981, p. 1204, Table 8), which shows that α_L value is a function of mean travel distance with proportionality constants ranging from 0.041 to 0.256. The authors state that this range of magnitude results mainly from differences in the relative hydraulic conductivity distributions interpreted from monitoring tracer movement in the field tracer tests. The authors obtained a total of nine proportionality constants. Of these values, three of them (0.256, 0.178, and 0.117) were determined from a single-well test and six of them (0.095, 0.041, 0.059, 0.093, 0.051, and 0.043) from a two-well test. The authors also state that this range in magnitude results mainly from differences in the relative hydraulic conductivity distributions interpreted from tracer movement in the field tracer tests.

The arithmetic average of the aforementioned nine proportionality constants is 0.1037, and based on this, the authors offered the following longitudinal dispersivity α_L equation:

$$\alpha_L = 0.1L \tag{4-128}$$

in which L is the mean travel distance or *scale length*. Regarding Eq. (4-128), the authors state that although the scale-dependent α_L expressions indicate that α_L would continuously increase in a linear fashion with mean travel distance, and this behavior is considered unlikely for large travel distances. A comparative exhibit of Eq. (4-128) with other equations is shown in Figure 4-11.

4.12.2.2 Neuman Method for Estimation of Longitudinal Dispersivity with Scale

In his paper, Neuman (1990) stated that a fractal model of log hydraulic conductivities is consistent with the scale effect exhibited by field- and laboratory-determined dispersivities since the 1970s. Neuman (1990) also stated that he interpreted the data for 134 longitudinal dispersivity (α_L) values without including the values obtained by calibrating numerical models compiled by Gelhar et al. (1985). Based on different interpretations, Neuman (1990) gives a total of four α_L equations [Neuman, 1990, Eqs. (22), (26), (29), and (32)] as a function of scale length L and stated that the semivariogram equation [Neuman, 1990, Eq. (29)] corresponds to

$$\alpha_L = 0.017L^{1.5} \tag{4-129}$$

which is recommended to be used by Neuman (1990) in practice (Figure 4-9).

Equation (4-129) is not empirical, and it has been derived theoretically in Neuman (1990) based on values corresponding to L less than 3,500 m. Equation (4-129) is valid for all L, but not to predict α_L values used in numerical models. It is only to predict single α_L values derived analytically on the assumption that hydraulic conductivity/transmissivity and α_L are uniform across the experimental plume. Such models are statistically unconditional (Shlomo P. Neuman, March 3, 2025, personal communication).

A comparative exhibit of Eq. (4-129) with other α_L equations is shown in Figure 4-11.

4.12.2.3 Xu and Eckstein Method for Estimation of Longitudinal Dispersivity with Scale

Xu and Eckstein (1995) generated regression equations from the longitudinal dispersivity α_L data included in Gelhar et al. (1992). The authors stated that they improved the correlation of the Gelhar et al. (1992) data by making use of all α_L values including the ones with low reliability.

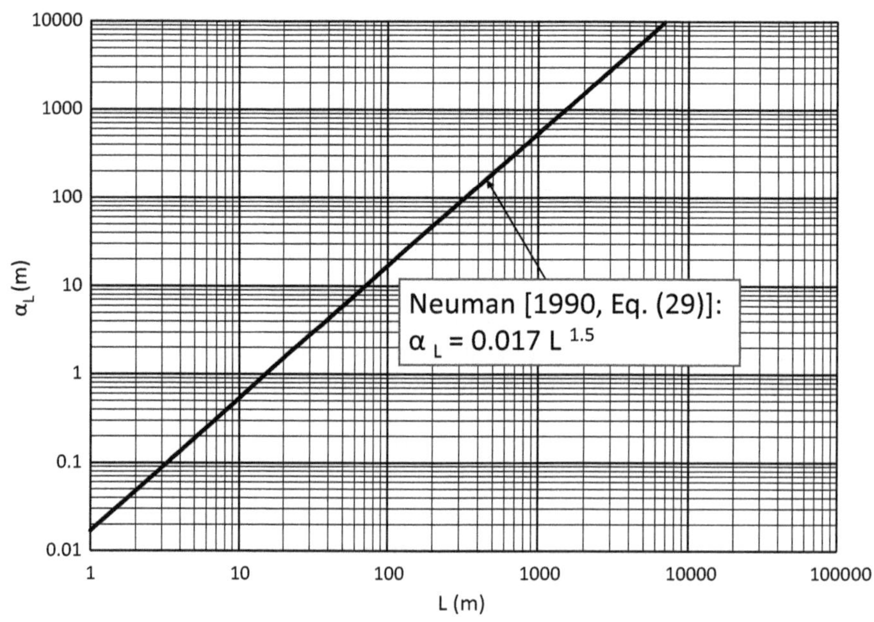

FIGURE 4-9
Longitudinal dispersivity α_L versus scale length L based on Eq. (4-129), which is Eq. (29) in Neuman (1990).

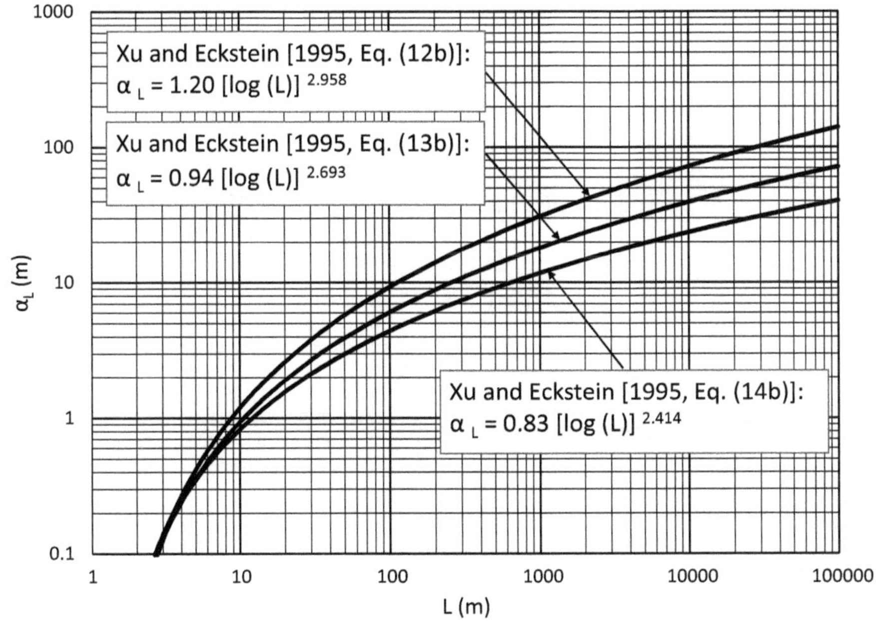

FIGURE 4-10
Longitudinal dispersivity α_L versus scale length L. based on Eqs. (4-130), (4-131), and (4-132), which are Eqs. (12b), (13b), and (14b) in Xu and Eckstein (1995), respectively.

For the purpose of comparison, Xu and Eckstein (1995) conducted three regressions with different weighting schemes: 1:1:1; 1:1.5:2; and 1:2:3. In these weighting schemes, the first number represents the weight value of the low reliability α_L points, the second number represents the weight value of the intermediate points, and the third number represents the weight value of the high points. The first scheme is equivalent to a regular regression line.

Based on the α_L values of Gelhar et al. (1992) aligning into a distinctly linear distribution on a $\log(\alpha_L)$ versus log[log (L)] diagram (Xu and Eckstein, 1995, p. 907, Figure 2), the authors generated the following empirical equations:

$$\alpha_L = 1.20\left[\log(L)\right]^{2.958} \quad \text{for } 1:1:1 \text{ scheme} \tag{4-130}$$

$$\alpha_L = 0.94\left[\log(L)\right]^{2.693} \quad \text{for } 1:1.5:2\text{s cheme} \tag{4-131}$$

$$\alpha_L = 0.83\left[\log(L)\right]^{2.414} \quad \text{for } 1:1:3 \text{ scheme} \tag{4-132}$$

The graphical forms of these equations are shown in Figure 4-10. Also, a comparative exhibit of these equations with other equations is shown in Figure 4-11.

4.12.2.4 Comparison of the Longitudinal Dispersivity Equations of Pickens and Grisak (1981), Neuman (1990), and Xu and Eckstein (1995)

Comparison of the aforementioned longitudinal dispersivity α_L equations of Pickens and Grisak (1981), Neuman (1990), and Xu and Eckstein (1995) presented in Sections 4.11.2.1–4.11.2.3, respectively, is given in Figure 4-11. The following characteristics can be noted from Figure 4-11:

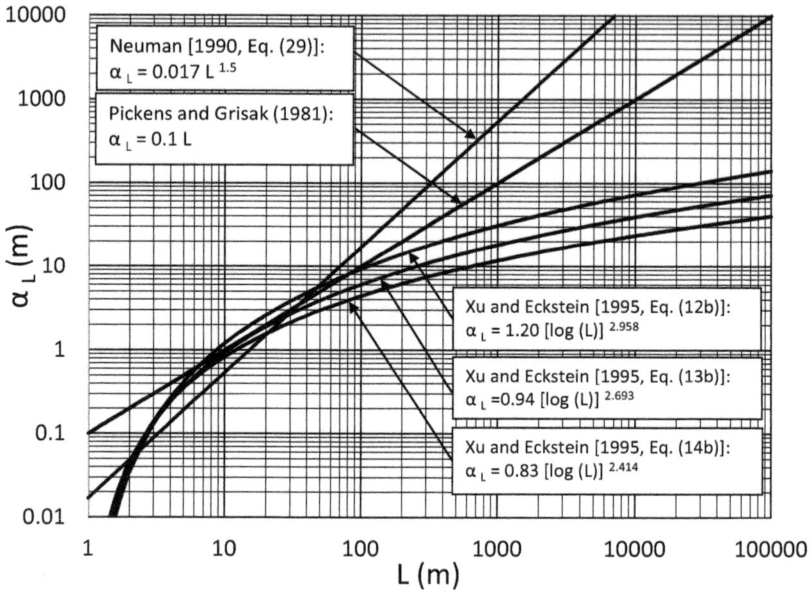

FIGURE 4-11
Longitudinal dispersivity α_L versus scale length L based on Pickens and Grisak (1981), Neuman (1990), and Xu and Eckstein (1995) equations.

1. For $L < 100$ m, the values of all α_L equations are relatively closer as compared with $L > 100$ m cases. But for $L > 100$ m, the differences become very significant depending on the scale length L. Some examples are given in the following items.
2. For $L = 100$ m, $\alpha_L = 10$ m [Pickens and Grisak (1981), Eq. (4-128)]; $\alpha_L = 17$ m [Neuman (1990), Eq. (4-129)]; and $\alpha_L = 10$ m, 6 m, and 4 m [Xu and Eckstein (1995), Eqs. (4-130), (4-131), and (4-132), respectively].
3. For $L = 1{,}000$ m, $\alpha_L = 100$ m [Pickens and Grisak (1981), Eq. (4-128)]; $\alpha_L = 538$ m [Neuman (1990), Eq. (4-129)]; and $\alpha_L = 31$ m, 18 m, and 12 m [Xu and Eckstein (1995), Eqs. (4-130), (4-131), and (4-132), respectively].
4. As mentioned in Section 4.12.2.2, Eq. (4-129) of Neuman (1990) is not an empirical equation; it has been derived theoretically in Neuman (1990) based on empirical values corresponding to L less than 3,500 m.
5. Equation (4-129) of Neuman (1990) is valid for all L, but not to predict α_L values used in numerical models, only to predict single α_L values derived analytically on the assumption that hydraulic conductivity/transmissivity and α_L are uniform across the experimental plume.

4.12.3 Longitudinal Dispersivity Estimation Method of Zech, Attinger, Bellin, Cvetkovic, Dagan, Dietrich, Fiori, and Teutsch [Zech et al. (2023) Method] Based on the Log Hydraulic Conductivity Variance (σ_f^2) and Longitudinal Correlation Scale (λ_f)

4.12.3.1 Introduction

As mentioned in Section 4.12.1, Zech et al. (2023) presented a method for the estimation of longitudinal dispersivity α_L with some guidelines based on the log hydraulic conductivity variance (σ_f^2) and longitudinal correlation scale (λ_f). This method is called the Zech et al. (2023) *Method*, which has eight authors: Alraune Zech, Sabine Attinger, Alberto Bellin, Vladimir Cvetkovic, Gedeon Dagan, Peter Dietrich, Aldo Fiori, and Georg Teutsch. As the heading of this section indicates, the method is based on log hydraulic conductivity variance (σ_f^2) and longitudinal correlation scale (λ_f), which are some well-established quantities of the stochastic subsurface hydrology. To simplify the presentation of the Zech et al. (2023) *Method*, first, some preliminaries regarding the stochastic subsurface hydrology (e.g., Freeze, 1975; Dagan, 1989; Gelhar, 1993; Rubin, 2003; Batu, 2006) are presented. Then, the method is described with some practical examples.

4.12.3.2 Some Preliminaries Regarding Stochastic Subsurface Hydrology

Most subsurface materials exhibit complex heterogeneity behavior, which significantly affects the spatial and temporal variations of solutes in these porous formations. Spatial variations in hydraulic conductivity in aquifer materials have been shown to have significant effects on the magnitude of dispersivities.

Generally, in aquifers, hydraulic conductivity follows a log-normal density function (Freeze, 1975). Consider a one-dimensional case with log hydraulic conductivities $f_1, f_2, \ldots, f_n$ at $x_1, x_2, \ldots, x_n$ distances from the origin given by

$$f_1\ (x_1) = \ln\left[K_1(x_1)\right], f_2\ (x_2) = \ln\left[K_2(x_2)\right], \ldots, f_n\ (x_n) = \ln\left[K_n(x_n)\right] \tag{4-133}$$

or in combined form

$$f_i\ (x_i) = \ln\left[K_i(x_i)\right], \quad i = 1, 2, \ldots, n \tag{4-134}$$

in which $K_i(x_i)$ is the hydraulic conductivity stochastic process and $f_i\ (x_i)$ is the log hydraulic conductivity, which is assumed to be statistically homogeneous (Figure 4-12). The parameter $f_i\ (x_i)$ is normally distributed. In the context of statistical properties, *statistically homogeneous* means that the properties are the same regardless of location, while *statistically isotropic* means that the properties are the same in all directions. Given the measurements of log hydraulic conductivity, the mean $\overline{f}$ is defined as

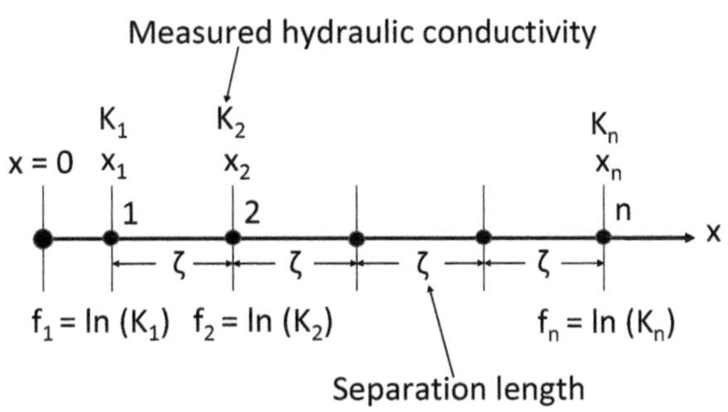

FIGURE 4-12
One-dimensional sequence of log hydraulic conductivity values $f_1, f_2, \ldots, f_n$ at $x_1, x_2, \ldots, x_n$ coordinates, respectively. [Adapted from Freeze et al. 1990, p. 430, Figure 13 (a).]

$$\bar{f} = \frac{f_1 + f_2 + \cdots + f_n}{n} = \frac{1}{n}\sum_{i=1}^{n} f_i \tag{4-135}$$

By showing

$$f_i' = f_i - \bar{f} \tag{4-136}$$

The *variance* $\sigma_{f'}^2$ is defined as [e.g., Spiegel, 1975, p. 160, Eq. (15)]

$$\sigma_f^2 = \frac{\left(f_1 - \bar{f}\right)^2 + \left(f_2 - \bar{f}\right)^2 + \cdots + \left(f_n - \bar{f}\right)^2}{n} = \frac{1}{n}\sum_{i=1}^{n}\left(f_i - \bar{f}\right)^2 \tag{4-137}$$

The variance σ_f^2 is one measure of the heterogeneity of an aquifer. Now, the expression of a common *covariance function* is given.

The covariance function is defined as [e.g., Freeze et al., 1990, p. 761; Domenico and Schwartz, 1998, p. 229, Eq. (10.25)]

$$R_f = \frac{1}{n}\sum_{i=1}^{n}\left(f_i - \bar{f}\right)\left(f_{i+1} - \bar{f}\right) \tag{4-138}$$

If the measurement f_i is made at position x_i and the measurement f_{i+k} at position x_{i+k}, then the *separation length* is defined by

$$\zeta = \left|x_{i+k} - x_i\right| \tag{4-139}$$

in which k is the lag. The covariance function may take a number of forms. One of the most commonly used functions is the three-dimensional statistically anisotropic exponential covariance function for the separation vector

$$\boldsymbol{\zeta} = \zeta_1\,\boldsymbol{i} + \zeta_2\,\boldsymbol{j} + \zeta_1\,\boldsymbol{k} \tag{4-140}$$

in which $\boldsymbol{i}$, $\boldsymbol{j}$, and $\boldsymbol{k}$ are the unit vectors in the x, y, and z coordinate directions, respectively. Its form is [e.g., Gelhar and Axness, 1990, p. 162, Eq. (4)]

$$R_f(\boldsymbol{\zeta}) = \sigma_f^2 \exp\left[-\left(\frac{\zeta_1^2}{\lambda_{1_f}} + \frac{\zeta_2^2}{\lambda_{2_f}} + \frac{\zeta_3^2}{\lambda_{3_f}}\right)^{\frac{1}{2}}\right] \tag{4-141}$$

in which λ_{1_f}, λ_{2_f}, and λ_{3_f} are the *correlation scales* (the distances to e^{-1} correlation, $e = 2.718282$) in the principal x, y, and z coordinate directions, respectively. Some authors call them the *correlation lengths* (e.g., Freeze et al., 1990, p. 753).

For the one-dimensional case along the x coordinate direction ($\zeta_1 \equiv \zeta, \zeta_2 = \zeta_3 = 0, \lambda_{1_f} \equiv \lambda_f$), Eq. (4-141) takes the form

$$R_f(\zeta) = \sigma_f^2 \exp\left(-\frac{|\zeta|}{\lambda_f}\right) \tag{4-142}$$

which is exactly the same as Gelhar [1993, p. 29, Eq. (2.2.6)] with the exception of notation. From Eq. (4-142), the *autocorrelation function* ρ_f is defined as [e.g., Freeze et al., 1990, p. 753, Eq. (13), with the exception of notation]

$$\rho_f = \frac{R_f(\zeta)}{\sigma_f^2} = \exp\left(-\frac{|\zeta|}{\lambda_f}\right) \tag{4-143}$$

in which λ_f is also called the *exponential decay parameter* by Freeze et al. (1990). It is a measure of the distance over which the f value is correlated; specifically, it is the distance over which ρ_f decays to a value of e^{-1} (Freeze et al., 1990, p. 753) in which $e = 2.718282$.

The geometric mean hydraulic conductivity K_g can be determined from Eq. (4-135) as

$$\bar{f} = \ln(K_g) \tag{4-144}$$

And solving it for K_g gives

$$K_g = e^{\bar{f}} \tag{4-145}$$

in which $e = 2.718282$.

Since $f_i - \bar{f} = \ln(K_i) - \ln(K_g) = \ln(K_i / K_g)$, both the variance σ_f^2 and standard deviation σ_f are dimensionless quantities.

Example 4-14

Measured hydraulic conductivity (K_i) values are given in Table 4-1 along the x-coordinate direction. The total length portion of the aquifer is 280.0 m and the measured K_i value locations are at every $\zeta = 20.0$ m separation length along the x-coordinate direction (Figure 4-12). In the y-coordinate direction, it is assumed that the same measured values are valid. Using the values in Table 4-1, determine (a) the arithmetic average $\bar{f}$ of the $\ln(K_i)$ values; (b) the variance (σ_f^2) and standard deviation (σ_f) of the hydraulic conductivity (K_i) data; (c) the covariance R_f; (d) the autocovariance ρ_f; (e) the λ_f correlation scale (or correlation length); and (f) write down the ρ_f autocovariance function and plot it as a function of ζ separation length.

TABLE 4-1

Calculation Details of Variance for the K_i Values of Example 4-14

i	x_i (m)	K_i (cm/s)	$f_i = \ln(K_i)$	$(f_i - \bar{f})$	$(f_i - \bar{f})^2$
1	20	0.000912	−6.999871	−0.748934	0.560902
2	40	0.002140	−6.146949	0.103987	0.010813
3	60	0.000875	−7.041287	−0.790350	0.624653
4	80	0.000221	−8.417348	−2.166411	4.693338
5	100	0.000492	−7.617032	−1.366095	1.866216
6	120	0.000503	−7.594920	−1.343984	1.806293
7	140	0.007060	−4.953310	1.297626	1.683834
8	160	0.003740	−5.588670	0.662267	0.438597
9	180	0.011600	−4.456750	1.794186	3.219105
10	200	0.023800	−3.738070	2.512867	6.314500
11	220	0.000767	−7.173024	−0.922087	0.850245
12	240	0.000627	−7.374564	−1.123627	1.262539
13	260	0.008850	−4.727338	1.523599	2.321353
14	280	0.003400	−5.683980	0.566957	0.321440
Sum			−87.513112		25.973828

Solution

a. The calculation details for variance (σ_f^2) are given in Table 4-1. The sum of values in the fourth column of Table 4-1 is

$$\sum_{i=1}^{14} f_i = -87.513$$

From Eq. (4-135), the arithmetic average of the f_i values is

$$\bar{f} = \frac{-87.513}{14} = -6.251$$

b. The sum of the last column of Table 4-1 is

$$\sum_{n=1}^{14} \left(f_i - \bar{f}\right)^2 = 25.9738$$

From Eq. (4-137), the variance is

$$\sigma_f^2 = \frac{25.9738}{14} = 1.86$$

The standard deviation is

$$\sigma_f = 1.36$$

c. The sum of the last column of Table 4-2 is

$$\sum_{i=1}^{13} \left(f_i - \bar{f}\right)\left(f_{i+1} - \bar{f}\right) = 9.030717$$

TABLE 4-2

Calculation Details of Covariance for the K_i Values of Example 4-14

i	x_i (m)	K_i (cm/s)	$f_i = \ln(K_i)$	$(f_i - \bar{f})$	$(f_{i+1} - \bar{f})$	$(f_i - \bar{f})$ $(f_{i+1} - \bar{f})$
1	20	0.000912	−6.999871	−0.748934	0.103987	−0.077879
2	40	0.002140	−6.146949	0.103987	−0.790350	−0.082186
3	60	0.000875	−7.041287	−0.790350	−2.166411	1.712223
4	80	0.000221	−8.417348	−2.166411	−1.366095	2.959524
5	100	0.000492	−7.617032	−1.366095	−1.343984	1.836010
6	120	0.000503	−7.594920	−1.343984	1.297626	−1.743989
7	140	0.007060	−4.953310	1.297626	0.662267	0.859375
8	160	0.003740	−5.588670	0.662267	1.794186	1.188230
9	180	0.011600	−4.456750	1.794186	2.512867	4.508552
10	200	0.023800	−3.738070	2.512867	−0.922087	−2.317082
11	220	0.000767	−7.173024	−0.922087	−1.123627	1.036082
12	240	0.000627	−7.374564	−1.123627	1.523599	−1.711957
13	260	0.008850	−4.727338	1.523599	0.566957	0.863815
14	280	0.003400	−5.683980	0.566957		
Sum						9.030717

From Eq. (4-138), the covariance is

$$R_f = \frac{9.030717}{13} = 0.695$$

d. From Eq. (4-143), the autocovariance is

$$\rho_f = \frac{R_f}{\sigma_f^2} = \frac{0.695}{1.86} = 0.374$$

e. From Eq. (4-143),

$$\ln(\rho_f) = -\frac{|\zeta|}{\lambda_f}$$

Solving for λ_f,

$$\lambda_f = -\frac{|\zeta|}{\ln(\rho_f)}$$

With $|\zeta| = 20.0$ m and $\rho_f = 0.374$,

$$\lambda_f = -\frac{20.0 \text{ m}}{\ln(0.374)} = 20.336 \text{ m}$$

f. With $\lambda_f = 20.336$ m correlation length, Eq. (4-143) becomes

$$\rho_f = \exp\left(-\frac{|\zeta|}{20.336 \text{ m}}\right)$$

And its graph is shown in Figure 4-13.

4.12.3.3 The Zech et al. (2023) Method for Estimation of Asymptotic Longitudinal Dispersivity ($\alpha_{L_{asymp}}$)

As mentioned in Section 4.12.3.1, Zech et al. (2023) presented a method for estimation of α_L with some guidelines based on the log hydraulic conductivity variance (σ_f^2) and longitudinal correlation scale (λ_f). This method is called the *Zech et al. (2023) method,* for which the following eight authors contributed: Alraune Zech, Sabine Attinger, Alberto Bellin, Vladimir Cvetkovic, Gedeon Dagan, Peter Dietrich, Aldo Fiori, and Georg Teutsch. The method is based on log hydraulic conductivity variance (σ_f^2) and longitudinal correlation scale (λ_f), which are described in Section 4.12.3.2. In the following sections, first, the equations of asymptotic value of the longitudinal dispersivity ($\alpha_{L_{asymp}}$) based on stochastic approaches are presented. Then, the Zech et al. (2023) method is presented. After that, comparative evaluations are made for the longitudinal dispersivity (α_L) estimation approaches between Neuman, Xu and Eckstein, and Zech et al. (2023) methods.

4.12.3.3.1 The Asymptotic Value of the Longitudinal Dispersivity ($\alpha_{L_{asymp}}$): Dagan Equation

For a large time ($t \rightarrow \infty$), Dagan (1988, p. 1495) determined the asymptotic longitudinal dispersion coefficient D_{xx} expression as

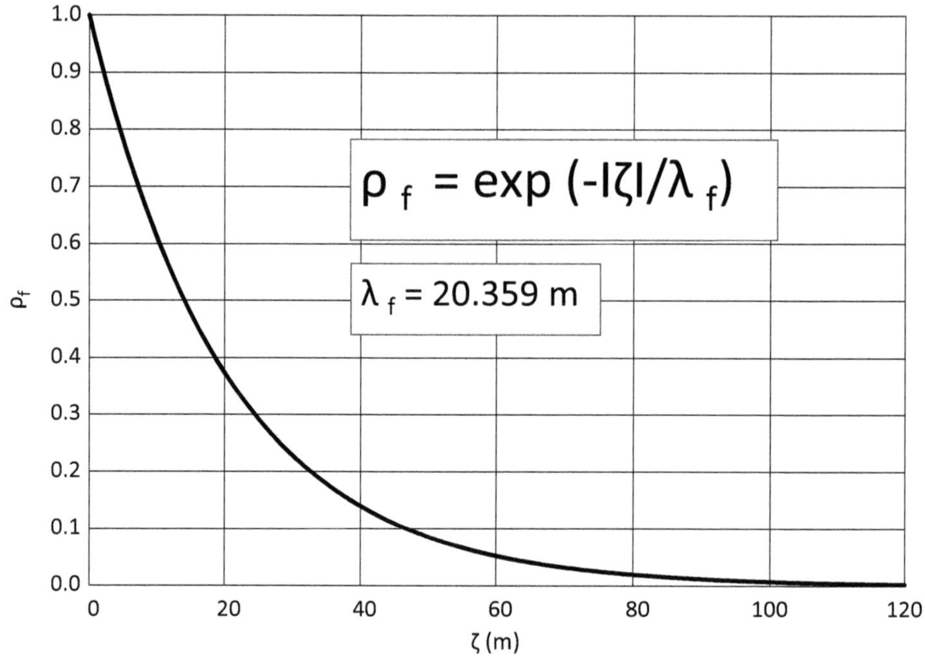

FIGURE 4-13
Autocorrelation function ρ_f versus separation length ζ for Example 4-14.

$$D_{xx_{asymp}} = \sigma_f^2 U \lambda_{f-h} \tag{4-146}$$

in which σ_f^2 is the variance, U is the ground water velocity along the x-coordinate direction, and λ_{f-h} is the longitudinal correlation scale along the x-coordinate direction as well. For σ_f^2 and λ_{f-h}, Dagan (1988) used the symbols σ_Y^2 and I_h, respectively.

In Section 4.10, D_{xx} is given by Eq. (4-121) as

$$D_{xx_{asymp}} \equiv D_{x_{asymp}} = \alpha_{L_{asymp}} U + D^* \tag{4-147}$$

Using Eq. (4-147) in Eq. (4-146) with the assumption that the effective molecular diffusion coefficient is negligible ($D^* = 0$), one gets

$$\alpha_{L_{asymp}} = \sigma_f^2 \lambda_{f-h} \tag{4-148}$$

which is exactly the same as Zech et al. (2023, p. 347 and p. 350) with the exception of notation. Eq. (4-148) is the *Dagan asymptotic longitudinal dispersivity equation.*

Zech et al. (2023, p. 347) state that the asymptotic $\alpha_{L_{asymp}}$ is typically reached after the plume has traveled a few correlation scales λ_{f-h}, which is typically up to 10 m depending on the particular site.

4.12.3.3.2 *The Asymptotic Value of the Longitudinal Dispersivity ($\alpha_{L_{asymp}}$): Gelhar and Axness Equation*

Based on the three-dimensional stochastic analysis of macrodispersion in aquifers, Gelhar and Axness [1983, p. 164, Eq. (33)] determined an expression for asymptotic longitudinal dispersivity as

$$\alpha_{L_{asymp}} = \frac{\sigma_f^2 \lambda_{f-h}}{\gamma^2} \tag{4-149}$$

in which (Gutjahr et al., 1978)

$$\gamma = 1 + \frac{\sigma_f^2}{6} \tag{4-150}$$

The Gelhar and Axness (1983) method is based on the solution of the stochastic differential equation for the solute concentration perturbation using Fourier–Stieltjes representations for the perturbed quantities.

Regarding Eq. (4-149), Dagan (1988, p. 1495) states that the asymptotic value [Eq. (4-148)] $D_{x_{asymp}} \to \sigma_f^2 U \lambda_{f-h}$, which is independent of natural logarithm base e, is the same as the one derived by Gelhar and Axness (1983), besides their factor γ, taken for consistency equal to unity. This means that the second term in Eq. (4-150) is neglected.

According to Zech et al. (2023, p. 350), weak-level heterogeneity, medium-level heterogeneity, and high-level heterogeneity are characterized by $\sigma_f^2 < 1$, $1 < \sigma_f^2 < 2$, and $\sigma_f^2 > 2$, respectively. For $\sigma_f^2 = 1$ and $\sigma_f^2 = 2$, from Eq. (4-150), $\gamma = 1.17$, and 1.33, respectively. This means that the relative errors of γ are 17% and 33% for $\sigma_f^2 = 1$ and $\sigma_f^2 = 2$, respectively.

Under the aforementioned conditions, Eq. (4-149) of Gelhar and Axness (1983) is written as

$$\alpha_{L_{asymp}} \cong \sigma_f^2 \lambda_{f-h} \tag{4-151}$$

4.12.3.3.3 The Asymptotic Value of the Longitudinal Dispersivity ($\alpha_{L_{asymp}}$): Neuman, Winter, and Newman Equation

Dagan (1988, p. 1495) also mentions the Neuman et al. (1987) paper as having another $\alpha_{L_{asymp}}$ expression. The only related expression of the authors is [Neuman et al., 1987, p. 457, Eq. (26) and p. 459, Eq. (64)]

$$\alpha_{L_{asymp}} = \frac{3}{8} \sigma_f^2 \lambda_{f-h} \tag{4-152}$$

which is significantly different from Dagan, and Gelhar and Axness equations, which are given by Eqs. (4-148) and (4-151), respectively.

4.12.3.3.4 Key Conclusions of Zech et al. (2023) for the Asymptotic Longitudinal Asymptotic Dispersivity ($\alpha_{L_{asymp}}$) Estimation

In Zech et al. (2023, p. 348), the authors first summarized the main findings of Zech et al. (2015) by starting from the compiled α_L data as a function of the scale length L by Gelhar et al. (1992). After reviewing the α_L data, the authors added field α_L data accumulated between 1992 and 2015 and removed the low reliability of α_L data. The main conclusions regarding α_L reached by Zech et al. (2023, p. 348) are as follows:

1. There is no justification for the assumed general scaling of α_L with L. It rather leads to inadvertently large predicted values of solute spreading.
2. For each aquifer, α_L is site specific being a function of parameters quantifying aquifer heterogeneity rather than the travel distance (Zech et al., 2015, p. 7671, Figure 4).

3. The local spatial evolution of α_L with solute travel distance shows a preasymptotic increase followed by stabilization at a constant value. The asymptotic α_L is typically reached after the plume traveled a few integral scale λ_f (Zech et al., 2023, p. 355, Figure A1). The magnitude of λ_f depends on the aquifer-specific level of heterogeneity, as quantified, for instance, by log hydraulic conductivity variance σ_f^2. This is in line with the theoretical predictions [e.g., outlined in Appendix A of Zech et al. (2023, pp. 354–356)] as well as shown for a few field cases where data were available (Zech et al., 2015, p. 7672, Figure 5).

Based on these conclusions and the analyses of the previous sections, the guidelines of the Zech et al. (2023) method for estimation of asymptotic longitudinal dispersivity ($\alpha_{L_{asymp}}$) are presented in the following section.

4.12.3.3.5 *Guidelines of the Zech et al. (2023) Method for Estimation of Asymptotic Longitudinal Dispersivity ($\alpha_{L_{asymp}}$)*

Regarding the guidelines of the Zech et al. (2023) method for estimation of longitudinal dispersivity based on Eq. (4-151), Zech et al. (2023, p. 350) make the following statements (with notation adaptation):

> We suggested using the first-order relationship $\alpha_L = \sigma_f^2 \lambda_{f-h}$ as a reasonable approximation of longitudinal macrodispersivity. However, the field characterization data needed to estimate the values of σ_f^2 and λ_{f-h} are generally scarce, especially for the preliminary transport prediction which is often of interest. Sometimes it is possible to estimate σ_f^2 from samples extracted along one or more wells. However, the correlation length λ_{f-h} is more difficult to estimate as it requires the availability of a few wells at different distances. Still, the value of σ_f^2 is indicative of the level of aquifer heterogeneity and may help to attribute it to one of the three groups of Table 1. In a rough division weak, medium and high heterogeneity are characterized by $\sigma_f^2 < 1$, $1 < \sigma_f^2 < 2$, $\sigma_f^2 > 2$, respectively.

The table mentioned in the quoted statements above is Table 4-3 for which the source data for the α_L values are given in Zech et al. (2023, pp. 357–358, Tables B1 and B2). The authors state that their tables include compiled α_L data till 2015 in addition of the α_L data in Gelhar et al. (1992).

In Table 4-3, Zech et al. (2023, p. 349) defined the heterogeneity levels as follows: (1) *Weak Heterogeneity*: Aquifer materials are sandy with some minor fraction of silt/clay and/or gravel; *(2) Medium Heterogeneity*: Aquifer materials ranging from gravel to sand with some silt/clay; and (3) *High Heterogeneity*: Aquifers having with various materials, from gravel to silt/clay, in similar proportions.

As can be seen from Table 4-3, the mean α_L increases with the level of heterogeneity, as expected.

Regarding the use of the guidelines of the Zech et al. (2023) method given in the quoted statements above, Zech et al. (2023, p. 351) make the following statements (with notation adaptation):

> In many common circumstances, neither estimates of σ_f^2 nor λ_{f-h} are available for the specific aquifer. However, the type of material (sand, gravel, silt/clay) and the proportions could be assessed. In such a case, we recommend the selection of the group based on the division of Section 'Amalgamation of Longitudinal Macrodispersivity Field Data' and adopting the appropriate range of values of Table 4-3. While this is an approximate procedure, it is surmised that it is more rational than the other aforementioned ones.

TABLE 4-3

Weighted Average Longitudinal Dispersivity α_L for Different Levels of Heterogeneities Based on Reported Field Data

Heterogeneity Level	Number of Sites	$\alpha_{L\text{-}mean}$ (M)	$\alpha_{L\text{-}SD}$ (M)
1—Weak heterogeneity	13	1.1	1.1
2—Medium heterogeneity	10	3.2	1.5
3—High heterogeneity	7	7.5	2.9

Source: After Zech et al. (2023, p. 350, Table 1).
$\alpha_{L\text{-}mean}$ (M) : Mean value of the longitudinal dispersivity.
$\alpha_{L\text{-}SD}$ (M) : Standard deviation of the longitudinal dispersivity.
1—Weak heterogeneity: Aquifer materials are sandy with some minor fraction of silt/clay and or gravel.
2—Medium heterogeneity: Aquifer materials ranging from gravel to sand with some silt/clay.
3—High heterogeneity: Aquifers having various materials, from gravel to silt/clay, in similar proportions.

4.12.4 Comparative Evaluation of the Methods of Estimation for the Longitudinal Dispersivity (α_L) Based on Pickens and Grisak (1981), Neuman (1990), Xu and Eckstein (1995), and Zech et al. (2023) for Some Selected Aquifers α_L Data

As mentioned in Section 4.12.2, the main feature of the methods of Pickens and Grisak (1981), Neuman (1990), and Xu and Eckstein (1995) is that the longitudinal dispersivity α_L depends on the scale length L.

On the other hand, as included in Section 4.12.3.3.4, one of the key conclusions of Zech et al. (2023, p. 348) is

> There is no justification for the assumed general scaling of α_L with L. It rather leads to in advertently large predicted values of solute spreading." The other key conclusion of Zech et al. (2023, p. 348) is: "The local spatial evolution of α_L with solute travel distance shows a preasymptotic increase followed by stabilization at a constant value. The asymptotic α_L is typically reached after the plume traveled a few integral scale λ_f (Zech et al., 2023, p. 355, Figure A1).

The purpose of this section is to make a comparative evaluation for the aforementioned methods of Pickens and Grisak (1981), Neuman (1990), and Xu and Eckstein (1995) for estimation of longitudinal dispersivity α_L depending on the scale length L along with the above-given quoted statement of Zech et al. (2023, p. 348) by using the corresponding data for the selected aquifers (Zech et al., 2015, p. 7672, Figure 5) which were provided to the author of this book by the first author of the Zech et al. (2023) method (Alraune Zech, March 3, 2025, personal communication).

The longitudinal dispersivity α_L versus scale length L data are given in Table 4-4 (Chalk River, Canada), Table 4-5 (Bordon, Canada), Table 4-6 (Cape Cod, Massachusetts, USA), Table 4-7 (Krauthausen, Germany), Table 4-8 (Corbas, France), and Table 4-9 (Horkheimer Insel, Germany). Their plotted version is shown in Figure 4-14, which corresponds to Zech et al. (2015, p. 7672, Figure 5).

TABLE 4-4

Longitudinal Dispersivity (α_L) Versus Scale (L) for the Chalk River (Canada) Site ($K_g = 0.2, \sigma_f = 0.23$)

L (m)	α_L (m)
266.0	0.55
116.2	0.51
101.6	0.23
94.8	0.34
93.4	0.21
80.0	0.34
71.5	0.19
25.8	0.19
5.1	0.24

Source: Zech et al. (2015, p. 7672, Figure 5 and personal communication, Alraune Zech, March 10, 2025).

TABLE 4-5

Longitudinal Dispersivity (α_L) versus Scale (L) for the Bordon (Canada) Site ($K_g = 0.05, \sigma_f = 0.24$)

L (m)	α_L (m)
90.0	0.5
59.0	0.42
42.0	0.33
39.0	0.27
35.0	0.27
24.0	0.33
8.0	0.23
6.0	0.24
4.0	0.27

Source: Zech et al. (2015, p. 7672, Figure 5 and personal communication, Alraune Zech, March 10, 2025).

TABLE 4-6

Longitudinal Dispersivity (α_L) versus Scale (L) for the Cape Cod (Massachusetts, USA) Site ($K_g = 1.3, \sigma_f = 0.24$)

L (m)	α_L (m)
212.0	0.96
198.8	0.94
182.6	0.92
164.2	0.99
149.0	0.76
135.3	0.89
115.3	0.85
101.3	0.93
90.0	0.9
78.7	0.85
65.6	0.9

(Continued)

TABLE 4-6 (*Continued*)

Longitudinal Dispersivity (α_L) versus Scale (L) for the Cape Cod (Massachusetts, USA) Site ($K_g = 1.3, \sigma_f = 0.24$)

L (m)	α_L (m)
51.7	0.83
39.4	0.66
26.1	0.67
17.1	0.59
7.4	0.45

Source: Zech et al. (2015, p. 7672, Figure 5 and personal communication, Alraune Zech, March 10, 2025).

TABLE 4-7

Longitudinal Dispersivity (α_L) versus Scale (L) for the Krauthausen (Germany) Site ($K_g = 1.4, \sigma_f = 1.08$)

L (m)	α_L (m)
239.0	2.67
202.0	2.71
170.0	3.64
131.0	1.63
99.0	1.72
68.0	3.25
56.0	3.36
45.0	2.91
29.0	2.86
15.0	2.97

Source: Zech et al. (2015, p. 7672, Figure 5 and personal communication, Alraune Zech, March 10, 2025).

TABLE 4-8

Longitudinal Dispersivity (α_L) versus Scale (L) for the Corbas (France) Site

L (m)	α_L (m)
150	10.5
50	4.6
25	2.4

Source: Zech et al. (2015, p. 7672, Figure 5 and personal communication, Alraune Zech, March 10, 2025).

In Figures 4-15–4-20, the six α_L longitudinal dispersivity datasets are compared individually with the methods of Pickens and Grisak (1981), Neuman (1990), and Xu and Eckstein (1995).

Regarding Figure 4-14, Zech et al. (2015, p. 7672) make the following statements (with notation adaptation):

> For all experiments in Figure 4-14, we observe a general trend of increasing dispersivity with distance. The studies of high reliability, in which a spatial moment analysis was performed, exhibit an asymptotic behavior. The sandy aquifers of low heterogeneity like

TABLE 4-9

Longitudinal Dispersivity (α_L) versus Scale (L) for the Horkheimer Insel (Germany) Site ($K_g = 14.0, \sigma_f = 2.34$)

L (m)	α_L (m)
56.4	6.0
44.5	7.09
41.0	5.92
34.1	4.0
32.7	5.0
17.1	2.5
12.2	2.0
9.9	1.02
8.9	1.38

Source: Zech et al. (2015, p. 7672, Figure 5 and personal communication, Alraune Zech, March 10, 2025).

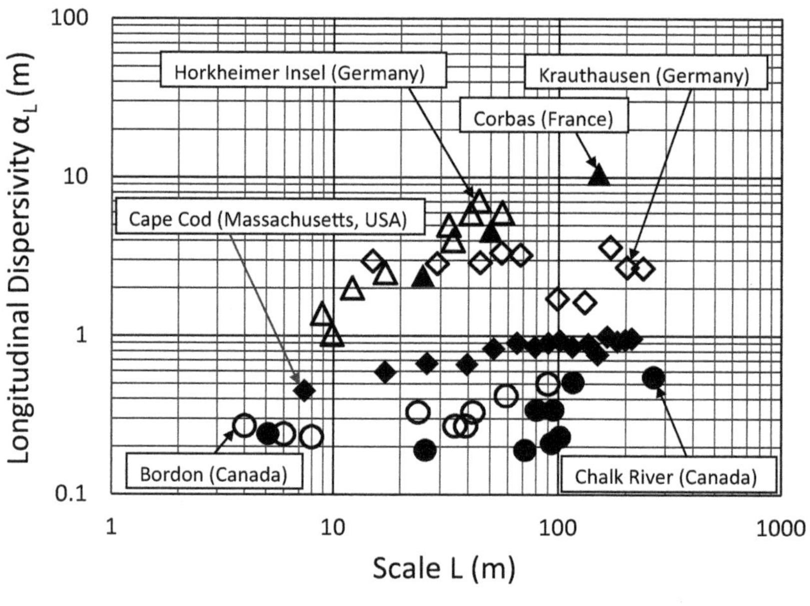

FIGURE 4-14
Longitudinal dispersivity α_L versus scale length L selected for a total of six aquifers. Adapted from Zech et al. (2015, p. 7672, Figure 5).

> at sites in Borden [Freyberg, 1986] ($\sigma_f^2 = 0.24$, $l = 2.7$ m), Chalk River [Moltyaner et al., 1993; Dagan and Neuman, 1997] ($\sigma_f^2 = 0.23$, $l = 1.5$ m) and Cape Cod [Garabedian et al., 1991] ($\sigma_f^2 = 0.24$, $l = 2.6$ m), exhibit an asymptotic value in the range of 1 m. The value at the more heterogeneous Krauthausen site [Vereecken et al., 2000] ($\sigma_f^2 = 1.08$, $l = 6.7$ m) is in the order of 4 m. For the highly heterogeneous sites of Horkheimer Island [Ptak and Teutsch, 1994] ($\sigma_f^2 = 2.34$, $l = 6 - 10$ m) and Corbas [Sauty, 1977], the travel distance of the tracer was apparently too short to arrive at constant asymptotic behavior.

In the above-given quoted statements, Zech et al. (2015) specifically state that the longitudinal dispersivity α_L versus scale length L data for Chalk River (Figure 4-15), Bordon (Figure 4-16), and Cape Cod (4-17) exhibit asymptotic behavior. However, by

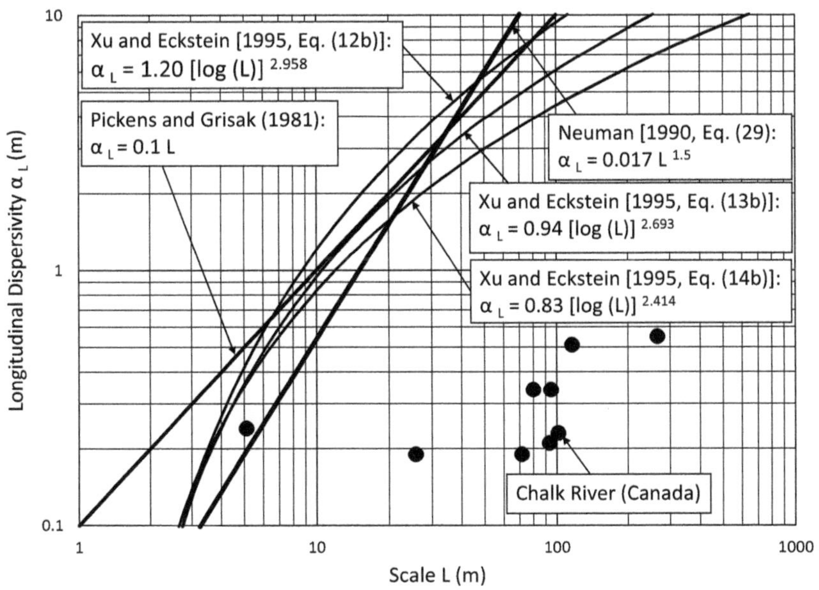

FIGURE 4-15
Comparison of the Chalk River (Canada) longitudinal dispersivity α_L values with the methods of Pickens and Grisak (1981), Neuman (1990), and Xu and Eckstein (1995).

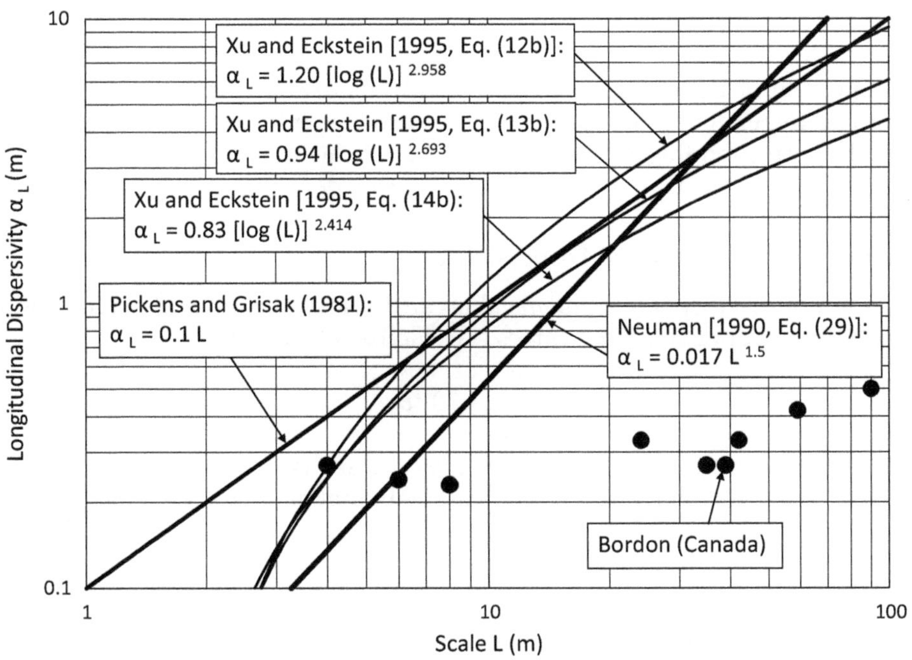

FIGURE 4-16
Comparison of the Bordon (Canada) longitudinal dispersivity α_L values with the methods of Pickens and Grisak (1981), Neuman (1990), and Xu and Eckstein (1995).

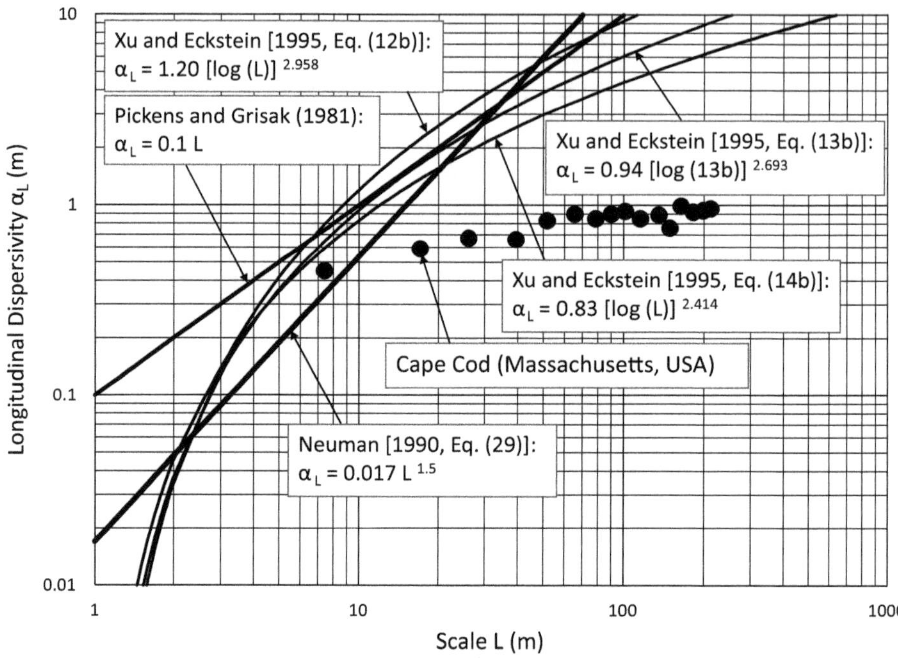

FIGURE 4-17
Comparison of the Cape Cod (Massachusetts, USA) longitudinal dispersivity α_L values with the methods of Pickens and Grisak (1981), Neuman (1990), and Xu and Eckstein (1995).

careful inspection of Figures 4-15–4-17, especially Figures 4-15 and 4-16, there is a clear scale length L dependency on the longitudinal dispersivity α_L rather than asymptotic behavior. The corresponding data for Figure 4-18 indicate that as L increases α_L decreases. Figures 4-19 and 4-20 clearly exhibit the scale length L dependency on the α_L values.

4.12.5 Estimation of Transverse Dispersivities

Transverse horizontal dispersivity (α_{TH}) values are typically an order of magnitude smaller than longitudinal dispersivity (α_L) values (e.g., Gelhar et al., 1992):

$$\alpha_{TH} \approx 10^{-1}\alpha_L \tag{4-153}$$

Transverse vertical dispersivity (α_{TV}) values are typically two orders of magnitude smaller than longitudinal dispersivity (α_L) values (e.g., Gelhar et al., 1992):

$$\alpha_{TV} \approx 10^{-2}\alpha_L \tag{4-154}$$

Zech et al. (2019) made a critical analysis of transverse dispersivity data and concluded that their work casts doubt on the commonly used rule of thumb without providing an alternative. They also state that there is no simple or inexpensive way of measuring transverse dispersion coefficients (α_{TH} and α_{TV}). The authors finally state that the status of transverse dispersion coefficients is a challenging task requiring further investigation of both theoretical and experimental natures.

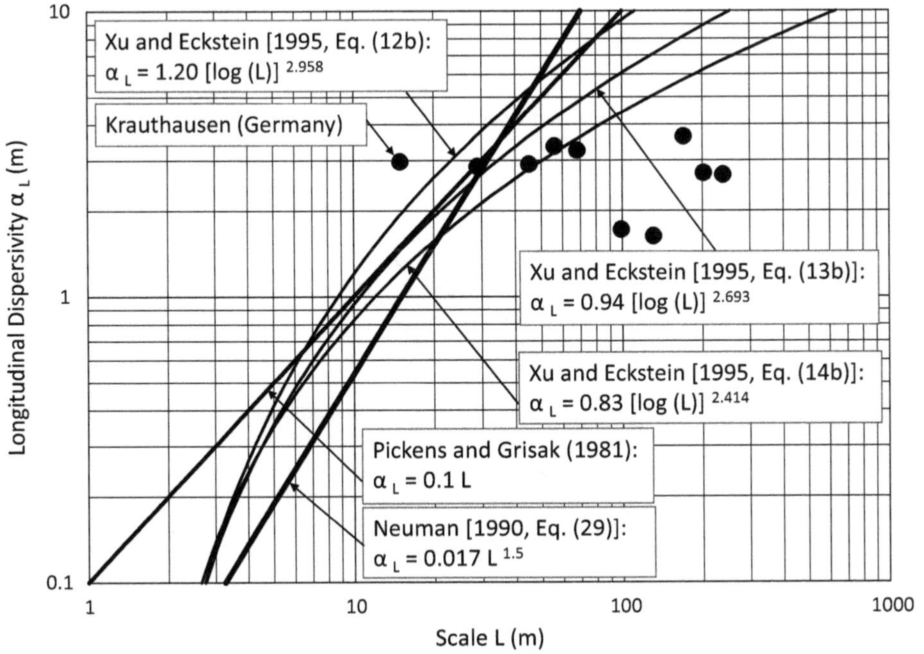

FIGURE 4-18
Comparison of the Krauthausen (Germany) longitudinal dispersivity α_L values with the methods of Pickens and Grisak (1981), Neuman (1990), and Xu and Eckstein (1995).

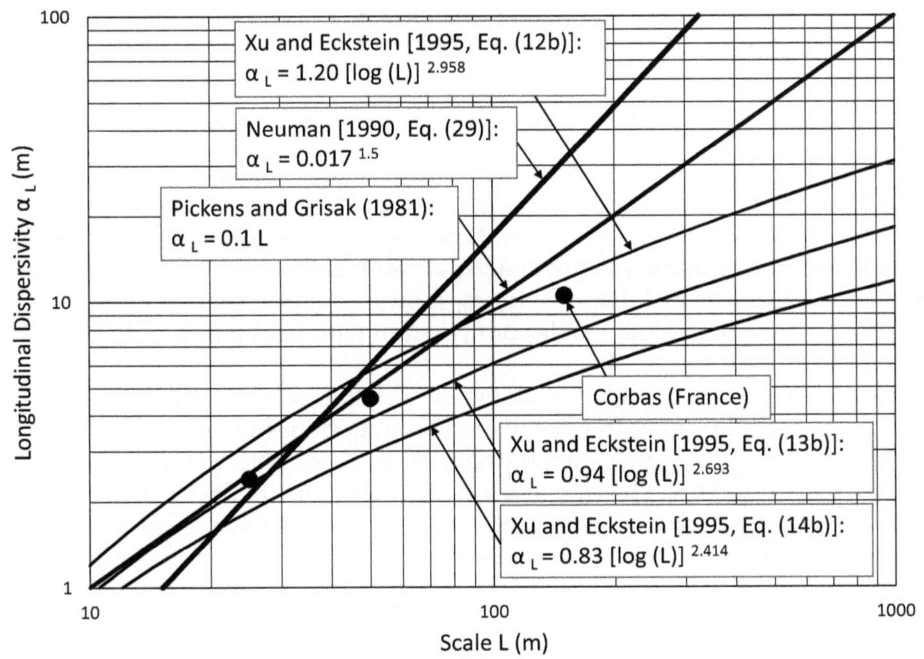

FIGURE 4-19
Comparison of the Corbas (France) longitudinal dispersivity α_L values with the methods of Pickens and Grisak (1981), Neuman (1990), and Xu and Eckstein (1995).

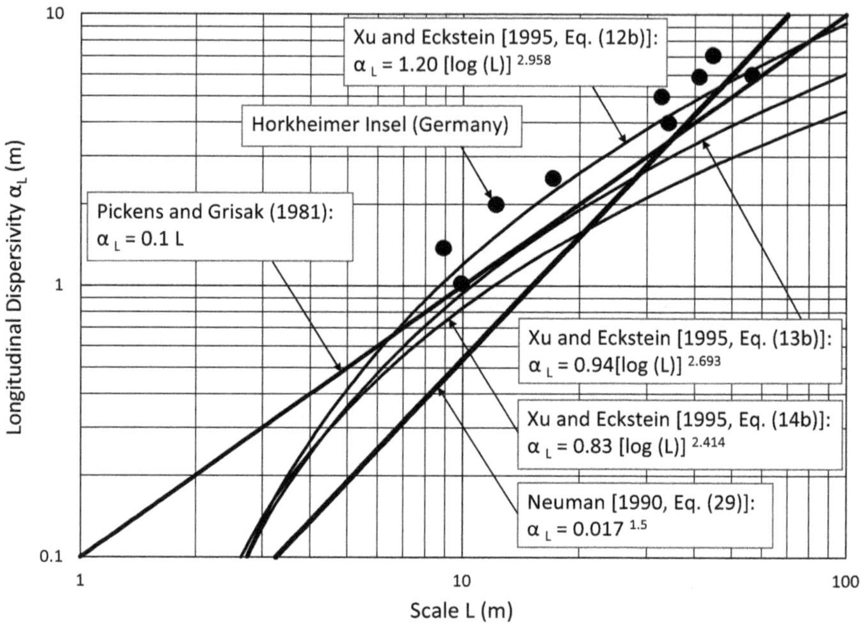

FIGURE 4-20
Comparison of the Horkheimer Insel (Germany) longitudinal dispersivity α_L values with the methods of Pickens and Grisak (1981), Neuman (1990), and Xu and Eckstein (1995).

Based on the above points, regarding α_{TH} and α_{TV} estimates, to the knowledge of the author of this book, Eqs. (4-153) and (4-154) may be used as transverse horizontal dispersivity and transverse vertical dispersivity estimates if no field-measured values are available.

4.12.6 Concluding Remarks for Longitudinal Dispersivity Estimation

Despite all theoretical efforts based on deterministic and stochastic approaches as well as field measurements during the past half-century, estimation of dispersivities for large field plumes is still a controversial issue and no concrete rules for use by practitioners have been established yet. Although many analyses based on stochastic differential equations indicate that the longitudinal dispersivity α_L shows asymptotic behavior with respect to scale length L, it appears that no viable confirmation with field-measured values has been established yet.

According to Gelhar et al. (1992, p. 1968, Figure 2), based on field tests and model calibration, with high reliability, longitudinal dispersivity α_L values range from 0.5 to 4 m. With intermediate reliability, α_L values range from 0.01 to 90 m. With low reliability, α_L values range from 0.03 to 6,000 m. The α_L values with respect to the level of heterogeneity in Table 4-3 proposed by Zech et al. (2023) are far below these values. In addition, Zech et al. (2023) suggest using the first-order relationship $\alpha_L = \sigma_f^2 \lambda_{f-h}$ as a reasonable approximation of longitudinal macrodispersivity, but the correlation length λ_{f-h} is more difficult to estimate as it requires the availability of a few wells at different distances.

In conclusion, it can be said that the aforementioned methods should be considered some guidelines along with professional judgment in estimating longitudinal dispersivity α_L, transverse horizontal dispersivity α_{TH}, and transverse vertical dispersivity α_{TV} values for real-world projects in practice.

4.13 Some One-Dimensional Analytical Solutions for Solute Transport in Porous Media

4.13.1 Introduction

In this section, some one-dimensional analytical solutions for solute transport under uniform flow conditions are presented. These solutions can be used to determine spatial and temporal solute concentration variations in support of two- and three-dimensional numerical solute transport modeling. Also these solutions can be used to determine the values of the retardation factor and degradation rates under field conditions. These solutions are based on both the first-type (Dirichlet) and third-type (Cauchy) or flux-type solute source conditions.

4.13.2 One-Dimensional First-Type Continuous Source Solutions

4.13.2.1 Transient Solution

4.13.2.1.1 Governing Differential Equation

The governing differential equation is given by Eq. (4-88) as

$$R_d \frac{\partial C}{\partial t} = D_x \frac{\partial^2 C}{\partial x^2} - U \frac{\partial C}{\partial x} - vR_d C \tag{4-155}$$

4.13.2.1.2 Initial and Boundary Conditions

The schematic of the problem is shown in Figure 4-21. The initial condition for $C(x,t)$ is

$$C(x,0) = 0 \quad x \geq 0 \tag{4-156}$$

The boundary conditions for $C(x,t)$ are

$$C(x,t) = C_M e^{\gamma t} \quad t \geq 0 \tag{4-157}$$

$$C(\infty,t) = 0 \quad t \geq 0 \tag{4-158}$$

$$\frac{\partial C(\infty,t)}{\partial x} = 0 \quad t \geq 0 \tag{4-159}$$

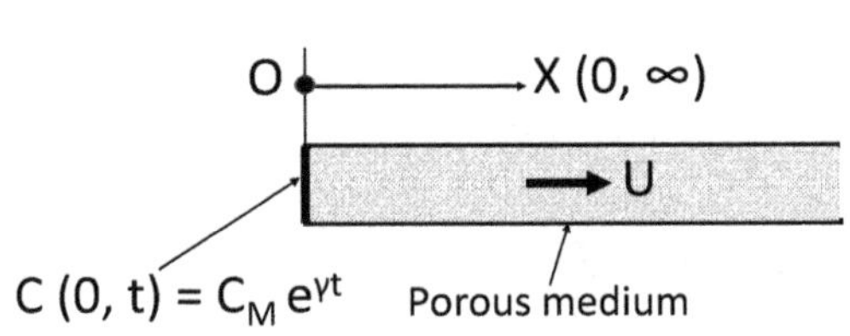

FIGURE 4-21
The schematic of a one-dimensional first-type continuous source solute transport model in a uniform porous medium flow field.

In Eq. (4-160), $\Upsilon\left(T^{-1}\right)$, and C_M is the source concentration when $\Upsilon = 0$, and $e(= 2.718281828\ldots)$ is the *Euler number.*

Equation (4-156) states that initially the concentration is zero everywhere in the transport domain. Eq (4-157) states that the source concentration is an exponentially varying time-dependent quantity. Eq (4-158) and (4-159) state that both the concentration C and its gradient $\partial C/\partial x$ with respect to x approach zero when the x coordinate tends to infinity.

4.13.2.1.3 Solution

The solution of Eq. (4-155) with the initial and boundary conditions given by Eqs. (4-156)–(4-159) is [Batu, 2006, p. 53, Eq. (3-12)]

$$C(x,t) = \frac{1}{2} C_M e^{\Upsilon t} \left\{ \exp\left[\frac{x\,(U-P)}{2D_x}\right] \operatorname{erfc}\left[\frac{R_d x - Pt}{2(D_x R_d t)^{\frac{1}{2}}}\right] + \exp\left[\frac{x\,(U+P)}{2D_x}\right] \operatorname{erfc}\left[\frac{R_d x + Pt}{2(D_x R_d t)^{\frac{1}{2}}}\right] \right\} \tag{4-160}$$

in which

$$P = \left[U^2 + 4D_x R_d (\nu + \Upsilon)\right]^{\frac{1}{2}} \tag{4-161}$$

$\operatorname{erfc}(u)$ is the complementary error function and is defined as

$$\operatorname{erfc}(u) = 1 - \frac{2}{\pi^{\frac{1}{2}}} \int_u^{\infty} u^{-\xi^2} d\xi \tag{4-162}$$

Equation (4-160) was derived as a special case of the general two-dimensional solution in Batu (1989). Full details of the general solution are also given in the book of Batu (2006, pp. 66–93) with verification against several numerical models.

Equation (4-160) is the same as the equation derived by Marino [1974, p. 1016, Eq. (35)] for $R_d = 1$ with the exception of notation and different variables.

4.13.2.1.3.1 Special Case for $\Upsilon = 0$ and $\nu = 0$ If $\Upsilon = 0$ and $\nu = 0$, Eq. (4-161) becomes $P = U$. Therefore, Eq. (4-160) becomes

$$C(x,t) = \frac{1}{2} C_M \left\{ \operatorname{erfc}\left[\frac{R_d x - Ut}{2(D_x R_d t)^{\frac{1}{2}}}\right] + \exp\left(\frac{Ux}{D_x}\right) \operatorname{erfc}\left[\frac{R_d x + Ut}{2(D_x R_d t)^{\frac{1}{2}}}\right] \right\} \tag{4-163}$$

It appears that Eq. (4-163) was derived by several researchers independently. According to Marino (1974), Eq. (4-163) was derived for $R_d = 1$ by Glueckauf et al. (1949). Bear (1972) states that Eq. (4-163) was first derived by Gröbner and Hofreiter (1949, 1950). On the other hand, according to van Genuchten and Parker (1984), Eq. (4-163) was first derived by Lapidus and Amundson (1952). Ogata and Banks (1961) present another derivation of Eq. (4-163) for

$R_d = 1$. With $R_d = 1$, Eq. (4-163) becomes equivalent to the equation included in Freeze and Cherry [1979, p. 391, Eq. (9.5)].

4.13.2.2 Steady-State Solution

4.13.2.2.1 Governing Differential Equation

Under steady-state solute transport conditions, $\partial C/\partial t$ is zero. Therefore, from Eq. (4-155), the governing differential equation becomes

$$D_x \frac{\partial^2 C}{\partial x^2} - U \frac{\partial C}{\partial x} - v R_d C = 0 \tag{4-164}$$

which is a one-dimensional steady-state solute transport differential equation.

4.13.2.2.2 Initial and Boundary Conditions

The schematic of the problem is shown in Figure 4-21. The boundary condition of $C(x)$ at the source is

$$C(0) = C_M \tag{4-165}$$

The boundary conditions for $C(x)$ are

$$C(\infty) = 0 \tag{4-166}$$

$$\frac{\partial C(\infty)}{\partial x} = 0 \tag{4-167}$$

Equation (4-165) states that the concentration at the source is C_M. Equations (4-166) and (4-167) state that both the concentration C and its gradient $\partial C / \partial x$ with respect to x approach zero when the x coordinate tends to infinity.

4.13.2.2.3 Solution

The solution of Eq. (4-164) with the boundary conditions is

$$C(x) = C_M \exp\left\{ \frac{Ux}{2D_x} \left[1 - \left(\frac{4D_x\, R_d v}{U^2} + 1 \right)^{\frac{1}{2}} \right] \right\} \tag{4-168}$$

which is exactly the same as Bear [1979, p. 269, Eq. (7-138)] with the exception of notation. If $v = 0$ in Eq. (4-168), it can easily be seen that $C(x) = C_M$, which means that under no degradation condition, the concentration becomes the same at every point on the x coordinate axis.

Equation (4-168) can also be written as

$$C(x) = C_M \exp\left[\frac{Ux}{2D_x} - x\left(\frac{vR_d}{D_x} + \frac{U^2}{4D_x^2} \right)^{\frac{1}{2}} \right] \tag{4-169}$$

which is exactly the same as the equation in Batu [2006, p. 59, Eq. (3-38)].

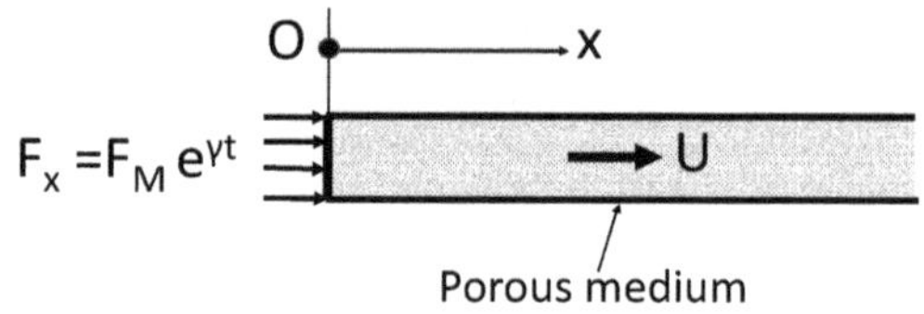

FIGURE 4-22
The schematic of a one-dimensional third-type continuous source solute transport model in a uniform porous medium flow field.

4.13.3 One-Dimensional Third-Type Continuous Source Solutions

4.13.3.1 Transient Solution

4.13.3.1.1 Governing Differential Equation

The governing differential equation is given by Eq. (4-88) as

$$R_d \frac{\partial C}{\partial t} = D_x \frac{\partial^2 C}{\partial x^2} - U \frac{\partial C}{\partial x} - \nu R_d C \tag{4-170}$$

4.13.3.1.2 Initial and Boundary Conditions

The schematic of the problem is shown in Figure 4-22. The initial condition for $C(x,t)$ is

$$C(x,0) = 0 \quad x \geq 0 \tag{4-171}$$

The boundary conditions are

$$F_x(0,t) = F_M e^{\Upsilon t} \quad t \geq 0 \tag{4-172}$$

$$C(\infty,t) = 0 \quad t \geq 0 \tag{4-173}$$

$$\frac{\partial C(\infty,t)}{\partial x} = 0 \quad t \geq 0 \tag{4-174}$$

In Eq. (4-172), $F_x(0,t)$ is the solute input flux at $x = 0$ (solute mass per unit area per unit time), $\Upsilon\left(T^{-1}\right)$ is a parameter, F_M is the solute flux when $\Upsilon = 0$, and $e(= 2.718281828\ldots)$ is the *Euler number.*

Equation (4-171) states that initially the concentration is zero everywhere in the transport domain. Eq (4-172) states that the solute source flux, which is time-dependent, has a specified value at $x = 0$. Therefore, it is the *third-type* or *flux-type source condition.* It is also called *Cauchy boundary condition.* Eqs (4-173) and (4-174) state that both the concentration C and its gradient $\partial C / \partial x$ with respect to x approach zero when the x coordinate tends to infinity.

The general form of the third-type source condition is given by Eq. (4-52a) and for the exponentially varying source flux case it takes the form

$$F_M e^{\Upsilon t} = \varphi_e U C - \varphi_e D_x \frac{\partial C}{\partial x} \tag{4-175}$$

4.13.3.1.3 Solution

The solution of Eq. (4-170) with the initial and boundary conditions given by Eqs. (4-171)–(4-174) is [Batu, 2006, p. 124, Eq. (3-230)]

$$C(x,t) = \frac{F_M}{\varphi_e}\frac{1}{D_x}\int_0^t F_1(\xi)\left[F_2(\xi) - F_3(\xi)F_4(\xi)\right]\exp\left[\gamma(t-\xi)\right]d\xi \tag{4-176}$$

in which

$$F_1(\xi) = \exp\left(\frac{Ux}{2D_x} - \nu\xi - \frac{U^2\xi}{4D_xR_d}\right) \tag{4-177}$$

$$F_2(\xi) = \left(\frac{D_x}{\pi R_d\xi}\right)^{\frac{1}{2}}\exp\left(-\frac{R_dx^2}{4D_x\xi}\right) \tag{4-178}$$

$$F_3(\xi) = \frac{U}{2R_d}\exp\left(\frac{Ux}{2D_x} + \frac{U^2\xi}{4R_dD_x}\right) \tag{4-179}$$

$$F_4(\xi) = \operatorname{erfc}\left[\frac{R_dx}{2(D_xR_d\xi)^{\frac{1}{2}}} + \frac{U}{2}\left(\frac{\xi}{R_dD_x}\right)^{\frac{1}{2}}\right] \tag{4-180}$$

4.13.3.1.3.1 Special Case for $\Upsilon = 0$ and $\nu = 0$ If $\gamma = 0$ and $F_M = \varphi_e UC_M$, the concentration $C(x,t)$ becomes volume-averaged (resident) concentration and is denoted by $C_r(x,t)$ which is defined in Section 4.3.2.

For $\gamma = 0$ and $\nu = 0$, with $C_r(x,t) \equiv C(x,t)$, Eq. (4-176) takes the form [Batu, 2006, p. 124, Eq. (3-237)]

$$\begin{aligned} C(x,t) = \frac{UC_M}{D_x}\int_0^t \exp\left(\frac{Ux}{2D_x} - \frac{U^2\xi}{4D_xR_d}\right)&\left\{\left(\frac{D_x}{\pi R_d\xi}\right)^{\frac{1}{2}}\exp\left(-\frac{R_dx^2}{4D_x\xi}\right)\right. \\ &\left. - \frac{U}{2R_d}\exp\left(\frac{Ux}{2D_x} + \frac{U^2\xi}{4R_dD_x}\right)\operatorname{erfc}\left[\frac{R_dx}{2(D_xR_d\xi)^{\frac{1}{2}}} + \frac{U}{2}\left(\frac{\xi}{R_dD_x}\right)^{\frac{1}{2}}\right]\right\}d\xi \end{aligned} \tag{4-181}$$

4.13.3.1.3.2 Solution of Lindstrom et al. (1967) for $\Upsilon = 0$ and $\nu = 0$ With *$\gamma = 0$ and $\nu = 0$*, Eq. (4-88) becomes

$$R_d\frac{\partial C}{\partial t} = D_x\frac{\partial^2 C}{\partial x^2} - U\frac{\partial C}{\partial x} \tag{4-182}$$

which is the differential equation Lindstrom et al. (1967) used. They used the initial and boundary conditions given by Eqs. (4-171), (4-173), and (4-174) with the following source condition instead of Eq. (4-175)

$$UC - D_x\frac{\partial C}{\partial x} = \frac{F_M}{\varphi_e} = UC_M \tag{4-183}$$

in which $F_M = \varphi_e UC_M$.

The boundary-value problem defined above was solved by Lindstrom et al. (1967), and the following equation was obtained [e.g., van Genuchten and Parker, 1984, p. 704, Eq. (5) with R_d]:

$$C(x,t)=\frac{1}{2}C_M\,\mathrm{erfc}\left[\frac{R_dx-Ut}{2(D_xR_dt)^{\frac{1}{2}}}\right]+\left(\frac{U^2t}{\pi R_dD_x}\right)^{\frac{1}{2}}\exp\left[-\frac{(R_dx-Ut)^2}{4D_xR_dt}\right]$$
$$-\frac{1}{2}C_M\left(1+\frac{Ux}{D_x}+\frac{U^2t}{D_xR_d}\right)\exp\left(\frac{Ux}{D_x}\right)\mathrm{erfc}\left[\frac{R_dx+Ut}{2(D_xR_dt)^{\frac{1}{2}}}\right] \quad (4\text{-}184)$$

which is also given in Batu [2006, p. 125, Eq. (3-239)] with $C_r(x,t)\equiv C(x,t)$. Although Eqs. (4-181) and (4-184) look different, they generate exactly the same numerical results, which are included in Batu [2006, p. 127, Table 3-5).

4.13.3.2 Steady-State Solution

4.13.3.2.1 Governing Differential Equation

Under steady-state solute transport conditions, $\partial C/\partial t$ is zero. Therefore, from Eq. (4-88), the governing differential equation becomes

$$D_x\frac{\partial^2C}{\partial x^2}-U\frac{\partial C}{\partial x}-\nu R_dC=0 \quad (4\text{-}185)$$

which is a one-dimensional steady-state solute transport differential equation.

4.13.3.2.2 Initial and Boundary Conditions

The schematic of the problem is shown in Figure 4-22. From Eqs. (4-52a) and (4-175), with $\Upsilon=0$, the boundary condition of $F_x\ (0)$ becomes

$$F_x\ (0)=F_M=\varphi_eUC-\varphi_eD_x\frac{\partial C}{\partial x} \quad (4\text{-}186)$$

The other boundary conditions for $C(x)$ at infinity are

$$C(\infty)=0 \quad (4\text{-}187)$$

$$\frac{\partial C(\infty)}{\partial x}=0 \quad (4\text{-}188)$$

Equation (4-186) states that the solute flux at the source is F_M. Equations (4-187) and (4-188) state that both the concentration C and its gradient $\partial C/\partial x$ with respect to x approach zero when the x coordinate tends to infinity.

4.13.3.2.3 Solution

The solution of Eq. (4-185) with the boundary conditions given by Eqs. (4-186)–(4-188) is [Batu, 2006, p. 126, Eq. (3-247)]

$$C(x)=\frac{F_M}{\varphi_e\left(\frac{U}{2}+D_xp_0\right)}\exp\left(\frac{Ux}{2D_x}-p_0x\right) \quad (4\text{-}189)$$

$$p_0 = \left(\frac{R_d}{D_x} v + \frac{U^2}{4D_x^2} \right)^{\frac{1}{2}} \tag{4-190}$$

Equation (4-189) is given as a special case of the general two- and three-dimensional analytical solutions [Batu, 2006; p. 143, Eq. (3-294); p. 167, Eq. (3-353)]. Eq. (4-189) is also given as a special case of the same general two-dimensional solute transport solution (Batu, 1993). These two- and three-dimensional analytical solute transport solutions have been verified extensively against some three-dimensional numerical models with excellent agreement (Batu, 1993, 2006).

With

$$F_x(0) = F_M = \varphi_e U C_M \tag{4-191}$$

Equation (4-189) becomes

$$C(x) = \frac{UC_M}{\left[\frac{U}{2} + D_x \left(\frac{R_d}{D_x} v + \frac{U^2}{4D_x^2} \right)^{\frac{1}{2}} \right]} \exp\left[\frac{Ux}{2D_x} - \left(\frac{R_d}{D_x} v + \frac{U^2}{4D_x^2} \right)^{\frac{1}{2}} x \right] \tag{4-192}$$

After some manipulations, finally one gets

$$C(x) = C_M \frac{\exp\left\{ \frac{Ux}{2D_x} \left[1 - \left(\frac{4D_x R_d v}{U^2} + 1 \right)^{\frac{1}{2}} \right] \right\}}{\left[\frac{1}{2} + \left(\frac{D_x R_d v}{U^2} + \frac{1}{4} \right)^{\frac{1}{2}} \right]} \tag{4-193}$$

Problems

4.1 In a solution, there is 75.86 mL(73.16 g) solvent and 18.42 mL(20.46 g) solute. Determine mass/mass concentration.

4.2 For Problem 4.1, determine the volume/volume concentration.

4.3 What is the molarity of a solution containing 85.32 g $KMnO_4$ (potassium permanganate) in 8 Lof solution.

4.4 What is the molarity of a solution containing 54.0 g CH_3OH in 500 cm^3 of solution.

4.5 Determine the molality of a solution of 72.44 g of glucose ($C_6H_{12}O_6$) by dissolving into 850 g of water.

4.6 If the measured K_d value of 1,2-dichlorobenzene is 0.45 cm^3/g and $\varphi = 0.26$, determine the value of R_d.

4.7 Describe the law regarding molecular diffusion by writing the related equation.

4.8 Write the solute flux equation under hydrodynamic dispersion conditions and describe the meaning of each term in it with an explanation.

4.9 Measured hydraulic conductivity (K_i) values are given in Table 1 of Problem 4.9 along the x-coordinate direction. The total length portion of the aquifer is 140.0 m and the measured K_i value locations are at every $\zeta = 10.0$ m separation length along (Figure 4-12) the x coordinate direction. In the y-coordinate direction, it is assumed that the same measured values are valid. Using the values in Table 1 of Problem 4.9, determine (a) the arithmetic average $\bar{f}$ of the $\ln(K_i)$ values; (b) the variance (σ_f^2) and standard deviation (σ_f) of the hydraulic conductivity (K_i) data; (c) the covariance R_f; (d) the autocovariance ρ_f; (e) the λ_f correlation scale (or correlation length); (f) write down the ρ_f autocovariance function and plot it as a function of ζ separation length.

TABLE 1 OF PROBLEM 4.9

K_i Values of x_i

L (m)	x_i (m)	K_i (cm/s)
1	10	0.007760
2	20	0.019700
3	30	0.005350
4	40	0.003660
5	50	0.005450
6	60	0.001760
7	70	0.086200
8	80	0.046300
9	90	0.045000
10	100	0.025200
11	110	0.005830
12	120	0.004160
13	130	0.066500
14	140	0.078200

5

Sufficiency Evaluation of the Inward Hydraulic Gradients in Contaminated Aquifers

5.1 Introduction

One of the most widely used remediation techniques for a contaminant plume in an aquifer is to take under control of existing contaminant plumes in aquifers. In such cases, a situation often can occur such that control of contaminant migration in an aquifer by the ground water flow toward the region of higher concentration is required. In other words, the goal is to generate inward flow gradients toward the center of the contaminant plume in order to reduce the concentrations below the imposed concentrations by regulatory requirements. The question is: Are the inward hydraulic gradients, generated by extraction wells or other means, sufficient to take under control of the plume? The purpose of this chapter is to provide answers to this question.

The pioneering study for the problem described above and its practical applications were presented by Al-Niami and Rushton (1977). Later on, additional contributions were made for different cases and for their verification against numerical solutions (Al-Niami and Rushton, 1978; Marino, 1978a; Marino, 1978b,b; Aral and Tang, 1992; Eldho, 1997).

5.2 Flow and Dispersion in Opposite Directions

5.2.1 Governing Differential Equation

The one-dimensional advection–dispersion partial differential equation is given by Eq. (4.57) in Chapter 4 as [e.g., Ogata, 1970, p. 112, Eq. (21)]

$$\frac{\partial C}{\partial t} = D\frac{\partial^2 C}{\partial x^2} - U\frac{\partial C}{\partial x} \tag{5-1}$$

in which C is the concentration, x is the Cartesian coordinate, U is the ground water velocity in x-direction, D is the dispersion coefficient in x-direction, and t is the time.

The velocity U is assumed to remain constant for all x values. Small changes in U will occur due to changes in density, but it is anticipated that this is unlikely to cause any significant errors.

5.2.2 Status of the Dispersion Coefficient for Flow against Dispersion

For the geometry shown in Figure 5-1, the dispersion coefficient D is assumed to be independent of the U velocity (Al-Niami and Rushton, 1977, p. 93; Aral and Tang, 1992,

DOI: 10.1201/9781003739128-5

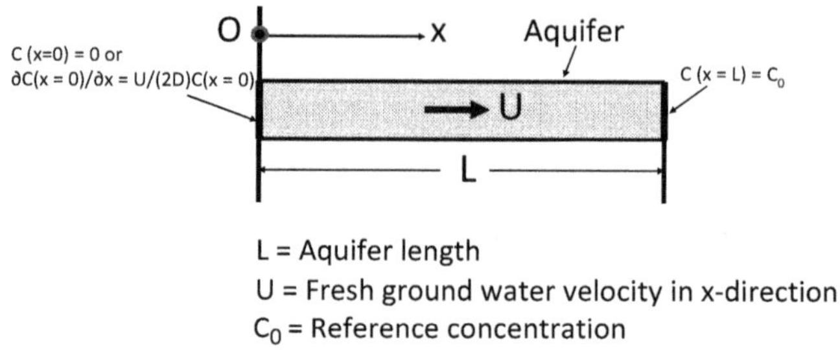

FIGURE 5-1
Schematic for one-dimensional flow against dispersion in aquifers. Adapted from Al-Niami and Rushton (1977, p. 90, Figure 1).

p. 206). With $D_{m_x} \equiv D_m$, The general form of the dispersion coefficient is [Chapter 4, Eq. (4.89)] $D = D^* + D_m = D_0\tau + \alpha_L U$, in which D_m is the mechanical dispersion coefficient and D^* is the effective molecular diffusion coefficient. Here, α_L is the longitudinal dispersivity, D_0 is the diffusion coefficient in pure water or diffusion coefficient in aqueous solution, and τ is the dimensionless *tortuosity factor*, approximately ranging between 0.3 and 0.7. Under these conditions, for flow against dispersion shown in Figure 5-1, $D \cong D^*$, which indicates that the possible control of dispersion in an aquifer by ground water flow is toward the region of higher solute concentration, namely, by molecular diffusion.

The values of the effective molecular diffusion coefficient D^* range between 1.0×10^{-10} and $1.0 \times 10^{-11}\ \mathrm{m^2/s}$ for nonreactive chemical species in clayey geological deposits. Values of coarse-grained unconsolidated materials can be somewhat higher than $1.0 \times 10^{-10}\ \mathrm{m^2/s}$ but are less than the coefficients for the chemical species in water, i.e., $< 2.0 \times 10^{-9}\ \mathrm{m^2/s}\left(0.0001728\ \mathrm{m^2/d}\right)$ (Freeze and Cherry, 1979, p. 393).

In their illustrative example, Al-Niami and Rushton (1977, p. 93) used $D = 0.5\ \mathrm{m^2/d}(5.787 \times 10^{-6}\ \mathrm{m^2/s})$. Aral and Tang (1992, p. 266) used the same value in their example. Eldho (1997, p. 265), in a similar example, used $D = 0.1\ \mathrm{cm^2/s}(0.864\ \mathrm{m^2/d})$. These values are significantly less than the dispersion coefficients under advection and dispersion effects used for mathematical modeling studies in aquifers, but much higher than the effective molecular diffusion coefficient $2.0 \times 10^{-9}\ \mathrm{m^2/s}$ value given above. However, these authors have not provided any rationales as to how they selected the values given above.

5.2.3 Description of the Problem

Consider the one-dimensional flow of fresh water through an aquifer having finite length L toward the region of high concentration shown in Figure 5-1. This moving fresh water tends to contain the dispersion of the higher concentration. As shown in Figure 5-1, the flow is toward $C(x = L) = C_0$ along the L length of the aquifer.

5.2.4 Initial and Boundary Conditions

The initial condition is

$$C(x, t = 0) = 0 \quad 0 \le x \le L \tag{5-2}$$

which states that initially the concentration is zero along the L length of the aquifer.

The boundary condition at $x = 0$ is

$$C(x=0,t)=0 \tag{5-3}$$

which states that the concentration is zero at $x = 0$ for all t. The boundary condition at $x = L$ is

$$C(x=L,t)=C_0 \tag{5-4}$$

which states that the concentration is C_0 at $x = L$. Eq(5-4) is a Dirichlet or first-type boundary condition.

The boundary condition given by Eq. (5-3) is set to zero for all times, which may not occur in practice. An alternative condition that allows some dispersion to occur at the boundary is a Cauchy or third-type boundary condition given as

$$\frac{\partial C(0,t)}{\partial x}=\frac{U}{2D}C(0,t) \tag{5-5}$$

for which derivation is given later on in Section 5.3.3 as Eq. (5-16). Eq(5-5) is based on the assumption that the quantity crossing the boundary is proportional to the concentration at the boundary.

5.3 Transient Solutions for Concentration and Comparison with Numerical Solutions

5.3.1 Solution Method

Al-Niami and Rushton (1977) solved the above-described boundary-value problem using the Laplace transformation solution method described in Carslaw and Jaeger (1959, pp. 297–319). Before that, they introduced a new function such that

$$C(x,t)=K(x,t)\exp\left(\frac{Ux}{2D}-\frac{U^2t}{4D}\right) \tag{5-6}$$

With this, Eq. (5-1) reduces to

$$\frac{\partial K}{\partial t}=D\frac{\partial^2 K}{\partial x^2} \tag{5-7}$$

From Eq. (5-6),

$$K(x,t)=C(x,t)\exp\left(-\frac{Ux}{2D}+\frac{U^2t}{4D}\right) \tag{5-8}$$

The initial condition given by Eq. (5-2) becomes

$$K(0,t=0)=0 \quad 0 \le x \le L \tag{5-9}$$

The boundary conditions given by Eqs. (5-3) and (5-4), respectively, take the forms

$$K(x=0,t)=0 \tag{5-10}$$

$$K(L,t)=C_0 \exp\left(-\frac{UL}{2D}+\frac{U^2t}{4D}\right) \tag{5-11}$$

5.3.2 Solution for $C(x=0,t)=0$ Boundary Condition

Using the Laplace transformation technique, Al-Niami and Rushton [1977, p. 90, Eq. (10), details in Appendix A] determined the following solution for the boundary condition given by Eq. (5-3):

$$C(x,t)=\frac{\sinh\left(\frac{Ux}{D}\right)}{\sinh\left(\frac{UL}{D}\right)}\exp\left[\frac{U(x-L)}{2D}\right] + \frac{2\pi D}{L^2}\exp\left[\frac{U(x-L)}{2D}-\frac{U^2t}{4D}\right]\sum_{n=1}^{\infty}\frac{(-1)^n\, n\exp\left[-D\left(\frac{n\pi}{L}\right)^2 t\right]}{\left[D\left(\frac{n\pi}{L}\right)^2+\frac{U^2}{4D}\right]}\sin\left(\frac{n\pi x}{L}\right) \tag{5-12}$$

Al-Niami and Rushton (1977, p. 90) state that Eq. (5-12) gives better results for $T = Dt/L^2 > 0.006$. For smaller values of time, they give an alternative expression [Al-Niami and Rushton, 1977, p. 96, Eq. (A.9)], which converges more rapidly. However, with the increased number of terms in the series, Eq. (5-12) can generate accurate results for all values of T.

Derivation of Eq. (5-12) is given in Example 5-1.

Example 5-1

Derive Eq. (5-12).

Solution

SHOWING THE CORRECTNESS OF THE $K(x,t)$ DIFFERENTIAL EQUATION

To simplify the solution process of Eq. (5-1), it is convenient to introduce another function $K(x,t)$ given by Eq. (5-6). With this transformation, Eq. (5-6) reduces to Eq. (5-7) and its correctness is shown below. From Eq. (566),

$$\frac{\partial C}{\partial t}=\frac{\partial K}{\partial t}\exp\left(\frac{Ux}{2D}-\frac{U^2t}{4D}\right)-K\frac{U^2}{4D}\exp\left(\frac{Ux}{2D}-\frac{U^2t}{4D}\right) \tag{E5-1-1}$$

$$\frac{\partial C}{\partial x}=\frac{\partial K}{\partial x}\exp\left(\frac{Ux}{2D}-\frac{U^2t}{4D}\right)+K\frac{U}{2D}\exp\left(\frac{Ux}{2D}-\frac{U^2t}{4D}\right) \tag{E5-1-2}$$

$$\frac{\partial^2 C}{\partial x^2} = \frac{\partial^2 K}{\partial x^2}\exp\left(\frac{Ux}{2D} - \frac{U^2 t}{4D}\right) + \frac{\partial K}{\partial x}\frac{U}{2D}\exp\left(\frac{Ux}{2D} - \frac{U^2 t}{4D}\right)$$
$$+ \frac{\partial K}{\partial x}\frac{U}{2D}\exp\left(\frac{Ux}{2D} - \frac{U^2 t}{4D}\right) + K\left(\frac{U}{2D}\right)^2 \exp\left(\frac{Ux}{2D} - \frac{U^2 t}{4D}\right) \quad \text{(E5-1-3)}$$

Using Eqs. (E5-1-2) and (E5-1-3),

$$D\frac{\partial^2 C}{\partial x^2} - U\frac{\partial C}{\partial x} = D\frac{\partial^2 K}{\partial x^2}\exp\left(\frac{Ux}{2D} - \frac{U^2 t}{4D}\right) + \frac{\partial K}{\partial x}\frac{U}{2}\exp\left(\frac{Ux}{2D} - \frac{U^2 t}{4D}\right)$$
$$+ \frac{\partial K}{\partial x}\frac{U}{2}\exp\left(\frac{Ux}{2D} - \frac{U^2 t}{4D}\right) + K\frac{U^2}{4D}\exp\left(\frac{Ux}{2D} - \frac{U^2 t}{4D}\right)$$
$$-U\frac{\partial K}{\partial x}\exp\left(\frac{Ux}{2D} - \frac{U^2 t}{4D}\right) - K\frac{U^2}{2D}\exp\left(\frac{Ux}{2D} - \frac{U^2 t}{4D}\right)$$
$$= D\frac{\partial^2 K}{\partial x^2}\exp\left(\frac{Ux}{2D} - \frac{U^2 t}{4D}\right) + K\frac{U^2}{4D}\exp\left(\frac{Ux}{2D} - \frac{U^2 t}{4D}\right) \quad \text{(E5-1-4)}$$
$$- K\frac{U^2}{2D}\exp\left(\frac{Ux}{2D} - \frac{U^2 t}{4D}\right)$$
$$= D\frac{\partial^2 K}{\partial x^2}\exp\left(\frac{Ux}{2D} - \frac{U^2 t}{4D}\right) - K\frac{U^2}{4D}\exp\left(\frac{Ux}{2D} - \frac{U^2 t}{4D}\right)$$

Substitution of Eqs. (E5-1-1) and (E5-1-4) into Eq. (5-1),

$$\frac{\partial K}{\partial t}\exp\left(\frac{Ux}{2D} - \frac{U^2 t}{4D}\right) - K\frac{U^2}{4D}\exp\left(\frac{Ux}{2D} - \frac{U^2 t}{4D}\right)$$
$$= D\frac{\partial^2 K}{\partial x^2}\exp\left(\frac{Ux}{2D} - \frac{U^2 t}{4D}\right) - K\frac{U^2}{4D}\exp\left(\frac{Ux}{2D} - \frac{U^2 t}{4D}\right) \quad \text{(E5-1-5)}$$

From here, one can write

$$\frac{\partial K}{\partial t} = D\frac{\partial^2 K}{\partial x^2} \quad \text{(E5-1-6)}$$

which is exactly the same as Eq. (5-7).

Initial and Boundary Conditions for $K(x,t)$

For $K(x,t)$, initial condition given by Eq. (5-2) becomes

$$K(x,t=0) = 0 \quad \text{(E5-1-7)}$$

The corresponding boundary condition given by Eq. (9-2) at $x = 0$ is

$$K(x=0,t) = 0 \quad \text{(E5-1-8)}$$

The boundary condition at $x = L$ given by Eq. (5-4) becomes

$$K(x=L,t)=C_0\exp\left(-\frac{UL}{2D}+\frac{U^2t}{4D}\right) \tag{E5-1-9}$$

LAPLACE DOMAIN DIFFERENTIAL EQUATION

Taking the Laplace transformation of Eq. (E5-1-6),

$$L\left\{\frac{\partial K}{\partial t}\right\}=L\left\{D\frac{\partial^2 K}{\partial x^2}\right\} \tag{E5-1-10}$$

The Laplace transform of the left term is

$$L\left\{\frac{\partial K}{\partial t}\right\}=su(x,s)-K(x,0) \tag{E5-1-11}$$

Using Eq. (E5-1-7),

$$L\left\{\frac{\partial K}{\partial t}\right\}=su(x,s) \tag{E5-1-12}$$

in which

$$K(x,0)=0 \tag{E5-1-13}$$

$$u(x,s)=L\{K(x,t)\} \tag{E5-1-14}$$

The Laplace transform of the left term of Eq. (E5-1-10) is

$$L\left\{D\frac{\partial^2 K}{\partial x^2}\right\}=D\frac{\partial^2 u}{\partial x^2} \tag{E5-1-15}$$

Substitution of Eqs. (E5-1-11) and (E5-1-15) into Eq. (E5-1-10),

$$\frac{\partial^2 u}{\partial x^2}-\frac{s}{D}u=0 \tag{E5-1-16}$$

which is the differential equation in the Laplace domain.

GENERAL SOLUTION OF THE LAPLACE DOMAIN DIFFERENTIAL EQUATION

Equation (E5-1-16) is a linear, homogeneous second-order differential equation. Its solution can be determined as follows (e.g., Spiegel, 1968, p. 105, 18.7): Its characteristic equation is

$$m^2-\frac{s}{D}=0 \tag{E5-1-17}$$

Its roots are

$$m_1=-\left(\frac{s}{D}\right)^{\frac{1}{2}} \quad m_2=\left(\frac{s}{D}\right)^{\frac{1}{2}} \tag{E5-1-18}$$

The solution is

$$u(x,s) = C_1 e^{-x\left(\frac{s}{D}\right)^{\frac{1}{2}}} + C_2 e^{x\left(\frac{s}{D}\right)^{\frac{1}{2}}} \qquad \text{(E5-1-19)}$$

BOUNDARY CONDITIONS IN THE LAPLACE DOMAIN

The Laplace domain forms of the boundary condition given by Eq. (E5-1-8) are

$$s(0,s) = 0 \qquad \text{(E5-1-20)}$$

The Laplace domain form of Eq. (E5-1-9) is

$$L\{K(L,t)\} = L\left\{C_0 \exp\left(-\frac{UL}{2D} + \frac{U^2 t}{4D}\right)\right\} = C_0 \exp\left(-\frac{UL}{2D}\right) L\left\{\exp\left(\frac{U^2 t}{4D}\right)\right\} \qquad \text{(E5-1-21)}$$

From here,

$$u(L,s) = C_0 \exp\left(-\frac{UL}{2D}\right)\frac{1}{(s-\alpha)} \qquad \text{(E5-1-22)}$$

in which

$$\alpha = \frac{U^2}{4D} \qquad \text{(E5-1-23)}$$

USAGE OF THE BOUNDARY CONDITIONS AND THE LAPLACE DOMAIN SOLUTION

Using Eq. (E5-1-20) in Eq. (E5-1-19),

$$C_1 + C_2 = 0 \quad C_1 = -C_2 \qquad \text{(E5-1-24)}$$

With this, Eq. (E5-1-19) becomes

$$u(x,s) = C_2\left[e^{x\left(\frac{s}{D}\right)^{\frac{1}{2}}} - e^{-x\left(\frac{s}{D}\right)^{\frac{1}{2}}}\right] \qquad \text{(E5-1-25)}$$

Using Eq. (E5-1-22),

$$u(L,s) = C_2\left[e^{L\left(\frac{s}{D}\right)^{\frac{1}{2}}} - e^{-L\left(\frac{s}{D}\right)^{\frac{1}{2}}}\right] = C_0 \exp\left(-\frac{UL}{2D}\right)\frac{1}{(s-\alpha)} \qquad \text{(E5-1-26)}$$

From here,

$$C_2 = \frac{C_0 \exp\left(-\frac{UL}{2D}\right)\frac{1}{(s-\alpha)}}{\left[e^{L\left(\frac{s}{D}\right)^{\frac{1}{2}}} - e^{-L\left(\frac{s}{D}\right)^{\frac{1}{2}}}\right]} \qquad \text{(E5-1-27)}$$

Using this in Eq. (E5-1-25),

$$u(x,s)=C_0\exp\left(-\frac{UL}{2D}\right)\frac{1}{(s-\alpha)}\frac{\left[e^{x\left(\frac{s}{D}\right)^{\frac{1}{2}}}-e^{-x\left(\frac{s}{D}\right)^{\frac{1}{2}}}\right]}{\left[e^{L\left(\frac{s}{D}\right)^{\frac{1}{2}}}-e^{-L\left(\frac{s}{D}\right)^{\frac{1}{2}}}\right]} \qquad \text{(E5-1-28)}$$

Using the formula

$$\sinh(x)=\frac{1}{2}\left(e^{x}-e^{-x}\right) \qquad \text{(E5-1-29)}$$

Equation (E5-1-28) takes the form

$$u(x,s)=C_0\exp\left(-\frac{UL}{2D}\right)\frac{1}{(s-\alpha)}\frac{\sinh\left[x\left(\frac{s}{D}\right)^{\frac{1}{2}}\right]}{\sinh\left[L\left(\frac{s}{D}\right)^{\frac{1}{2}}\right]} \qquad \text{(E5-1-30)}$$

which is the Laplace domain solution. Eq(E5-1-30) is exactly the same as Al-Niami and Rushton [1977, p. 95, Eq. (A-4)] with the exception of notation.

INVERSE LAPLACE TRANSFORM

The Complex Inversion Formula: If $f(s)=L\{F(t)\}$, then $L^{-1}\{f(s)\}$ is given by [e.g., Spiegel, 1965, p. 201, Eq. (1)]

$$F(x,t)=\frac{1}{2\pi i}\int_{\Upsilon-i\infty}^{\Upsilon+i\infty} e^{st} f(x,s)ds \quad t>0 \qquad \text{(E5-1-31)}$$

and $F(x,t)=0$ for $t<0$. This result is called the *complex integral formula*. It is also known as *Bromwich's integral formula*. Eq (E5-1-31) provides a direct means for obtaining the inverse Laplace transform of a given function $f(s)$ like Eq. (E5-1-30).

Usage of the Complex Inverse Formula and Cauchy's Theorem: Using Eq. (E5-1-30) in Eq. (E5-1-31)

$$K(x,t)=\frac{1}{2\pi i}C_0\exp\left(-\frac{UL}{2D}\right)\int_{\Upsilon-i\infty}^{\Upsilon+i\infty}\frac{e^{st}}{(s-\alpha)}\frac{\sinh(\mu x)}{\sinh(\mu L)}ds \qquad \text{(E5-1-32)}$$

in which

$$\mu=\left(\frac{s}{D}\right)^{\frac{1}{2}} \qquad \text{(E5-1-33)}$$

It follows from Bromwich's integral formula and the residue theorem [e.g., Carslaw and Jaeger, 1959, p. 312; Spiegel, 1965, pp. 201–202, Eqs. (1) and (4)]

$$K(x,t)=\frac{1}{2\pi i}\exp\left(-\frac{UL}{2D}\right)\sum\left[\text{residues of } e^{st}\frac{\sinh(\mu x)}{(s-\alpha)\sinh(\mu L)}\right.$$
$$\left.\text{at poles of}\frac{\sinh(\mu x)}{(s-\alpha)\sinh(\mu L)}\right] \tag{E5-1-34}$$

Determination of the Residue at the Simple Pole $s=\alpha$: $u(x,s)$ given by Eq. (E5-1-34) has a residue at the simple pole $s=\alpha$, which makes $s-\alpha$ zero. Therefore, one pole of the integrand is $s=\alpha$ with residue [e.g., Spiegel, 1965, pp. 205.206, Problem 5]

$$\text{Residue at }(s=\alpha)=\lim_{s\to\alpha}(s-\alpha)\frac{e^{st}}{(s-\alpha)}\frac{\sinh(\mu x)}{\sinh(\mu L)}=\frac{e^{\alpha t}\sinh(\mu x)}{\sinh(\mu L)} \tag{E5-1-35}$$

Using Eq. (E5-1-33), it takes the form

$$\text{Residue at }(s=\alpha)=\frac{e^{\alpha t}\sinh\left[\left(\frac{\alpha}{D}\right)^{\frac{1}{2}}x\right]}{\sinh\left[\left(\frac{\alpha}{D}\right)^{\frac{1}{2}}L\right]} \tag{E5-1-36}$$

Other Poles: The term $\sinh(\mu L)$ in the denominator of Eq. (E5-1-34) becomes zero for

$$\mu_n=\frac{in\pi}{L}\quad n=0,1,2,\ldots, \tag{E5-1-37}$$

because

$$\sinh(\mu_n L)=\sinh(in\pi)=i\sin(n\pi)=0\quad n=0,1,2,\ldots, \tag{E5-1-38}$$

From Eq. (E5-1-33),

$$\mu_n=\left(\frac{s_n}{D}\right)^{\frac{1}{2}} \tag{E5-1-39}$$

Then,

$$s_n=D\mu_n^2=D\left(\frac{in\pi}{L}\right)^2=-D\left(\frac{n\pi}{L}\right)^2\quad n=0,1,2,\ldots, \tag{E5-1-40}$$

$$\text{Residues at } s_n=\lim_{s\to s_n}\left[\frac{(s-s_n)}{\sinh(\mu_n L)}\right]\left[\frac{e^{s_n t}\sinh(\mu_n x)}{(s_n-\alpha)}\right] \tag{E5-1-41}$$

Since the first bracketed term becomes 0/0 at $s = s_n$ and μ_n values, the rule of L'Hospital must be applied. Therefore,

$$\text{Residues at } s_n = \frac{\lim\limits_{s\to s_n} \frac{1}{1!}\frac{d}{ds}\left[(s-s_n)e^{s_n t}\sinh(\mu_n x)\right]}{(s_n-\alpha)\lim\limits_{s\to s_n}\frac{1}{1!}\frac{d}{d\mu_n}\left[\sinh(\mu_n L)\right]\frac{d\mu_n}{ds_n}} = \frac{e^{s_n t}\sinh(\mu_n x)}{(s_n-\alpha)\frac{d}{d\mu_n}\left[\sinh(\mu_n L)\right]\frac{d\mu_n}{ds_n}} \qquad \text{(E5-1-42)}$$

From Eq. (E5-1-40),

$$D\mu_n^2 = s_n \quad 2\mu_n D d\mu_n = ds_n \quad \frac{d\mu_n}{ds_n} = \frac{1}{2\mu_n D} \qquad \text{(E5-1-43)}$$

Then, Eq. (E5-1-42) becomes

$$\text{Residues at } s_n = \frac{(1)e^{s_n t}\sinh(\mu_n x)}{(s_n-\alpha)L\cosh(\mu_n L)\frac{1}{2\mu_n D}} \qquad \text{(E5-1-44)}$$

From Eq. (E5-1-37),

$$\mu_n L = in\pi \qquad \text{(E5-1-45)}$$

Therefore,

$$\cosh(\mu_n L) = \cosh(in\pi) = \cos(n\pi) = (-1)^n \qquad \text{(E5-1-46)}$$

Also, from Eq. (E5-1-45) and the first expression of Eq. (E5-1-43),

$$s_n = D\mu_n^2 = D\left(\frac{in\pi}{L}\right)^2 = -D\left(\frac{n\pi}{L}\right)^2 \qquad \text{(E5-1-47)}$$

Using Eqs. (E5-1-46) in Eq. (E5-1-44),

$$\text{Residues at } s_n = \frac{2D}{L}\frac{\mu_n}{(s_n-\alpha)}(-1)^n e^{s_n t}\sinh(\mu_n x) \qquad \text{(E5-1-48)}$$

Using the residue theorem [e.g., Carslaw and Jaeger, 1959, p. 312; Spiegel, 1965, p. 202, Eq. (4)] with Eqs. (E5-1-36) and (E5-1-48), Eq. (E5-1-34) can be written as

$$K(x,t) = \frac{1}{2\pi i}e^{-\frac{UL}{2D}}(2\pi i)\left\{\frac{e^{\alpha t}\sinh\left[\left(\frac{\alpha}{D}\right)^{\frac{1}{2}}x\right]}{\sinh\left[\left(\frac{\alpha}{D}\right)^{\frac{1}{2}}L\right]} + \frac{2D}{L}\sum_{n=1}^{\infty}\frac{(-1)^n\mu_n e^{s_n t}\sinh(\mu_n x)}{(s_n-\alpha)}\right\} \qquad \text{(E5-1-49)}$$

Inserting this in Eq. (5-6),

$$C(x,t)=\exp\left(\frac{Ux}{2D}-\frac{U^2t}{4D}-\frac{UL}{2D}\right)\left\{\frac{e^{\alpha t}\sinh\left[\left(\frac{\alpha}{D}\right)^{\frac{1}{2}}x\right]}{\sinh\left[\left(\frac{\alpha}{D}\right)^{\frac{1}{2}}L\right]}+\frac{2D}{L}\sum_{n=1}^{\infty}\frac{(-1)^n\mu_n e^{s_n t}\sinh(\mu_n x)}{(s_n-\alpha)}\right\} \tag{E5-1-50}$$

Using Eqs. (E5-1-23), (E5-1-37), (E5-1-47), and

$$\sinh\left(\frac{in\pi}{L}x\right)=i\sin\left(\frac{n\pi}{L}x\right) \tag{E5-1-51}$$

$$C(x,t)=\exp\left[\frac{U(x-L)}{2D}-\frac{U^2t}{4D}\right]\left\{\frac{\exp\left(-\frac{U^2t}{4D}\right)\sinh\left(\frac{Ux}{D}\right)}{\sinh\left(\frac{UL}{D}\right)}+\frac{2D}{L}\sum_{n=1}^{\infty}\frac{(-1)^n\left(\frac{in\pi}{L}\right)\exp\left[-D\left(\frac{n\pi}{L}\right)^2t\right]i\sin\left(\frac{n\pi}{L}x\right)}{\left[-D\left(\frac{n\pi}{L}\right)^2-\alpha\right]}\right\} \tag{E5-1-52}$$

After some manipulations,

$$C(x,t)=\frac{\sinh\left(\frac{Ux}{D}\right)}{\sinh\left(\frac{UL}{D}\right)}\exp\left[\frac{U(x-L)}{2D}\right]+\frac{2\pi D}{L^2}\exp\left[\frac{U(x-L)}{2D}-\frac{U^2t}{4D}\right]\sum_{n=1}^{\infty}\frac{(-1)^n n\exp\left[-D\left(\frac{n\pi}{L}\right)^2t\right]}{\left[D\left(\frac{n\pi}{L}\right)^2+\frac{U^2}{4D}\right]}\sin\left(\frac{n\pi x}{L}\right) \tag{E5-1-53}$$

Equation (E5-1-53) is exactly the same as Eq. (5-12).

5.3.3 Solution for $\frac{\partial C(0,t)}{\partial x}=\frac{U}{2D}C(0,t)$ Boundary Condition

The boundary condition given by Eq. (5-5) can be derived from Eq. (5-6):

$$\frac{\partial C(x,t)}{\partial x}=\frac{\partial[K(x,t)]}{\partial x}\exp\left(\frac{Ux}{2D}-\frac{U^2t}{4D}\right)+K(x,t)\left(\frac{U}{2D}\right)\exp\left(\frac{Ux}{2D}-\frac{U^2t}{4D}\right) \tag{5-13}$$

Using Eq. (5-6),

$$\frac{\partial C(x,t)}{\partial x}=\frac{\partial\left[K(x,t)\right]}{\partial x}\exp\left(\frac{Ux}{2D}-\frac{U^2t}{4D}\right)+\left(\frac{U}{2D}\right)C(x,t) \tag{5-14}$$

At $x=0$,

$$\frac{\partial C(x=0,t)}{\partial x}=\frac{\partial\left[K(x=0,t)\right]}{\partial x}\exp\left(-\frac{U^2t}{4D}\right)+\left(\frac{U}{2D}\right)C(x=0,t) \tag{5-15}$$

The first term on the right-hand side can be negligibly small compared with the second term. Therefore, it becomes

$$\frac{\partial C(x=0,t)}{\partial x}=\frac{U}{2D}C(x=0,t) \tag{5-16}$$

which is exactly the same as Eq. (5-5).

With this boundary condition and the initial condition given by Eq. (5-2) and the other boundary condition given by Eq. (5-4), Al-Niami and Rushton [1977, p. 91, Eq. (12), details in Appendix B] determined the following solution:

$$C(x,t)=\frac{\cosh\left(\frac{Ux}{2D}\right)}{\cosh\left(\frac{UL}{2D}\right)}\exp\left[\frac{U(x-L)}{2D}\right]$$

$$-\frac{\pi D}{L^2}\exp\left[\frac{U(x-L)}{2D}-\frac{U^2t}{4D}\right]\sum_{n=1}^{\infty}\frac{(-1)^n(2n+1)\exp\left[-D\frac{(2n+1)^2}{4L^2}\pi^2t\right]}{\left[D\frac{(2n+1)^2}{4L^2}\pi^2+\frac{U^2}{4D}\right]}\cos\left[\frac{(2n+1)\pi}{2L}x\right] \tag{5-17}$$

Equation (5-17) is also exactly the same as Marino [1978a, p. 153, Eq. (15)]. Derivation of Eq. (5-17) is given in Example 5-2.

Al-Niami and Rushton (1977, p. 91) state that Eq. (9.17) gives better results for $T=Dt/L^2>0.2$. For $T<0.2$, they give an alternative expression [Al-Niami and Rushton, 1977, p. 97, Eq. (B-6)]. However, with the increased number of terms in the series, Eq. (5-17) can generate accurate results for all values of T.

Example 5-2

Derive Eq. (5-17).

Solution

Governing Differential Equation for $K(x,t)$

With

$$C(x,t)=K(x,t)\exp\left(\frac{Ux}{2D}-\frac{U^2t}{4D}\right) \tag{E5-2-1}$$

Equ (5-1) reduces to

$$\frac{\partial K}{\partial t} = D\frac{\partial^2 K}{\partial x^2} \tag{E5-2-2}$$

whose correctness is shown in Example 5-1.

INITIAL AND BOUNDARY CONDITIONS FOR $K(x,t)$

For $K(x,t)$, initial condition given by Eq. (5-2) becomes

$$K(x,t=0) = 0 \quad 0 \le x \le L \tag{E5-2-3}$$

The corresponding boundary condition given by Eq. (5-5) at $x = 0$ is

$$\frac{\partial C(0,t)}{\partial x} = \frac{U}{2D}C(0,t) \tag{E5-2-4}$$

From Eq. (E5-2-1),

$$\frac{\partial C(x,t)}{\partial x} = \frac{\partial K(x,t)}{\partial x}\exp\left(\frac{Ux}{2D} - \frac{U^2t}{4D}\right) + K(x,t)\left(\frac{U}{2D}\right)\exp\left(\frac{Ux}{2D} - \frac{U^2t}{4D}\right) \tag{E5-2-5}$$

At $x = 0$,

$$\frac{\partial C(0,t)}{\partial x} = \frac{\partial K(0,t)}{\partial x}\exp\left(-\frac{U^2t}{4D}\right) + K(0,t)\left(\frac{U}{2D}\right)\exp\left(-\frac{U^2t}{4D}\right) \tag{E5-2-6}$$

From Eq. (E5-2-1), at $x = 0$,

$$C(0,t) = K(0,t)\exp\left(-\frac{U^2t}{4D}\right) \tag{E5-2-7}$$

Introducing Eqs. (E5-2-6) and (E5-2-7) into Eq. (E5-2-4),

$$\frac{\partial K(0,t)}{\partial x}\exp\left(-\frac{U^2t}{4D}\right) + K(0,t)\left(\frac{U}{2D}\right)\exp\left(-\frac{U^2t}{4D}\right) = \frac{U}{2D}K(0,t)\exp\left(-\frac{U^2t}{4D}\right) \tag{E5-2-8}$$

From here, one can write

$$\frac{\partial K(0,t)}{\partial x} = 0 \tag{E5-2-9}$$

From, Eq. (E5-2-1), the boundary condition at $x = L$ given by Eq. (5-4) becomes

$$K(x = L,t) = C_0 \exp\left(-\frac{UL}{2D} + \frac{U^2t}{4D}\right) \tag{E5-2-10}$$

DIFFERENTIAL EQUATION IN THE LAPLACE DOMAIN

The differential equation of Eq. (E5-2-2) in the Laplace domain is given by Eq. (E5-1-16) of Example 5-1 as

$$\frac{\partial^2 u}{\partial x^2} - \frac{s}{D}u = 0 \tag{E5-2-11}$$

GENERAL SOLUTION OF THE LAPLACE DOMAIN DIFFERENTIAL EQUATION

The general solution of Eq. (E5-2-11) is given by Eq. (E5-1-19) in Example 5-1 as

$$u(x,s) = C_1 e^{-x\left(\frac{s}{D}\right)^{\frac{1}{2}}} + C_2 e^{x\left(\frac{s}{D}\right)^{\frac{1}{2}}} \tag{E5-2-12}$$

USAGE OF THE BOUNDARY CONDITIONS AND THE LAPLACE DOMAIN SOLUTION

The Laplace transform of Eq. (E5-2-9) is

$$\frac{\partial u(0,s)}{\partial x} = 0 \tag{E5-2-13}$$

Using Eq. (E9-2-12),

$$\frac{\partial u(x,s)}{\partial x} = C_1\left[-\left(\frac{s}{D}\right)^{\frac{1}{2}}\right]e^{-x\left(\frac{s}{D}\right)^{\frac{1}{2}}} + C_2\left[\left(\frac{s}{D}\right)^{\frac{1}{2}}\right]e^{x\left(\frac{s}{D}\right)^{\frac{1}{2}}} = 0 \tag{E5-2-14}$$

From here,

$$C_1 = C_2 \tag{E5-2-15}$$

Then, Eq. (E5-2-12) becomes

$$u(x,s) = C_2\left[e^{-x\left(\frac{s}{D}\right)^{\frac{1}{2}}} + e^{x\left(\frac{s}{D}\right)^{\frac{1}{2}}}\right] \tag{E5-2-16}$$

The Laplace transform of Eq. (E5-2-10) is

$$L\{K(L,t)\} = L\left\{C_0 \exp\left(-\frac{UL}{2D} + \frac{U^2 t}{4D}\right)\right\} = C_0 \exp\left(-\frac{UL}{2D}\right)L\left\{\exp\left(\frac{U^2 t}{4D}\right)\right\} \tag{E5-2-17}$$

From here,

$$u(L,s) = C_0 \exp\left(-\frac{UL}{2D}\right)\frac{1}{(s-\alpha)} \tag{E5-2-18}$$

in which

$$\alpha = \frac{U^2}{4D} \tag{E5-2-19}$$

At $x = L$, Eq. (E5-2-16) becomes

$$u(L,s) = C_2\left[e^{-L\left(\frac{s}{D}\right)^{\frac{1}{2}}} + e^{L\left(\frac{s}{D}\right)^{\frac{1}{2}}}\right] \tag{E5-2-20}$$

From Eqs. (E5-2-18) and (E5-2-19), it becomes

$$C_0 \exp\left(-\frac{UL}{2D}\right)\frac{1}{(s-\alpha)} = C_2\left[e^{-L\left(\frac{s}{D}\right)^{\frac{1}{2}}} + e^{L\left(\frac{s}{D}\right)^{\frac{1}{2}}}\right] \tag{E5-2-21}$$

Solving for C_2,

$$C_2 = \frac{C_0 \exp\left(-\frac{UL}{2D}\right)\frac{1}{(s-\alpha)}}{\left[e^{-L\left(\frac{s}{D}\right)^{\frac{1}{2}}} + e^{L\left(\frac{s}{D}\right)^{\frac{1}{2}}}\right]} \tag{E5-2-22}$$

Using this in Eq. (E5-2-16),

$$u(x,s) = C_0 \frac{\exp\left(-\frac{UL}{2D}\right)}{(s-\alpha)} \frac{\left[e^{-x\left(\frac{s}{D}\right)^{\frac{1}{2}}} + e^{x\left(\frac{s}{D}\right)^{\frac{1}{2}}}\right]}{\left[e^{-L\left(\frac{s}{D}\right)^{\frac{1}{2}}} + e^{L\left(\frac{s}{D}\right)^{\frac{1}{2}}}\right]} \tag{E5-2-23}$$

With the formula

$$\cosh(x) = \frac{1}{2}\left(e^{x} + e^{-x}\right) \tag{E5-2-24}$$

Equation (E5-2-23) takes the form

$$u(x,s) = f(x,s) = C_0 \frac{\exp\left(-\frac{UL}{2D}\right)}{(s-\alpha)} \frac{\cosh\left[x\left(\frac{s}{D}\right)^{\frac{1}{2}}\right]}{\cosh\left[L\left(\frac{s}{D}\right)^{\frac{1}{2}}\right]} \tag{E5-2-25}$$

which is the Laplace domain solution. Equation (E5-2.25) is exactly the same as Al-Niami and Rushton [1977, p. 96, Eq. (B-2)] and Marino [1978a, p. 157, Eq. (A-5)] with the exception of notation.

INVERSE LAPLACE TRANSFORM

The Complex Inversion Formula: If $f(s)=L\{F(t)\}$, then $L^{-1}\{f(s)\}$ is given by [e.g., Spiegel, 1965, p. 201, Eq. (1)]

$$F(x,t)=\frac{1}{2\pi i}\int_{\Upsilon-i\infty}^{\Upsilon+i\infty} e^{st} f(x,s)ds \quad t>0 \tag{E5-2-26}$$

and $F(x,t)=0$ for $t<0$. This result is called the *complex integral formula*. It is also known as *Bromwich's integral formula*. Eq (E5-2-26) provides a direct means for obtaining the inverse Laplace transform of a given function $f(s)$ like Eq. (E5-2-23).

Usage of the Complex Inverse Formula and Cauchy's Theorem: Using Eq. (E5-2-25) in Eq. (E5-2-26)

$$K(x,t)=\frac{1}{2\pi i}C_0\exp\left(-\frac{UL}{2D}\right)\int_{\Upsilon-i\infty}^{\Upsilon+i\infty}\frac{e^{st}}{(s-\alpha)}\frac{\cosh(\mu x)}{\cosh(\mu L)}ds \tag{E5-2-27}$$

in which

$$\mu=\left(\frac{s}{D}\right)^{\frac{1}{2}} \tag{E5-2-28}$$

It follows from the Bromwich's integral formula and the residue theorem [e.g., Carslaw and Jaeger, 1959, p. 312; Spiegel, 1965, pp. 201–202, Eqs. (1) and (4)],

$$K(x,t)=\frac{1}{2\pi i}\exp\left(-\frac{UL}{2D}\right)\sum\left[\text{residues of } e^{st}\frac{\cosh(\mu x)}{(s-\alpha)\cosh(\mu L)}\right.$$
$$\left.\text{at poles of }\frac{\cosh(\mu x)}{(s-\alpha)\cosh(\mu L)}\right] \tag{E5-2-29}$$

Determination of the Residue at the Simple Pole $s=\alpha$: $u(x,s)$ given by Eq. (E5-2-25) has a residue at the simple pole $s=\alpha$, which makes $s-\alpha$ zero. Therefore, one pole of the integrand is $s=\alpha$ with residue [e.g., Spiegel, 1965, pp. 205–206, Problem 5]

$$\text{Residue at }(s=\alpha)=\lim_{s\to\alpha}(s-\alpha)\frac{e^{st}}{(s-\alpha)}\frac{\cosh(\mu x)}{\cosh(\mu L)}=\frac{e^{\alpha t}\cosh(\mu x)}{\cosh(\mu L)} \tag{E5-2-30}$$

Using Eq. (E5-2-28), it takes the form

$$\text{Residue at }(s=\alpha)=\frac{e^{\alpha t}\cosh\left[\left(\frac{\alpha}{D}\right)^{\frac{1}{2}}x\right]}{\cosh\left[\left(\frac{\alpha}{D}\right)^{\frac{1}{2}}L\right]} \tag{E5-2-31}$$

Other Poles: The term $\cosh(\mu L)$ in the denominator of Eq. (E5-2-29) becomes zero for

$$\mu_n = \frac{(2n+1)\pi i}{2L} \quad n = 0,1,2,\ldots, \tag{E5-2-32}$$

because

$$\cosh(\mu_n L) = \cosh\left[\frac{(2n+1)\pi i L}{2L}\right] = \cos\left[\frac{(2n+1)\pi}{2}\right] \tag{E5-2-33}$$

Therefore, for $n = 0,1,2,\ldots$, $\cosh(\mu_n L) = \cos\left[\frac{(2n+1)\pi}{2}\right]$ becomes zero. From Eq. (E5-2-28),

$$\mu_n = \left(\frac{s_n}{D}\right)^{\frac{1}{2}} \tag{E5-2-34}$$

Then,

$$s_n = D\mu_n^2 = D\left[\frac{(2n+1)\pi i}{2L}\right]^2 = -D\frac{(2n+1)^2\pi^2}{4L^2} \quad n = 0,1,2,\ldots, \tag{E5-2-35}$$

$$\text{Residues at } s_n = \lim_{s \to s_n}\left[\frac{(s - s_n)}{\cosh(\mu_n L)}\right]\left[\frac{e^{s_n t}\cosh(\mu_n x)}{(s_n - \alpha)}\right] \tag{E522-36}$$

Since the first bracketed term becomes $0/0$ at $s = s_n$ and μ_n values, the rule of L'Hospital must be applied. Therefore,

$$\begin{aligned}\text{Residues at } s_n &= \frac{\lim\limits_{s \to s_n}\frac{1}{1!}\frac{d}{ds}\left[(s - s_n)e^{s_n t}\cosh(\mu_n x)\right]}{(s_n - \alpha)\lim\limits_{s \to s_n}\frac{1}{1!}\frac{d}{d\mu_n}\left[\cosh(\mu_n L)\right]\frac{d\mu_n}{ds_n}} \\ &= \frac{e^{s_n t}\cosh(\mu_n x)}{(s_n - \alpha)\frac{d}{d\mu_n}\left[\cosh(\mu_n L)\right]\frac{d\mu_n}{ds_n}}\end{aligned} \tag{E5-2-37}$$

From Eq. (E522-34),

$$D\mu_n^2 = s_n \quad 2\mu_n D d\mu_n = ds_n \quad \frac{d\mu_n}{ds_n} = \frac{1}{2\mu_n D} \tag{E5-2-38}$$

Then, Eq. (E5-2-36) becomes

$$\text{Residues at } s_n = \frac{(1)e^{s_n t}\cosh(\mu_n x)}{(s_n - \alpha)L\sinh(\mu_n L)\frac{1}{2\mu_n D}} \tag{E5-2-39}$$

From Eq. (E5-2-32),

$$\mu_n L = \frac{(2n+1)\pi i}{2} \tag{E5-2-40}$$

Therefore,

$$\sinh(\mu_n L) = \sinh\left[\frac{(2n+1)\pi i}{2}\right] = \mathrm{i}\sin\left[\frac{(2n+1)\pi}{2}\right] \tag{E5-2-41}$$

Also, from Eq. (E5-2-32) and the first expression of Eq. (E5-2-38),

$$s_n = D\mu_n^2 = D\left[\frac{(2n+1)\pi i}{2L}\right]^2 = -D\left[\frac{(2n+1)\pi}{2L}\right]^2 \tag{E5-2-42}$$

Using Eqs. (E5-2-32) and (E5-2-42) in Eq. (E5-2-38),

$$\text{Residues at } s_n = \sum_{n=1}^{\infty} \frac{e^{-D\left[\frac{(2n+1)\pi}{2L}\right]^2 t} \cosh\left[\frac{(2n+1)\pi i}{2L}x\right]}{\left\{-D\left[\frac{(2n+1)\pi}{2L}\right]^2 - \alpha\right\} L\sinh\left[\frac{(2n+1)\pi i}{2}\right] \frac{1}{2\left[\frac{(2n+1)\pi i}{2L}\right]D}} \tag{E5-2-43}$$

Since

$$\cosh\left[\frac{(2n+1)\pi i}{2L}x\right] = \cos\left[\frac{(2n+1)\pi}{2L}x\right] \tag{E5-2-44}$$

$$\sinh\left[\frac{(2n+1)\pi i}{2}\right] = i\sin\left[\frac{(2n+1)\pi}{2}\right] = i(-1)^n \tag{E5-2-45}$$

It takes the form

$$\text{Residues at } s_n = \sum_{n=1}^{\infty} \frac{e^{-D\left[\frac{(2n+1)\pi}{2L}\right]^2 t} \cos\left[\frac{(2n+1)\pi}{2L}x\right]}{\left\{-D\left[\frac{(2n+1)\pi}{2L}\right]^2 - \alpha\right\} Li(-1)^n \frac{1}{2\left[\frac{(2n+1)\pi i}{2L}\right]D}} \tag{E5-2-46}$$

After some manipulations,

$$\text{Residues at } s_n = -\frac{\pi D}{L^2} \sum_{n=1}^{\infty} \frac{(-1)^n (2n+1) e^{-D\left[\frac{(2n+1)\pi}{2L}\right]^2 t}}{\left\{D\left[\frac{(2n+1)\pi}{2L}\right]^2 + \frac{U^2}{4D}\right\}} \cos\left[\frac{(2n+1)\pi}{2L}x\right] \tag{E5-2-47}$$

Using the residue theorem [e.g., Carslaw and Jaeger, 1959, p. 312; Spiegel, 1965, p. 202, Eq. (4)] with Eqs. (E5-2-30) and (E5-2-46), Eq. (E5-2-29) can be written as

$$K(x,t)=\frac{1}{2\pi i}e^{-\frac{UL}{2D}}(2\pi i)\left\{\frac{e^{\alpha t}\cosh\left[\left(\frac{\alpha}{D}\right)^{\frac{1}{2}}x\right]}{\cosh\left[\left(\frac{\alpha}{D}\right)^{\frac{1}{2}}L\right]}\right.$$

(E5-2-48)

$$\left.-\frac{\pi D}{L^2}\sum_{n=1}^{\infty}\frac{(-1)^n(2n+1)e^{-D\left[\frac{(2n+1)\pi}{2L}\right]^2 t}}{\left\{D\left[\frac{(2n+1)\pi}{2L}\right]^2+\frac{U^2}{4D}\right\}}\cos\left[\frac{(2n+1)\pi}{2L}x\right]\right\}$$

Inserting this in Eq. (E5-2-1),

$$C(x,t)=e^{\left(\frac{Ux}{2D}-\frac{U^2t}{4D}\right)}e^{\left(-\frac{UL}{2D}\right)}\left\{\frac{e^{\alpha t}\cosh\left[\left(\frac{\alpha}{D}\right)^{\frac{1}{2}}x\right]}{\cosh\left[\left(\frac{\alpha}{D}\right)^{\frac{1}{2}}L\right]}-\right.$$

(E5-2-49)

$$\left.-\frac{\pi D}{L^2}\sum_{n=1}^{\infty}\frac{(-1)^n(2n+1)e^{-D\left[\frac{(2n+1)\pi}{2L}\right]^2 t}}{\left\{D\left[\frac{(2n+1)\pi}{2L}\right]^2+\frac{U^2}{4D}\right\}}\cos\left[\frac{(2n+1)\pi}{2L}x\right]\right\}$$

Using Eq. (E5-2-19) and after some manipulations,

$$C(x,t)=\frac{\cosh\left(\frac{Ux}{2D}\right)}{\cosh\left(\frac{UL}{2D}\right)}\exp\left[\frac{U(x-L)}{2D}\right]$$

$$-\frac{\pi D}{L^2}\exp\left[\frac{U(x-L)}{2D}-\frac{U^2t}{4D}\right]\sum_{n=1}^{\infty}\frac{(-1)^n(2n+1)\exp\left[-D\frac{(2n+1)^2}{4L^2}\pi^2t\right]}{\left[D\frac{(2n+1)^2}{4L^2}\pi^2+\frac{U^2}{4D}\right]}\cos\left[\frac{(2n+1)\pi}{2L}x\right]$$

(E5-2-50)

Equation (E5-2-50) is exactly the same as Eq. (5-17).

5.3.4 Comparison with Numerical Solutions

Al-Niami and Rushton (1978) also presented a finite-difference scheme based on the displaced boundary technique in which only the portion where there is a rapid change in concentration is modeled and comparisons with the analytical solutions are reasonably good.

Using the boundary element method, Eldho (1997, p. 265, Figure 2) compared the solution of Marino's solution [Marino, 1978a, p. 153, Eq. (15)], which is also given by Eq. (5-17) based on Al-Niami and Rushton [1977, p. 91, Eq. (12)]. Comparisons presented in Eldho (1997, p. 265, Figure 2) are reasonably good.

5.4 Steady-State Solutions for Concentration

Al-Niami and Rushton (1977) pointed out that one important feature of the flow against dispersion problem is that at large times a balance is reached between the dispersion and the flow. Therefore, the dispersion is contained. The steady-state solutions corresponding to the boundary conditions given by Eqs. (5-3) and (5-5) at $x = 0$ are presented below.

5.4.1 Steady-State Solution for the Boundary Condition $C(x = 0, t = \infty) = 0$

The steady-state solution is [Al-Niami and Rushton, 1977, p. 91, Eq. (13)]

$$\frac{C(x,t=\infty)}{C_0} = \frac{e^{\frac{Ux}{2D}} - 1}{e^{\frac{UL}{2D}} - 1} \tag{5-18}$$

which is derived in Example 5-3.

Example 5-3

Derive Eq. (5-18).

Solution

The corresponding transient solution for the boundary condition given by Eq. (5-3) is given by Eq. (5-12). For $t = \infty$, the second term of Eq. (5-12) becomes zero. Therefore, the solution takes the form

$$\frac{C(x,t=\infty)}{C_0} = \frac{\sinh\left(\frac{Ux}{2D}\right)}{\sinh\left(\frac{UL}{2D}\right)} \exp\left[\frac{U(x-L)}{2D}\right] \tag{E5-3-1}$$

Using

$$\sinh\left(\frac{Ux}{2D}\right) = \frac{1}{2}\left(e^{\frac{Ux}{2D}} - e^{-\frac{Ux}{2D}}\right) \tag{E5-3-2}$$

$$\sinh\left(\frac{UL}{2D}\right)=\frac{1}{2}\left(e^{\frac{UL}{2D}}-e^{-\frac{UL}{2D}}\right) \quad \text{(E5-3-3)}$$

Equation (E5-3-1) becomes

$$\frac{C(x,t=\infty)}{C_0}=\frac{\left(e^{\frac{Ux}{2D}}-e^{-\frac{Ux}{2D}}\right)}{\left(e^{\frac{UL}{2D}}-e^{-\frac{UL}{2D}}\right)}\left(e^{\frac{Ux}{2D}}\right)\left(e^{-\frac{UL}{2D}}\right) \quad \text{(E5-3-4)}$$

After some manipulations, one gets

$$\frac{C(x,t=\infty)}{C_0}=\frac{e^{\frac{Ux}{2D}}-1}{e^{\frac{UL}{2D}}-1} \quad \text{(E5-3-5)}$$

which is exactly the same as Eq. (5-12).

5.4.2 Steady-State Solution for the Boundary Condition $\frac{\partial C(0,t=\infty)}{\partial x}=\frac{U}{2D}C(0,t=\infty)$

The steady-state solution corresponding to the boundary condition given by Eq. (5-5) is [Al-Niami and Rushton, 1977, p. 91, Eq. (14)]

$$\frac{C(x,t=\infty)}{C_0}=\frac{e^{\frac{Ux}{2D}}+1}{e^{\frac{UL}{2D}}+1} \quad \text{(5-19)}$$

which is derived in Example 5-4.

Example 5-4

Derive Eq. (5-19).

Solution

The corresponding transient solution for the boundary condition given by Eq. (5-5) is given by Eq. (5-17). For $t=\infty$, the second term of Eq. (5-17) becomes zero. Therefore, the solution becomes

$$\frac{C(x,t=\infty)}{C_0}=\frac{\cosh\left(\frac{Ux}{2D}\right)}{\cosh\left(\frac{UL}{2D}\right)}\exp\left[\frac{U(x-L)}{2D}\right] \quad \text{(E5-4-1)}$$

Using

$$\cosh\left(\frac{Ux}{2D}\right)=\frac{1}{2}\left(e^{\frac{Ux}{2D}}+e^{-\frac{Ux}{2D}}\right) \quad \text{(E5-4-2)}$$

$$\cosh\left(\frac{UL}{2D}\right)=\frac{1}{2}\left(e^{\frac{UL}{2D}}+e^{-\frac{UL}{2D}}\right) \quad \text{(E5-4-3)}$$

Equation (E5-4-1) becomes

$$\frac{C(x,t=\infty)}{C_0}=\frac{\left(e^{\frac{Ux}{2D}}+e^{-\frac{Ux}{2D}}\right)}{\left(e^{\frac{UL}{2D}}+e^{-\frac{UL}{2D}}\right)}\left(e^{\frac{Ux}{2D}}\right)\left(e^{-\frac{UL}{2D}}\right) \qquad \text{(E5-4-4)}$$

After some manipulations, one gets

$$\frac{C(x,t=\infty)}{C_0}=\frac{e^{\frac{Ux}{2D}}+1}{e^{\frac{UL}{2D}}+1} \qquad \text{(E5-4-5)}$$

which is exactly the same as Eq. (5-19).

5.4.3 Comparison with Numerical Solutions

Using the boundary element method, Eldho (1997, p. 267, Figure 7) compared the solution of Al-Niami and Rushton [1977, p. 91, Eq. (13)], which is also given by Eq. (5-18). Comparisons presented in Eldho (1997, p. 267, Figure 7) are reasonably good.

5.5 Potential Applications to Remediation Projects

5.5.1 Evaluation of the Steady-State Concentration Expressions

Al-Niami and Rushton (1977) evaluated Eqs. (5-18) and (5-19) graphically in two alternative forms, and they are presented below.

Figure 5-2 shows the variation of C/C_0 with x/L when dispersion is balanced by ground water flow. Each curve refers to a constant value of $\mu = UL/(2D)$. Full lines for the boundary condition are given by Eq. (5-3); broken lines for the boundary condition are given by Eq. (5-5). For values of $\mu = uL/(2D)$ greater than approximately 7, the two different boundary conditions give identical results.

Figure 5-3 shows equal C/C_0 curves with logarithmic $\mu = UL/(2D)$ and $(L-x)/L$ coordinates. The lower portions of the curves are straight lines since $e^{\frac{Ux}{2D}}$ and $e^{\frac{UL}{2D}}$ are large compared to unity. As shown in Figure 5-3, within the vicinity of the $x=0$ boundary in Figure 5-1, different C/C_0 values can occur if the two different boundary conditions apply. Figure 5-4 is an extended version of Figure 5-3 such that the maximum value of $\mu = UL/(2D)$ is 10,000.

5.5.2 Application of the Steady-State Concentration Expressions

Applications of the steady-state concentration expressions given by Eqs. (5-18) and (5-19) are presented below with some examples.

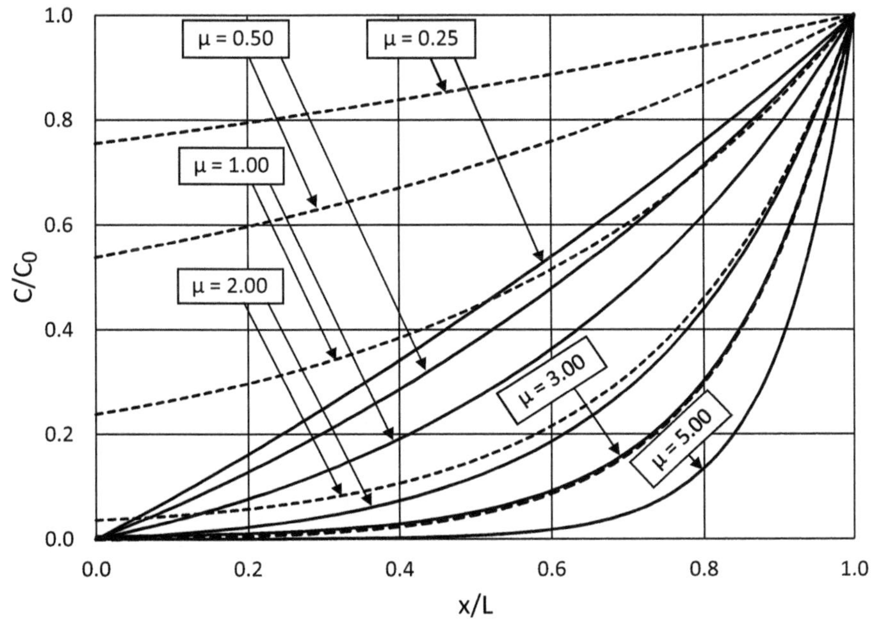

FIGURE 5-2
Variation of C/C_0 with x/L when dispersion is balanced by ground water flow. Each curve refers to a constant value of $\mu = UL/(2D)$. Full lines for the boundary condition given by Eq. (5-3); broken lines for the boundary condition given by Eq. (5-5). Adapted from Al-Niami and Rushton (1977 p. 92, Figure 2).

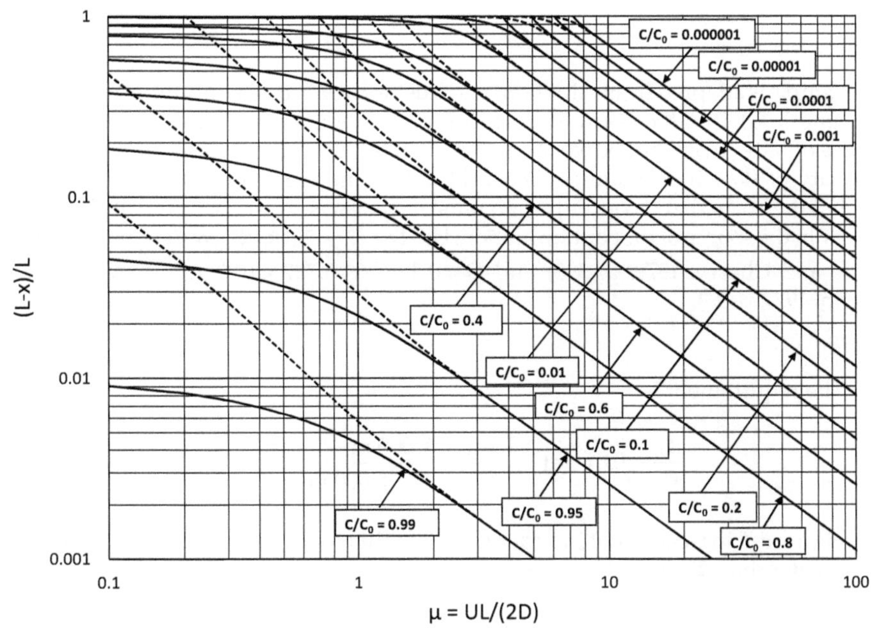

FIGURE 5-3
Variation of $(L-x)/L$ with $\mu = UL/(2D)$ when dispersion is balanced by ground water flow. Each curve refers to a constant value of C/C_0. Full lines for the boundary condition given by Eq. (5-3); broken lines for the boundary condition given by Eq. (5-5). Adapted from Al-Niami and Rushton (1977, p. 92, Figure 3).

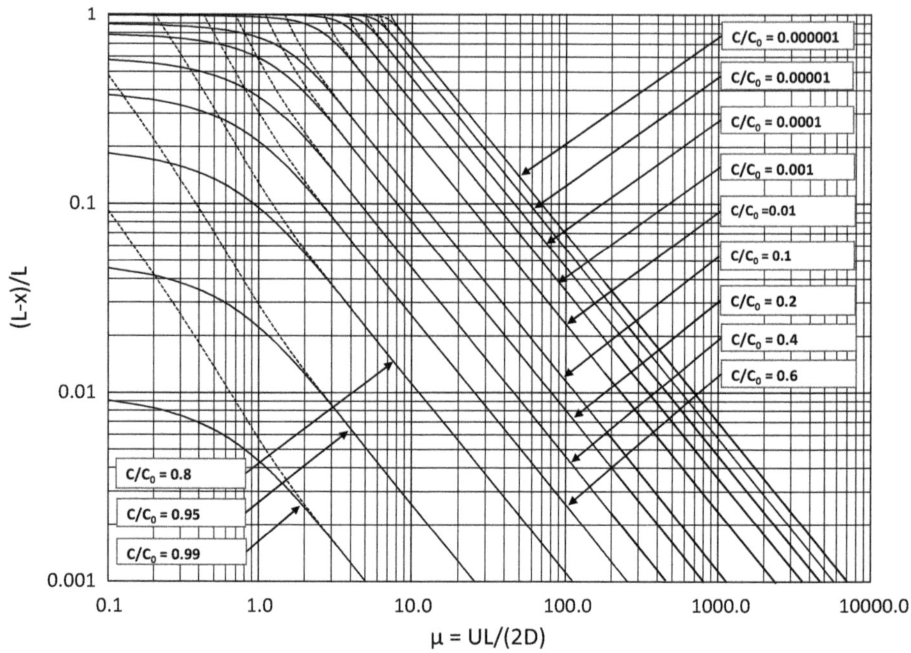

FIGURE 5-4
Variation of $(L-x)/L$ with $\mu = UL/(2D)$ when dispersion is balanced by ground water flow. Each curve refers to a constant value of C/C_0. Full lines for the boundary condition given by Eq. (5-3); broken lines for the boundary condition given by Eq. (5-5).

Example 5-5

This example is adapted from Al-Niami and Rushton (1977, p. 93). (a) Suppose that an aquifer of length 1,000 m, the normalized concentration must never exceed $C/C_0 = 0.01$ at a distance of 100 m from the source of contaminant (Figure 5-5). (a) Determine the required U velocity of flow in the aquifer. Also (b) determine the required U velocity of flow such that the normalized concentration must never exceed $C/C_0 = 0.0001$ at a distance of 100 m from the source of the contamination. Use $D = D^* = 2.0\times10^{-9}\ \text{m}^2/\text{s}\left(0.0001728\ \text{m}^2/\text{d}\right)$ (Freeze and Cherry, 1979, p. 393) and $D = 0.5\ \text{m}^2/\text{d}(5.787\times10^{-6}\ \text{m}^2/\text{s})$ (Al-Niami and Rushton, 1977, p. 93).

Solution

a. The requirement is

$$\frac{C}{C_0} < 0.01 \quad \text{at} \quad \frac{L-x}{L} = \frac{1{,}000\ \text{m} - 900\ \text{m}}{1{,}000\ \text{m}} = 0.1$$

Figure 5-6 indicates that in order to satisfy this condition, $\mu = UL/(2D)$ must be greater than 23, namely,

$$U > \frac{2\mu D}{L} = \frac{2(23)D}{1{,}000\ \text{m}} = 0.046D$$

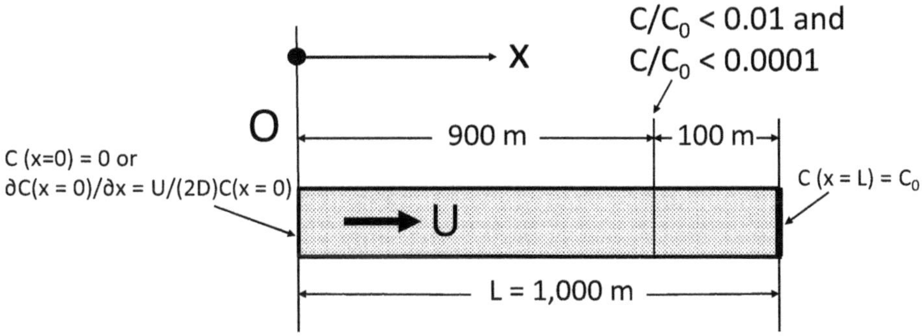

FIGURE 5-5
Schematic for Example 5-5.

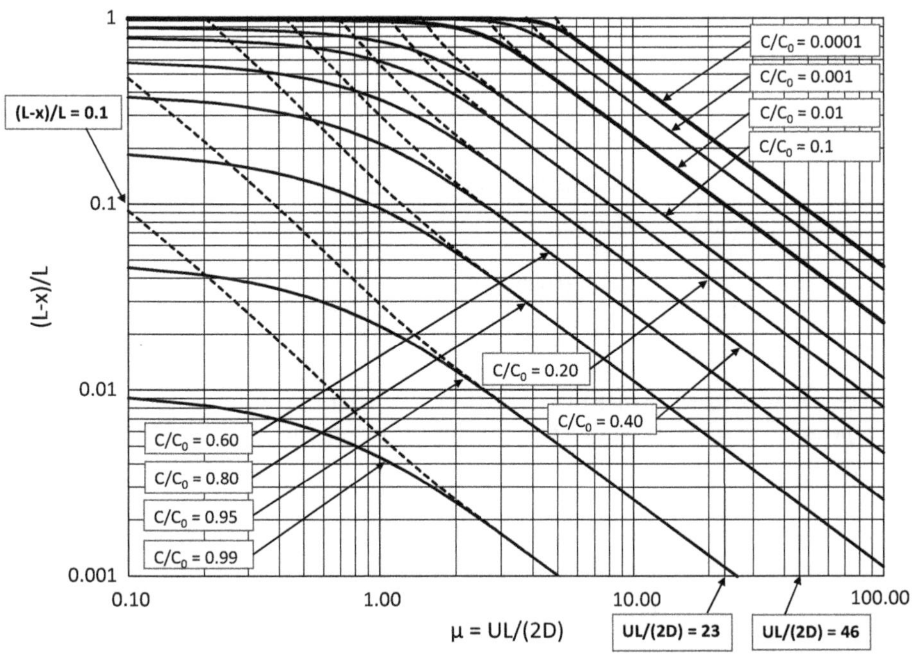

FIGURE 5-6
Application of Figure 5-3 for Example 5-5.

$$D = D^* = 2.0 \times 10^{-9}\ \text{m}^2/\text{s} = 0.0001728\ \text{m}^2/\text{d}:$$

$$U > \frac{2(23)\left(0.0001728\ \dfrac{\text{m}^2}{\text{d}}\right)}{1{,}000\ \text{m}} = 7.95 \times 10^{-6}\ \frac{\text{m}}{\text{d}}$$

This means that in order to satisfy the $C/C_0 < 0.01$ condition, the velocity in the aquifer (U) must be greater than 7.95×10^{-6} m/d.

$$D = 0.5\ \text{m}^2/\text{d} = 5.787 \times 10^{-6}\ \text{m}^2/\text{s}:$$

$$U > \frac{2(23)\left(0.5\ \frac{\text{m}^2}{\text{d}}\right)}{1{,}000\ \text{m}} = 0.023\ \frac{\text{m}}{\text{d}}$$

This means that in order to satisfy the $C/C_0 < 0.01$ condition, the velocity in the aquifer (U) must be greater than 0.023 m/d.

b. The requirement is

$$\frac{C}{C_0} < 0.0001 \quad \text{at} \quad \frac{L - x}{L} = \frac{1{,}000\ \text{m} - 900\ \text{m}}{1{,}000\ \text{m}} = 0.1$$

Figure 5-6 indicates that in order to satisfy this condition, $\mu = UL/(2D)$ must be greater than 46, namely,

$$U > \frac{2\mu D}{L} = \frac{2(46)D}{1{,}000\ \text{m}} = 0.092D$$

$$D = D^* = 2.0 \times 10^{-9}\ \text{m}^2/\text{s}\left(0.0001728\ \text{m}^2/\text{d}\right):$$

$$U > \frac{2(46)\left(0.0001728\ \frac{\text{m}^2}{\text{d}}\right)}{1{,}000\ \text{m}} = 15.90 \times 10^{-6}\ \frac{\text{m}}{\text{d}}$$

This means that in order to satisfy the $C/C_0 < 0.0001$ condition, the velocity in the aquifer (U) must be greater than 15.90×10^{-6} m/d.

$$D = 0.5\ \text{m}^2/\text{d}(5.787 \times 10^{-6}\ \text{m}^2/\text{s}):$$

$$U > \frac{2\mu D}{L} = \frac{2(46)\left(0.5\ \frac{\text{m}^2}{\text{d}}\right)}{1{,}000\ \text{m}} = 0.046\ \frac{\text{m}}{\text{d}}$$

This means that in order to satisfy the $C/C_0 < 0.0001$ condition, the velocity in the aquifer (U) must be greater than 0.046 m/d.

Example 5-6

In Figure 5-7, (a) suppose that an aquifer of length 1,000 m, the normalized concentration must never exceed $C/C_0 = 0.000001$ at a distance of 10 m from the source of the contaminant. Determine the required U velocity of flow in the aquifer. Also (b) determine the required U velocity of flow such that the normalized concentration must never exceed $C/C_0 = 0.000001$ at a distance of 1 m from the source of the contaminant. Use $D = D^* = 2.0 \times 10^{-9}\ \text{m}^2/\text{s}\left(0.0001728\ \text{m}^2/\text{d}\right)$ (Freeze and Cherry, 1979, p. 393) and $D = 0.5\ \text{m}^2/\text{d}(5.787 \times 10^{-6}\ \text{m}^2/\text{s})$ (Al-Niami and Rushton, 1977, p. 93).

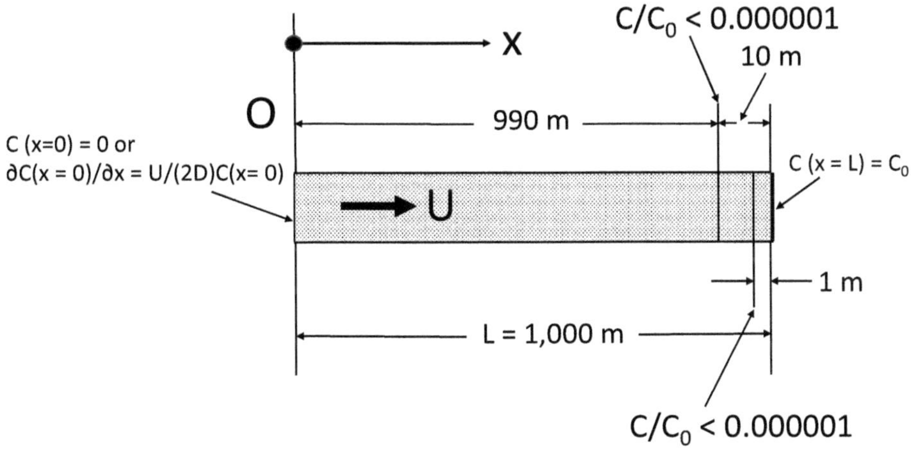

FIGURE 5-7
Schematic for Example 5-6.

Solution

a. The requirement is

$$\frac{C}{C_0} < 0.000001 \quad \text{at} \quad \frac{L-x}{L} = \frac{1{,}000\text{ m} - 990\text{ m}}{1{,}000\text{ m}} = 0.01$$

Figure 5-8 indicates that in order to satisfy this condition, $\mu = UL/(2D)$ must be greater than 700, namely,

$$U > \frac{2\mu D}{L} = \frac{2(700)D}{1{,}000\text{ m}} = 1.4D$$

$$D = D^* = 2.0 \times 10^{-9}\text{ m}^2/\text{s}\left(0.0001728\text{ m}^2/\text{d}\right):$$

$$U > \frac{2\mu D}{L} = \frac{2(700)\left(0.0001728\,\dfrac{\text{m}^2}{\text{d}}\right)}{1{,}000\text{ m}} = 2.42 \times 10^{-4}\,\frac{\text{m}}{\text{d}}$$

This means that in order to satisfy the $C/C_0 < 0.000001$ condition, the velocity in the aquifer (U) must be greater than 2.42×10^{-4} m/d.

$$D = 0.5\text{ m}^2/\text{d}(5.787 \times 10^{-6}\text{ m}^2/\text{s}):$$

$$U > \frac{2\mu D}{L} = \frac{2\left(0.5\,\dfrac{\text{m}^2}{\text{day}}\right)(700)}{1{,}000\text{ m}} = 0.7\,\frac{\text{m}}{\text{day}}$$

This means that in order to satisfy the $C/C_0 < 0.000001$ condition, the velocity in the aquifer (U) must be greater than 0.7 m/d.

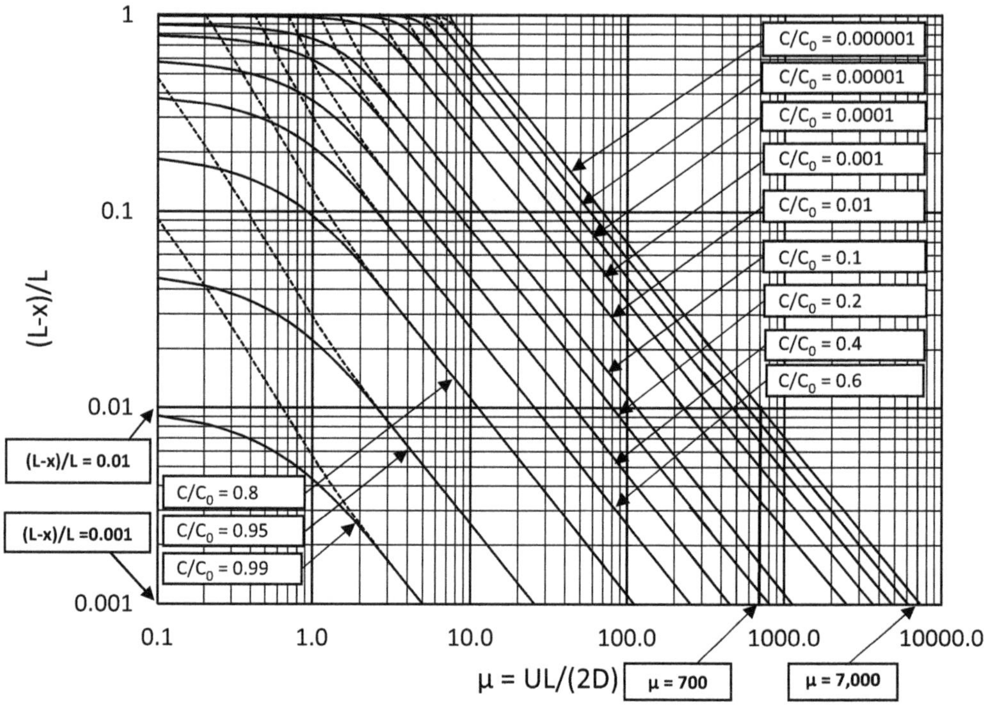

FIGURE 5-8
Application of Figure 5-4 for Example 5-6.

b. The requirement is

$$\frac{C}{C_0} < 0.000001 \quad \text{at} \quad \frac{L-x}{L} = \frac{1{,}000\ \text{m} - 999\ \text{m}}{1{,}000\ \text{m}} = 0.001$$

Figure 5-8 indicates that in order to satisfy this condition, $\mu = UL/(2D)$ must be greater than 7,000, namely,

$$U > \frac{2\mu D}{L} = \frac{2(7{,}000)D}{1{,}000\ \text{m}} = 14D$$

$$D = D^* = 2.0 \times 10^{-9}\ \text{m}^2/\text{s} = 0.0001728\ \text{m}^2/\text{d}:$$

$$U > \frac{2\mu D}{L} = \frac{2(7{,}000)\left(0.0001728\ \dfrac{\text{m}^2}{\text{d}}\right)}{1{,}000\ \text{m}} = 2.42 \times 10^{-3}\ \frac{\text{m}}{\text{d}}$$

This means that in order to satisfy the $C/C_0 < 0.000001$ condition, the velocity in the aquifer (U) must be greater than 2.42×10^{-3} m/d.

$$D = 0.5\ \text{m}^2/\text{d} = 5.787\times 10^{-6}\ \text{m}^2/\text{s}:$$

$$U > \frac{2\mu D}{L} = \frac{2(7{,}000)\left(0.5\,\dfrac{\text{m}^2}{\text{d}}\right)}{1{,}000\ \text{m}} = 7\,\frac{\text{m}}{\text{d}}$$

This means that in order to satisfy the $C/C_0 < 0.000001$ condition, the velocity in the aquifer (U) must be greater than 7 m/d.

Example 5-7

In Figure 5-9, the aquifer of length 50 m, the normalized concentration must never exceed $C/C_0 = 0.00001$ at a distance of 0.1 m from the source of the contaminant. Determine the required U velocity of flow in the aquifer. Use $D = D^* = 2.0\times 10^{-9}\ \text{m}^2/\text{s}\left(0.0001728\ \text{m}^2/\text{d}\right)$ (Freeze and Cherry, 1979, p. 393) and $D = 0.5\ \text{m}^2/\text{d}(5.787\times 10^{-6}\ \text{m}^2/\text{s})$ (Al-Niami and Rushton, 1977, p. 93).

Solution

The requirement is

$$\frac{C}{C_0} < 0.00001 \quad \text{at} \quad \frac{L-x}{L} = \frac{50\ \text{m} - 49.9\ \text{m}}{50\ \text{m}} = 0.002$$

Figure 5-10 indicates that in order to satisfy this condition, $\mu = UL/(2D)$ must be greater than 3,000, namely,

$$U > \frac{2\mu D}{L} = \frac{2(3{,}000)D}{50\ \text{m}} = 120D$$

$$D = D^* = 2.0\times 10^{-9}\ \text{m}^2/\text{s}\left(0.0001728\ \text{m}^2/\text{d}\right):$$

$$U > \frac{2\mu D}{L} = \frac{2(3{,}000)\left(0.0001728\,\dfrac{\text{m}^2}{\text{day}}\right)}{50\ \text{m}} = 0.021\,\frac{\text{m}}{\text{day}}$$

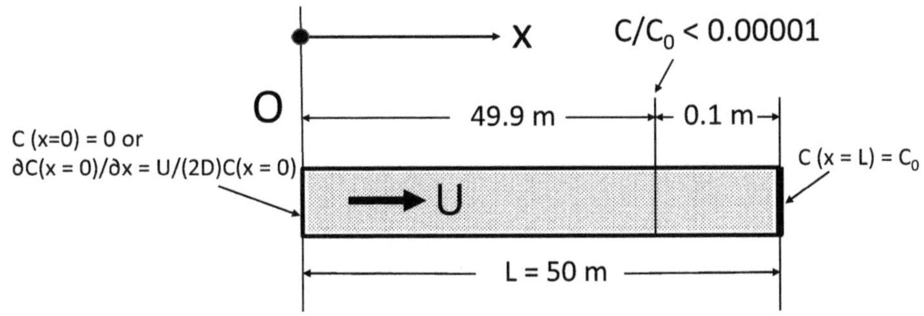

FIGURE 5-9
Schematic for Example 5-7.

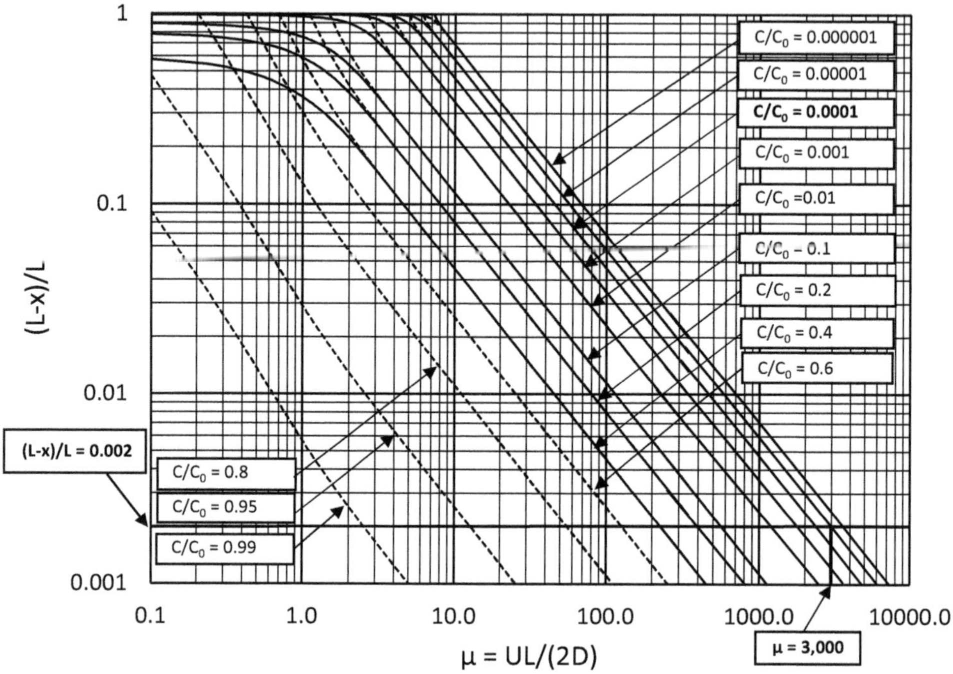

FIGURE 5-10
Application of Figure 5-4 for Example 5-7.

This means that in order to satisfy the $C/C_0 < 0.00001$ condition, the velocity in the aquifer (U) must be greater than 0.021 m/d.

$$D = 0.5\ \text{m}^2/\text{d}(5.787 \times 10^{-6}\ \text{m}^2/\text{s}):$$

$$U > \frac{2\mu D}{L} = \frac{2(3{,}000)\left(0.5\,\frac{\text{m}^2}{\text{d}}\right)}{1{,}000\ \text{m}} = 3\,\frac{\text{m}}{\text{d}}$$

This means that in order to satisfy the $C/C_0 < 0.00001$ condition, the velocity in the aquifer (U) must be greater than 3 m/d.

Problems

5.1 Explain the difference between the transient solutions given by Eqs. (5-12) and (5-17) by writing the corresponding equations.

5.2 Explain the status of the steady-state solutions given by Eqs. (5-18) and (5-19) and their application areas.

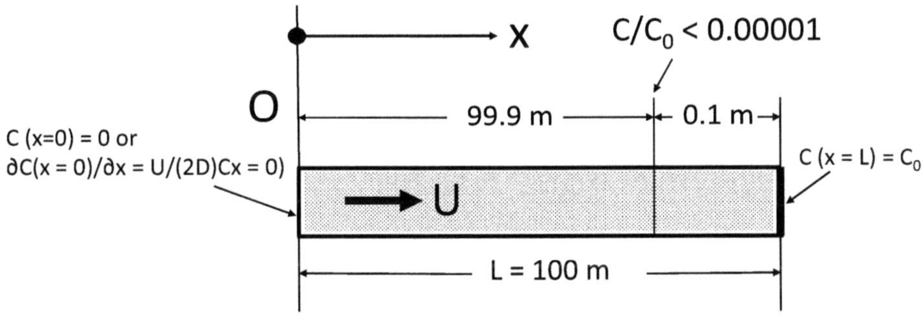

5.3 Suppose that an aquifer of length 100 m (Figure Problem 5.3), the normalized concentration must never exceed $C/C_0 = 0.00001$ at a distance of 0.1 m from the source of the contaminant. Determine the required U velocity of flow in the aquifer. Use $D = D^* = 2.0 \times 10^{-9}\ \text{m}^2/\text{s}(0.0001728\ \text{m}^2/\text{d})$ (Freeze and Cherry, 1979, p. 393) and $D = 0.5\ \text{m}^2/\text{d}(5.787 \times 10^{-6}\ \text{m}^2/\text{s})$ (Al-Niami and Rushton, 1977, p. 93).

6

Estimation of Degradation Rates of Solute Contaminants in Porous Media

6.1 Introduction

6.1.1 Terminology for Mass Loss during the Transport Process

Concepts regarding the degradation of materials began with decaying radioactive materials, and the term "radioactive decay" has emerged (e.g., Bear, 1972, 1979; Mercer et al., 1982). As the decay concept evolved, eventually simple differential equations with a decay constant were developed to quantify the decay phenomenon. As the name indicates, the term "radioactive decay" has been used specifically for radioactive materials. Later on, the term "decay" entered in the ground water literature as a more general term. During the past decades, some authors preferred to use the term "degradation" instead of "decay." Some authors use the term "weathering" for the combined effects of the degradation phenomena. It is necessary to point out that the term "degradation" is a general term and represents the effects of all mass loss in the ground water environment including biodegradation and volatilization. Therefore, in this book, the term "degradation" is used in a broader sense.

6.1.2 Literature for Degradation Rate Estimates

During the past half-century, some methods were developed to estimate the degradation rate constant for various contaminant constituents under laboratory and field conditions using measured dissolved solute concentrations (Buscheck and Alcantar, 1995; Chapelle et al., 1995; Zhang and Heathcote, 2003; Batu, 2010; Batu et al., 2013). These methods are based on the steady-state, one-dimensional solutions of the advective–dispersive solute transport differential equation with degradation and sorption effects using the first-type (Dirichlet-type) source condition with the exception of Batu (2010) for which the third-type (Cauchy-type) source condition is used. As will be pointed out later on, the first-type source solutions do not satisfy the mass balance at the inlet; whereas, the third-type source solution exactly satisfies the mass balance at the inlet. As a result, methods based on the first-type source solutions are associated with some errors in estimating degradation rates depending on the longitudinal dispersivity and the retardation factor.

As mentioned in Section 4.3, distinguishing between volume-averaged or resident concentrations (C_r) and flux-averaged or flowing concentrations (C_f) is widely accepted and has been used for decades. Distinctions between C_r and C_f have been made by scientists in the field of petroleum and chemical engineering (Brigham, 1974; Kreft and Zuber, 1978, 1979). Parker and van Genuchten (1984) showed that it is of fundamental importance to make such a distinction in order to stipulate boundary conditions appropriate for specific experimental solute detection models.

DOI: 10.1201/9781003739128-6

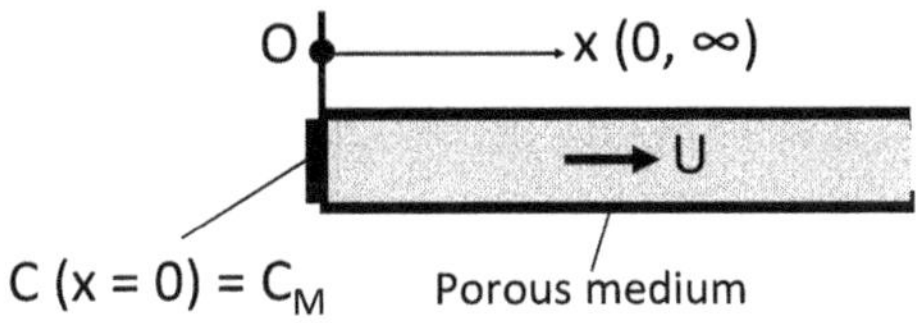

FIGURE 6-1
The geometry of one-dimensional steady-state solute transport in a porous medium having the first-type source condition.

6.2 One-Dimensional Steady-State First-Type Source Solution

6.2.1 Problem Description

The geometry of one-dimensional steady-state solute transport with the first-type source is shown in Figure 6-1, in which $U(L/T)$ is the ground water velocity (the ratio between the Darcy flux q and effective porosity φ_e) and $C_M\left(M/L^3\right)$ is the source concentration. As shown in Figure 6-1, the x coordinate tends to infinity.

6.2.2 Governing Differential Equation and Boundary Conditions

It has been shown that the solute transport differential equation in porous media for the flux- and volume-averaged (resident) concentrations is the same (Batu and van Genuchten, 1990). For the flux-averaged concentration case, the one-dimensional steady-state solute transport differential equation is [e.g., Batu, 2006, p. 58, Eq. (3-31)]

$$D_x \frac{\partial^2 C_f}{\partial x^2} - U \frac{\partial C_f}{\partial x} - R_d \nu C_f = 0 \tag{6-1}$$

in which $D_x(L^2/T)$ is the hydrodynamic dispersion coefficient, $C_f\left(M/L^3\right)$ is the flux-averaged concentration, $U(L/T)$ is the ground water velocity (the ratio between the Darcy flux q and effective porosity φ_e), R_d (unitless) is the retardation factor, and $\nu(1/T)$ is the degradation parameter, and $x(L)$ is the Cartesian coordinate. The expression for D_x is

$$D_x = D^* + \alpha_L U \tag{6-2}$$

in which $D^*\left(L^2/T\right)$ is the effective molecular diffusion coefficient and $\alpha_L\,(L)$ is the longitudinal dispersivity.

The solution for $C_f(x)$ is only dependent on the x coordinate. Therefore, the boundary condition at the source is

$$C_f(x=0) = C_M \tag{6-3}$$

The boundary conditions at infinity ($x = \infty$) are as follows:

$$C_f(x=\infty) = 0 \tag{6-4a}$$

$$\frac{\partial C_f(x=\infty)}{\partial x} = 0 \tag{6-4b}$$

6.2.3 Solution for Flux-Averaged Concentration

The solution of the problem described by Eqs. (6-1)–(6-4) is

$$C_f(x) = C_M \exp\left\{\frac{Ux}{2D_x}\left[1-\left(\frac{4D_xR_d\nu}{U^2}+1\right)^{\frac{1}{2}}\right]\right\} \tag{6-5}$$

which was first given by Bear [1979, p. 269, Eq. (7-138)]. The same solution is also given by Batu [2006, p. 59, Eq. (3-38)] as a special case of general two- and three-dimensional analytical solutions (Batu, 1996), and these solutions have been verified against some analytical and numerical solutions with excellent agreement (Batu, 1989, 2006). The degradation parameter estimation method developed by Buscheck and Alcantar (1995) is based on Eq. (6-5), and others (Chapelle et al., 1995; McNab and Dooher, 1998; Zhang and Heathcote, 2003) have used the same equation as well.

6.2.4 Advective–Dispersive Flux for Flux-Averaged Concentration

The advective–dispersive flux (F_{x_f}) is the only criterion needed to determine the concentration variation solution to satisfy mass balance. The advective–dispersive flux expression is [e.g., Freeze and Cherry, 1979, p. 550, Eq. (A10.4); Batu, 2006, p. 57, Eq. (3-24)]

$$F_{x_f}(x) = \varphi_e UC(x) - \varphi_e D_x \frac{\partial C_f(x)}{\partial x} \tag{6-6}$$

in which φ_e is the effective porosity and the other parameters are defined previously. The dimension of F_{x_f} is $M/L^2/T$, which means that mass loading is per unit time per unit area. According to Eq. (6-3), the concentration at $x = 0$ is C_M and the input flux is

$$F_{x_f}(x=0) = \varphi_e UC_M \tag{6-7}$$

Equation (6-7) is the definition of the flux-averaged concentration at the source. From Eqs. (6-5), (6-6), and (6-7), the advective–dispersive flux for the first-type source can be expressed as

$$F_{x_f}(x) = \phi_e UC_M \exp\left\{\frac{Ux}{2D_x}\left[1-\left(\frac{4D_xR_d\nu}{U^2}+1\right)^{\frac{1}{2}}\right]\right\} - \frac{1}{2}\phi_e UC_M\left[1-\left(\frac{4D_xR_d\nu}{U^2}+1\right)^{\frac{1}{2}}\right] \exp\left\{\frac{Ux}{2D_x}\left[1-\left(\frac{4D_xR_d\nu}{U^2}+1\right)^{\frac{1}{2}}\right]\right\} \tag{6-8}$$

At $x = 0$, Eq. (6-8) gives

$$F_{x_f}(x=0) = \phi_e UC_M\left[\frac{1}{2}+\left(\frac{D_xR_d\nu}{U^2}+\frac{1}{4}\right)^{\frac{1}{2}}\right] \tag{6-9}$$

If the degradation parameter (v) becomes zero, Eq. (6-9) becomes

$$F_{x_f} = \varphi_e U C_M \tag{6-10}$$

which is exactly the same as Eq. (6-7). From Eq. (6-9), it is interesting to observe that the advective–dispersive flux is the function of the degradation parameter (v) as well as the flow and solute transport parameters.

6.3 One-Dimensional Steady-State Third-Type Source Solution

6.3.1 Problem Description

The geometry of one-dimensional steady-state solute transport with the third-type source is shown in Figure 6-2, in which $U\left(L/T\right)$ is the ground water velocity (the ratio between the Darcy flux q and effective porosity φ_e) and $F_M\left(M/L^2/T\right)$ is solute the source flux. As shown in Figure 6-2, the x coordinate tends to infinity.

6.3.2 Governing Differential Equation and Boundary Conditions

As mentioned above, the flux-averaged and volume-averaged (resident) concentrations satisfy the same differential equation. Therefore, for volume-averaged concentrations case, the one-dimensional steady-state differential equation is (van Genuchten and Parker, 1984; Parker and van Genuchten, 1984)

$$D_x \frac{\partial^2 C_r}{\partial x^2} - U \frac{\partial C_r}{\partial x} - R_d v C_r = 0 \tag{6-11}$$

in which $C_r(M/L^3)$ is the volume-averaged concentration and the rest of parameters have been defined previously. The solution for $C_r(x)$ is dependent only on the x coordinate. Therefore, the boundary condition at the source is

$$F_{x_r}(x = 0) = F_M \tag{6-12}$$

in which $F_M(M/L^2/T)$ is the solute input flux at the source. Using Eq. (6-6), Eq. (6-12) can be expressed as

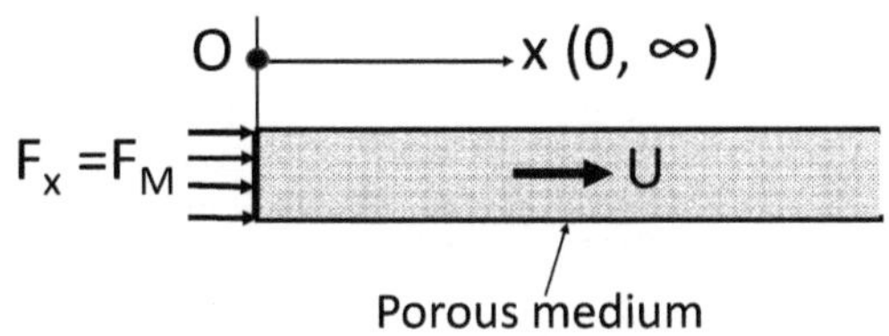

FIGURE 6-2
The geometry of one-dimensional steady-state solute transport in a porous medium having the third-type source condition.

$$F_{x_r}(x=0) = F_{M_r} = \left[\varphi_e U C_r - \varphi_e D_x \frac{\partial C_r}{\partial x}\right]_{x=0^+} \tag{6-13}$$

Combining Eqs. (6-12) and (6-13) gives the equivalent forms of van Genuchten and Parker [1984, page 703, Eq. (3)] and Parker and van Genuchten [1984, page 867, Eq. (7)]. As these authors stated, $x = 0^+$ indicates evaluation at the inlet boundary just inside the profile. It must be pointed out that F_M is the mass input at the inlet and its expression is given by Eq. (6-17).

The boundary conditions at infinity ($x = \infty$) are as follows:

$$C_r(x=\infty) = 0 \tag{6-14a}$$

$$\frac{\partial C_r(x=\infty)}{\partial x} = 0 \tag{6-14b}$$

6.3.3 Solution for Volume-Averaged Concentration Variation

The solution of the above-defined problem was given by Gershon and Nir (1969). The solution of the same but more general case was given as a special case of a more general one-dimensional analytical solute transport solution (van Genuchten, 1981). Although the Gershon and Nir (1969) paper is cited in van Genuchten (1981), it appears that the author was not aware that his special case solution was derived previously by Gershon and Mir. The solution of this simple one-dimensional boundary-value problem is also given in Batu (2006) based on Batu (1993) as a special case of the two- and three-dimensional analytical solutions without being aware of the aforementioned solutions. Therefore, the solution of the above-defined problem is [Batu, 2006, p. 126, Eq. (3-247)]

$$C_r(x) = \frac{F_M}{\varphi_e\left(\frac{U}{2} + D_x p_0\right)} \exp\left(\frac{Ux}{2D_x} - p_0 x\right) \tag{6-15}$$

in which

$$p_0 = \left(\frac{R_d}{D_x}\nu + \frac{U^2}{4D_x^2}\right)^{\frac{1}{2}} \tag{6-16}$$

With

$$F_{x_r}(x=0) = F_M = \varphi_e U C_M \tag{6-17}$$

Eq.(6-15) becomes

$$C_r(x) = C_M \frac{\exp\left\{\frac{Ux}{2D_x}\left[1 - \left(\frac{4D_x R_d \nu}{U^2} + 1\right)^{\frac{1}{2}}\right]\right\}}{\left[\frac{1}{2} + \left(\frac{D_x R_d \nu}{U^2} + \frac{1}{4}\right)^{\frac{1}{2}}\right]} \tag{6-18}$$

Equation (6-15) is given as a special case of the general two- and three-dimensional analytical solutions [Batu, 2006; page 143, Eq. (3-294); page 167, Eq. (3-353)]. Equation (6-15) is also given as a special case of the same general two-dimensional solute transport solution (Batu, 1993) and these two- and three-dimensional analytical solute transport solutions have been verified extensively against some three-dimensional numerical models with excellent agreement (Batu, 1993, 2006).

6.3.4 Advective–Dispersive Flux for Volume-Averaged Concentrations

From Eqs. (6-13) and (6-15), the convective–dispersive flux expression is [Batu, 2006, page 126, Eq. (3-249)]

$$F_{x_r}(x) = \varphi_e U C_r(x) - \frac{D_x F_M}{\left(\frac{U}{2} + D_x p_0\right)}\left(\frac{U}{2D_x} - p_0\right)\exp\left(\frac{Ux}{2D_x} - p_0 x\right) \tag{6-19}$$

in which $C_r(x)$ is given by Eq. (6-15). After simplifications, Eq. (6-19) takes the form

$$F_{x_r} = \varphi_e U C_M \exp\left\{\frac{Ux}{2D_x}\left[1 - \left(\frac{4D_x R_d \nu}{U^2} + 1\right)^{\frac{1}{2}}\right]\right\} \tag{6-20}$$

At $x = 0$, Eq. (6-20) becomes

$$F_{x_r}(x = 0) = \varphi_e U C_M \tag{6-21}$$

It is important to point out that unlike the first-type source case, the convective–dispersive flux at $x = 0$ for the resident concentration is not a function of the degradation parameter (ν) and the flow and solute transport parameters. In other words, Eq. (6-21) is the exact expression of the solute input flux at the source.

6.4 The Relationship between the First- and Third-Type Source Analytical Solutions

Combination of the flux-averaged concentration defined by Eq. (6-7) with the volume-averaged concentration defined by Eq. (6-13) gives (Kreft and Zuber, 1979; Parker and van Genuchten, 1984)

$$C_f(x) = C_r(x) - \frac{D_x}{U}\frac{\partial C_r(x)}{\partial x} \tag{6-22}$$

which is valid for nonzero ground water velocity (U) values. Flux-averaged concentrations may be interpreted physically as representing the mean of the microscopic fluid concentrations weighted by their respective microscopic fluid velocities. The discrepancy between C_f and C_r increases with the dispersion coefficient D_x (Kreft and Zuber, 1978). In the special case of $U = 0$ with mixing solely by diffusion, C_f has no physical relevance and is mathematically undefined (Parker and van Genuchten, 1984).

Equation (6-22) has some important ramifications, and it has been used in two ways in the literature. The first one is to come up with some physical explanations for the volume-averaged and flux-averaged concentrations (e.g., Parker and van Genuchten, 1984, pp. 867–868). The second one is to determine the third-type source solution from the first-type source solution and vice versa (Kreft and Zuber, 1979). Application of this approach is also given in Batu and van Genuchten (1990, p. 347). It is necessary to point out that Eq. (6-22) has nothing to do with the mass balance conservation regarding the first-type source solution.

Equation (6-22) gives the flux-averaged concentration (C_f) in terms of the volume-averaged concentration (C_r). As mentioned above, Eq. (6-22) can be used to derive the first-type solution from the third-type solution. Also, the third-type solution may be derived from the first-type solution by using the inverse of Eq. (6-22) [van Genuchten et al., 1984, p. 340, Eqs. (74a) and (74b)]. By applying this method [Batu and van Genuchten, 1990, p. 347, Eq. (44)], one can write

$$C_r(x) = \frac{U}{D_x}\exp\left(\frac{Ux}{D_x}\right)\int_x^{\infty}\exp\left(-\frac{U\xi}{D_x}\right)C_f(\xi)\,d\xi \tag{6-23}$$

in which $C_f(\xi)$ is given by Eq. (6-5) by replacing ξ with x. Using Eq. (6-5) in Eq. (6-23) and performing the integration, one gets exactly the same solution for the third-type source as given by Eq. (6-18). It must be emphasized that the relationships given by Eqs. (6-22) and (6-23) can only hold for the simplified solute transport cases, and the necessary conditions are that the ground water flow is steady unidirectional and the flow direction is perpendicular to the input boundary at $x = 0$. Further information regarding this subject with two- and three-dimensional first-type and third-type analytical solutions can be found in Batu (2006).

6.5 Comparative Evaluation of the First- and Third-Type Source Solutions for Mass Balance at the Inlet

The flux- and volume-averaged solutions will be compared using the following dimensionless parameters:

$$X = \frac{Ux}{2D_x} \qquad N = \frac{D_x R_d \nu}{U^2} \tag{6-24}$$

With these dimensionless parameters, the flux-averaged concentration expression given by Eq. (6-5) becomes

$$C_f(X) = C_M \exp\left\{X\left[1-(4N+1)^{\frac{1}{2}}\right]\right\} \tag{6-25}$$

The advective–dispersive flux expression given by Eq. (6-8) takes the form

$$F_{X_f}(X) = \frac{1}{2}\varphi_e U C_M\left[1+(4N+1)^{\frac{1}{2}}\right]\exp\left\{X\left[1-(4N+1)^{\frac{1}{2}}\right]\right\} \tag{6-26}$$

With the dimensionless parameters defined by Eq. (6-24), the volume-averaged concentration expression given by Eq. (6-18) becomes

$$C_r(X) = C_M \frac{\exp\left\{X\left[1-(4N+1)^{\frac{1}{2}}\right]\right\}}{\left[\frac{1}{2}+\left(N+\frac{1}{4}\right)^{\frac{1}{2}}\right]} \tag{6-27}$$

The advective–dispersive flux expression given by Eq. (6-20) takes the form

$$F_{X_r}(X) = \varphi_e U C_M \exp\left\{X\left[1-(4N+1)^{\frac{1}{2}}\right]\right\} \tag{6-28}$$

Review of the concentration and convective–dispersive flux expressions for the first-type and third-type sources reveals some interesting features, and these are presented below.

When $X = 0$, the first-type source solution, Eq. (6-25), always gives $C_f(X=0) = C_M$ without depending on the degradation parameter (N). For the third-type source case, Eq. (6-27) becomes

$$C_r(X=0) = \frac{C_M}{\left[\frac{1}{2}+\left(N+\frac{1}{4}\right)^{\frac{1}{2}}\right]} \tag{6-29}$$

which indicates that $C_r(X=0)$ at the source depends on N and $C_r(X=0)$ becomes equal to C_M if $N=0$. In other words, for degrading solutes, $C_r(X=0)$ is always less than C_M. This situation at first may sound somewhat striking, but the review of the flux component expressions can clarify the physical reality of the source conditions. Eq(6-28) indicates that the third-type source flux for $X = 0$ is

$$F_{X_r}(X=0) = \varphi_e U C_M \tag{6-30}$$

without depending on the values of N. And this is nothing but the source input flux condition as given by Eq. (6-17). In other words, for the third-type source case, the mass balance constraint is exactly satisfied at the inlet location. However, for the first-type source, this is not the case. For $X = 0$, Eq. (6-26) becomes

$$F_{X_f}(X=0) = \frac{1}{2}\varphi_e U C_M\left[1+(4N+1)^{\frac{1}{2}}\right] \tag{6-31}$$

which indicates that the mass balance constraint is not satisfied for degraded solutes at the inlet location. In other words, the input flux $F_{X_f}(X=0)$ increases as the value of N increases and this is an obvious mass balance violation. Only when $N = 0$, the mass balance constraint given by Eq. (6-17) or (6-30) is satisfied.

Ground water practitioners mostly take samples from flowing ground water in the field. However, concentration measurements are made in the laboratory under no flow conditions ($U = 0$) and the measured concentrations are volume-averaged concentrations (C_r)

not flux-averaged concentrations (C_f). As mentioned previously, in the special case of $U = 0$, mixing occurs solely by molecular diffusion, C_f has no physical meaning and is mathematically undefined (Parker and van Genuchten, 1984, p. 867). Using the measured volume-averaged concentration at the source (C_M), the calculated concentration at $x = 0$ by Eq. (6-18) will be less than the laboratory-measured C_M value depending on the values of N. If the ground water velocity (U) is greater than zero, this is not unexpected.

Based on the above-mentioned facts, measurement of volume-averaged concentrations (C_r) is straightforward in principle by core sampling. However, the measurement of flux-averaged concentrations (C_f) is generally more difficult. Parker and van Genuchten (1984, p. 1159) state that the least ambiguous procedure for determining C_f at any arbitrary location within the porous medium is to calculate $C_f(x,t)$ from $C_r(x,t)$ data either through definition of C_f as the ratio of solute flux to hydraulic flux or through some other expression relating C_f and C_r [e.g., Eq. (6-22)].

In the first-type and third-type steady-state solute transport analytical solutions, the critical parameter is N, which is defined by the second expression in Eq. (6-24). The dimensionless parameter N can only become zero if the degradation parameter (ν) is zero. If $N = 0$, the concentration variations become the same and uniform from $X = 0$ to infinity for both first-type and third-type source cases. Likewise, for $N = 0$; at $X = 0$, the advective–dispersive fluxes for the first-type and third-type source cases become the same and uniform as well by exactly satisfying the mass balance constraint at the inlet location. As the value of N increases, the deviations between the first-type and third-type source solution concentrations increase. For $N = 1$, Figures 6-3 and 6-4 show comparisons of normalized source concentrations and normalized advective–dispersive fluxes, respectively. Likewise, for $N = 5$, Figures 6-5 and 6-6 show comparisons of normalized source concentrations and normalized advective–dispersive fluxes, respectively. These figures show that the discrepancies between the first-type and third-type solutions increase as the values of

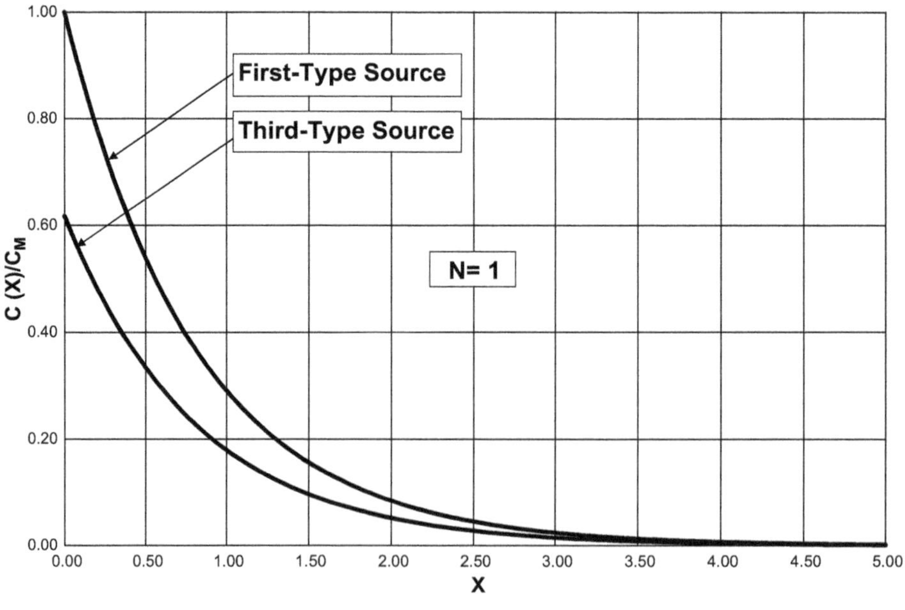

FIGURE 6-3
Comparison of first- and third-type normalized source concentrations for $N = 1$.

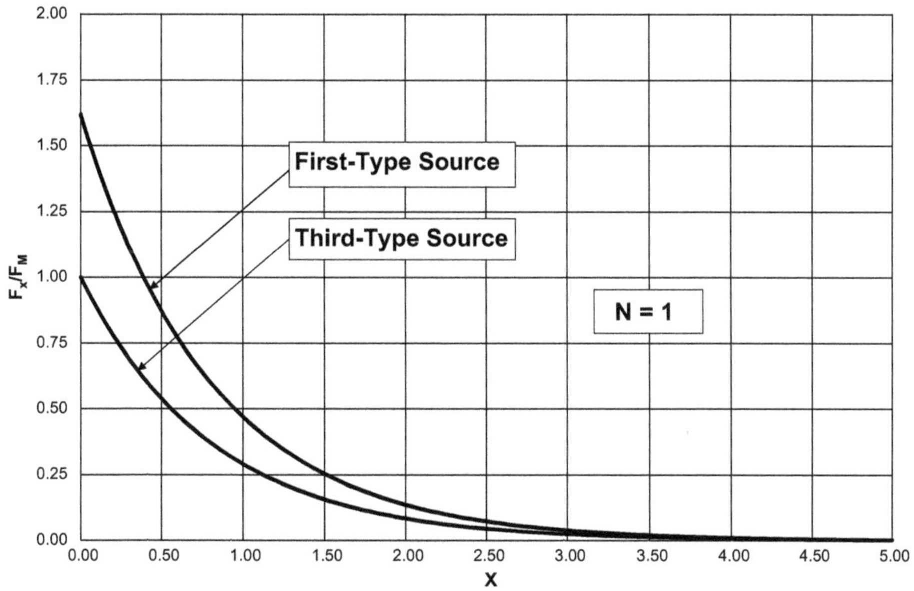

FIGURE 6-4
Comparison of first- and third-type normalized source advective–dispersive fluxes $N = 1$.

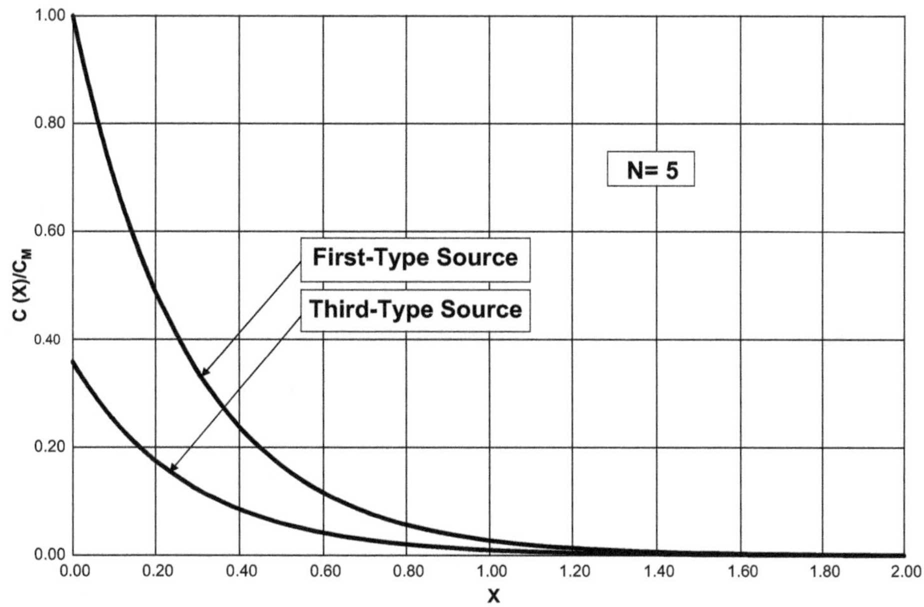

FIGURE 6-5
Comparison of first- and third-type normalized source concentrations for $N = 5$.

N increases. It must be emphasized that for the third-type source case, the ratio of F_{X_r}/F_M is always equal to unity (Figures 6-4 and 6-6) at the source ($X = 0$) for every N value, which means that the mass balance constraint as given by Eq. (6-30) is exactly satisfied at

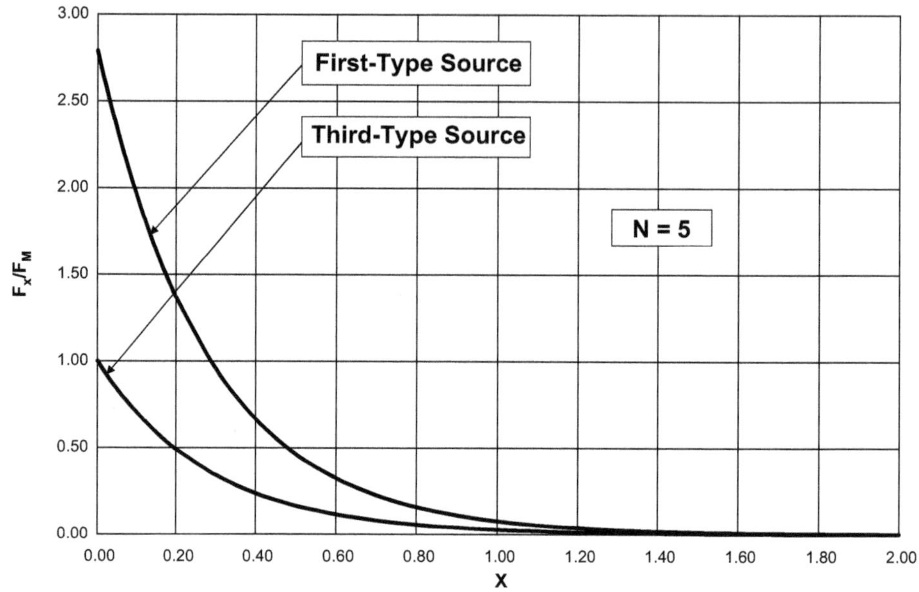

FIGURE 6-6
Comparison of first- and third-type normalized source advective–dispersive fluxes $N = 5$.

the inlet location. However, this is not the case for the first-type source solution since the ratio F_{X_f}/F_M increases as the values of N increase. For example, for $N = 1$, $F_{X_f}/F_M = 1.62$ (Figure 6-4) and for $N = 5$ (Figure 6-6), $F_{X_f}/F_M = 2.79$. Here, $F_{X_f}/F_M = 2.79$ means that at the source the flux is $F_{X_f}(X = 0) = 2.79\varphi_e UC_M$, which means that the flux is 2.79 times higher than the actual flux, a significant mass balance violation. Strictly speaking, the first-type source solution given by Eq. (6-5) fails to satisfy the mass balance constraint at the inlet location ($x = 0$).

6.6 Evaluation of the First-Type Source Solution-Based Method for Estimating Degradation Parameter

6.6.1 Literature for Degradation Rate Estimates Based on the First-Type Source Solution

As mentioned in Section 6.1.2, during the past half-century, methods (Buscheck and Alcantar, 1995; Chapelle et al., 1995; Zhang and Heathcote, 2003) were developed to estimate the degradation rate constant for various contaminant constituents using the first-type one-dimensional steady-state solute transport solution in porous media, which is given by Eq. (6-5). Based on the analyses in previous sections, it has been proven that the first-type source solutions do not satisfy the mass balance at the inlet. As a result, methods based on the first-type solutions are associated with some errors in estimating degradation rates.

Since the aforementioned methods are based mainly on the pioneering method of Buscheck and Alcantar (1995), in this section, this method will be described.

6.6.2 Buscheck and Alcantar Method Degradation Rate Estimation Method Based on the First-Type Source Solution

6.6.2.1 Equations for Transient and Steady-State Plumes

For transient plumes (the authors call them as shrinking plumes), Buscheck and Alcantar (1995) state that dissolved hydrocarbon plumes may decrease in size, as observed by declining contaminant concentrations in monitoring wells; therefore, exponential regression methods can be used to evaluate whether concentration versus time data fit a first-order decay observed for petroleum hydrocarbons under certain conditions. Then, they give the solution for the first-order decay as

$$C_f(t) = C_i e^{-kt} \tag{6-32}$$

in which $C_f(t)(M/L^3)$ is the concentration as a function of time t (T), C_i (M/L^3) is the initial concentration at $t = 0$, and k $(1/T)$ is the first-order attenuation rate.

For steady-state plumes (the authors call them as stable plumes), Buscheck and Alcantar (1995) state that Kemblowski et al. (1987) recast (altering or presenting it in a different way) Eq. (6-32) for concentration as a function of distance as

$$C_f(x) = C_M e^{-k\frac{x}{u}} \tag{6-33}$$

in which $C_M(M/L^3)$ is the source concentration.

6.6.2.2 Degradation Rate Expression from the Simplified First-Type Source Solution

The Buscheck and Alcantar (1995) method is based on Eq. (6-5) under the condition that the effective molecular diffusion coefficient D^* in Eq. (6-2) is negligible as compared with the mechanical dispersion coefficient $\alpha_L U$. With this,

$$R_d = \frac{U}{U_c} \tag{6-34}$$

in which $U(L/T)$ is the ground water velocity (the ratio between the Darcy flux q and effective porosity φ_e) and $U_c(L/T)$ is the contaminant velocity. Under these conditions, Eq. (6-5) takes the form

$$C_f(x) = C_M \exp\left\{\frac{x}{2\alpha_L}\left[1 - \left(\frac{4v\alpha_L}{U_c} + 1\right)^{\frac{1}{2}}\right]\right\} \tag{6-35}$$

which is exactly the same as Buscheck and Alcantar [1995, Eq. (6)] with the exception of notation (Figure 6-1). The authors state that Eqs. (6-33) and (6-35) take the form

$$C_f(x) = C_M \exp(mx) \tag{6-36}$$

in which

$$m = \frac{1}{2\alpha_L}\left[1-\left(\frac{4v\alpha_L}{U_c}+1\right)^{\frac{1}{2}}\right] \tag{6-37}$$

In Eq. (6-36), m is the slope of the log-linear data. Buscheck and Alcantar state that the term $-k/U$ and Eq. (6-37) both describe the slope of the log-linear data and can be equated to solve for the decay rate v. Therefore,

$$-\frac{k}{U} = \frac{1}{2\alpha_L}\left[1-\left(\frac{4v\alpha_L}{U_c}+1\right)^{\frac{1}{2}}\right] \tag{6-38}$$

which can be written as

$$-\frac{2\alpha_L k}{U} = 1-\left(\frac{4v\alpha_L}{U_c}+1\right)^{\frac{1}{2}} \tag{6-39}$$

From here,

$$\left(\frac{4v\alpha_L}{U_c}+1\right)^{\frac{1}{2}} = 1+\frac{2\alpha_L k}{U} \tag{6-40}$$

Taking the square of both sides

$$\frac{4v\alpha_L}{U_c}+1 = \left(1+\frac{2\alpha_L k}{U}\right)^2 \tag{6-41}$$

And solving it for v,

$$v = \frac{U_c}{4\alpha_L}\left[\left(1+2\alpha_L\frac{k}{U}\right)^2-1\right] \tag{6-42}$$

which is exactly the same as Buscheck and Alcantar [1995, Eq. (8)] with the exception of notation.

6.6.2.3 Degradation Estimation Method of Buscheck and Alcantar Based on the First-Type Source Solution

Buscheck and Alcantar (1995) presented the application of their Eq. (6-42) with an example. They regressed the values of source concentration (C_M) and k/U using the measured concentration values at several monitoring wells in distance (x) from the source versus concentration (logarithmic) coordinate system. As will be shown later on, the first-type source solution generates a mass balance violation at the inlet. As a result, the first-type source solution-based method overestimates the degradation parameter depending on the values of α_L and R_d.

6.6.3 Evaluation of the First-Type Source-Solution Based Method

As mentioned previously, Buscheck and Alcantar (1995) proposed the steady-state first-type source solution, Eq. (6-5), in estimating the degradation parameters and later many authors and institutions adopted this approach and published papers and reports on the same subject. However, Eq. (6-5) fails to satisfy the mass balance constraint depending on the value of the degradation parameter N. The parameter N, defined by the second expression of Eq. (6-24), must be determined to use Eq. (6-5). Comparing Eqs. (6-25) and (6-27) indicates that if N is negligible as compared with "1/4" in the denominator of Eq. (6-27), the first-type and third-type source solutions produce approximately the same results. This situation can be shown with an example in Chapelle et al. (1996, page 696, Figure 6): The given parameters are $D_x = 5\ \text{ft}^2/\text{d}$, $U = 2\ \text{ft}/\text{d}$, $R_d = 1$, and $v = 0.04\ \text{d}^{-1}$. Therefore, $N = D_x R_d v / U^2 = 0.05$. With this value, the denominator of Eq. (6-27) becomes 1.05. This means that $C_r(x) = 0.95\ C_f(x)$. This may be assumed as a negligible error depending on the site-specific application. If the values of N are significantly high, the error introduced by the first-type source solution increases in accordance with the increased values of N. The values of N increase for solutes in low-permeable porous materials with high sorption capacity and high degradation rates. Therefore, it is recommended that the first-type source-based model should not be used to determine the degradation parameters if N is not negligible as compared with "1/4." More specifically, the first-type source solution given by Eq. (6-5) should be used with caution in determining the degradation parameters based on measured concentrations. Based on the limitation of the first-type source solution; instead, it is recommended that the third-type source model given by Eq. (6-18) be used in estimating the degradation parameters because the third-type source solution exactly satisfies the mass balance constraint at the inlet location while the first-type source solution does not. The mass balance constraints are further elaborated below.

It can easily be shown that the $C_f(x)$ expression, Eq. (6-5), and F_{x_f} expression, Eq. (6-8), exactly satisfy the mass balance expression

$$\frac{\partial F_{x_f}}{\partial x} + \varphi_e R_d v C_f = 0 \tag{6-43}$$

in terms of flux-averaged concentration. With the F_{x_f} expression, Eq. (6-43) takes the same form as Eq. (6-1). In other words, Eq. (6-43) satisfies one of the mass balance constraints, but it does not mean that Eq. (6-5) is the correct solution to be used in estimating the values of degradation rates. Moreover, satisfying mass balance is not unexpected because the solute transport differential equation can be formulated in terms of both C_f and C_r as shown by Eqs. (6-1) and (6-11) (Batu and van Genuchten, 1990). As a result, the mass balance equation in terms of C_r and F_{x_r}

$$\frac{\partial F_{x_r}}{\partial x} + \varphi_e R_d v C_r = 0 \tag{6-44}$$

is also exactly satisfied by the $C_r(x)$, Eq. (6-18), and $F_{x_r}(x)$, Eq. (6-19), expressions. Eq(6-43) does not mean that the $C_f(x)$ solution is appropriate with the inlet condition of the problem. It is known that the solute flux at $x = 0$ is equal to $\varphi_e U C_M$, which is given by Eq. (6-21). As shown previously, the $C_r(x)$ solution exactly satisfies the flux expression given by Eq. (6-21). However, as indicated by Eq. (6-9) or (6-31), the first-type source solution violates this condition such that the solute flux is also dependent on N, which is defined by the second expression of Eq. (6-24). With the fact that degradation may occur beyond the inlet location ($x > 0$), the dependency of F_{xf} on the degradation parameter (N) at $x = 0$ is also a clear violation of the mass balance constraint at the inlet location. Therefore, the appropriate method for estimating the degradation parameter (ν) values based on measured concentrations is to use the solution based on the volume-averaged concentration (C_r) given by Eq. (6-18).

This subject has been further elaborated by V. Batu, M.Th. van Genuhcten, and J.C. Parker in Batu et al. (2013) in responding some comments for Batu (2010). Regarding the use of the third-type source condition, the authors make the following statements (Batu et al., 2013, p. 12):

> In conclusion, we remain solidly behind the statements in Batu (2010) that third-type solutions should be used in transport studies that focus on resident concentrations since they guarantee solute flux continuity across inlet boundaries. On the other hand, first-type boundary conditions satisfy mass balance criteria only if the modeled concentrations are meant to represent flux concentrations; they do not satisfy mass balance constraints for resident concentrations.

6.6.4 Comparative Evaluation of the First-Type Source Condition with an Example

The following parameters are used in the example: $C_M = 1{,}000\ \text{mg/m}^3 = 1\ \text{mg/L}$, $\varphi_e = 0.25$, $U = 0.10\ \text{m/d}$, $\alpha_L = 2\ \text{m}$, $D^* = 2.0 \times 10^{-9}\ \text{m}^2/\text{s} = 1.728 \times 10^{-4}\ \text{m}^2/\text{d}$, $\nu = 0.01\ \text{d}^{-1}$, and $R_d = 10$. From Eq. (6-21), the solute flux at the source can be calculated as $F_M = 25\ \text{mg/d/m}^2$. This is the input for the third-type source solution. Instead, for the first type-source solution, C_M is the input. Using the above parameter values, the second expression of Eq. (6-24)

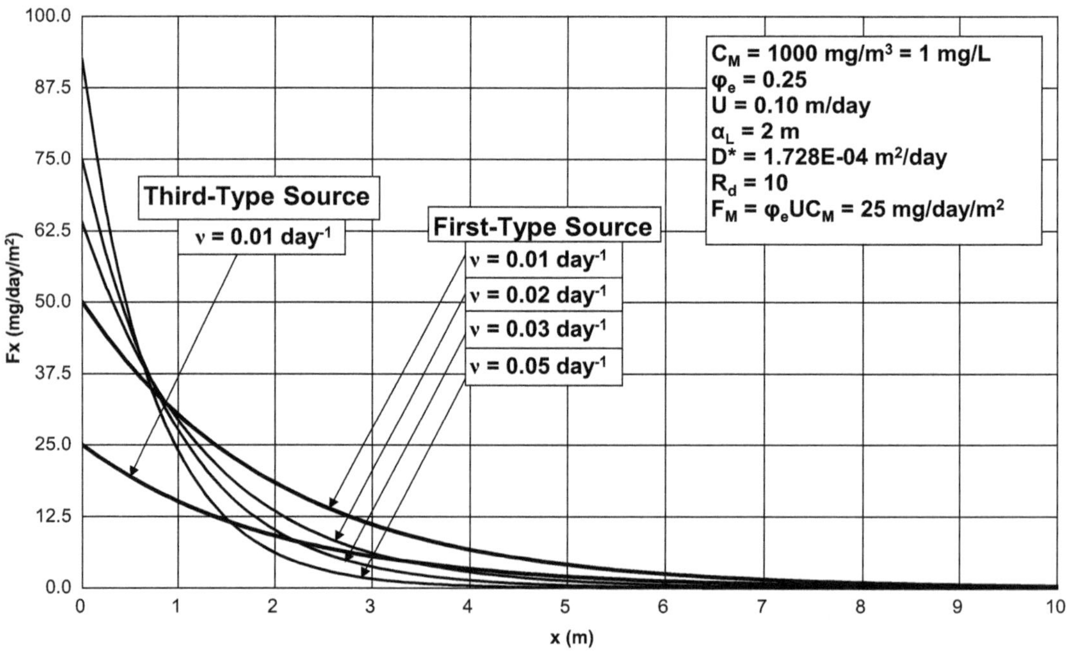

FIGURE 6-7
Comparison of the first-type and third-type source concentration profile for $\nu = 0.01\ \text{d}^{-1}$ and first-type source concentration profiles for $\nu = 0.02$, 0.03, and 0.05 d^{-1}. Adapted from Batu (2010, p. 566, Figure 3).

gives $N = 2$. Figure 6-7, which presents the concentration (both C_f and C_r) variation with the distance (x) for the first-type and third-type source cases for $v = 0.01\ \mathrm{d}^{-1}$, shows that the concentration profiles are significantly different and there is no match between the two curves with the exception of large distance (x) values. The most striking situation is that the first-type source solution overpredicts the concentration variation. For comparison purposes, the concentration profiles for $v = 0.02$, 0.03, and $0.05\ \mathrm{d}^{-1}$ for the first-type solution are also included in Figure 6-7. Some segments of these curves corresponding to the increased x values are relatively closer to the third-type source concentrations profile, but they correspond to higher v values. In other words, the first-type source solution results overpredict the v values. This situation is the result of the mass balance violation of the first-type source solution at the inlet location and its ramifications can be seen explicitly from Figure 6-8, which presents the advective–dispersive flux variations (both F_{x_f} and F_{x_r}) with the distance (x) for the first-type and third-type source cases for $v = 0.01\ \mathrm{d}^{-1}$. As can be seen from Figure 6-8, the advective–dispersive flux profiles are significantly different and there is no match between the two curves with the exception of large distance (x) values. Furthermore, for the third-type source solution, the flux at $x = 0$ is $F_{x_r} = 25\ \mathrm{mg/d/m^2}$ and the mass balance constraint at the inlet location is exactly satisfied. This value is significantly different from the first-type source solution flux ($F_{x_f} = 50\ \mathrm{mg/d/m^2}$). Therefore, this is a significant mass balance violation of the first-type source solution. For comparison purposes, the convective–dispersive flux profiles for $v = 0.02$, 0.03, and $0.05\ \mathrm{d}^{-1}$ for the first-type source solution are also presented.

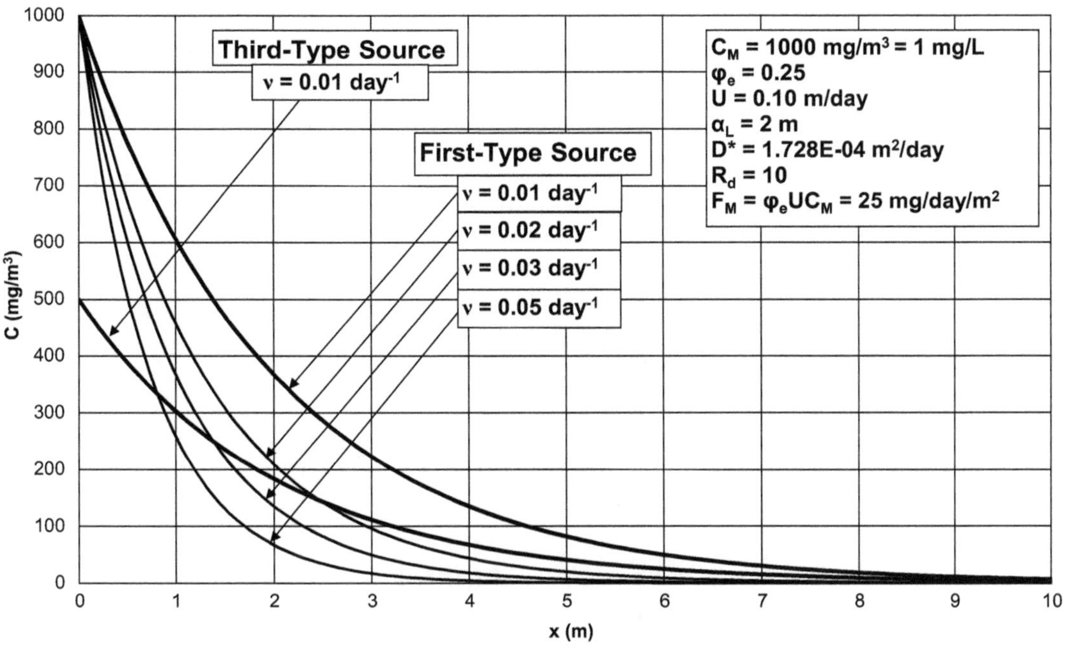

FIGURE 6-8
Comparison of the first-type and third-type source advective–dispersive flux profiles for $v = 0.01\ \mathrm{d}^{-1}$ and first-type advective–dispersive flux profiles for $v = 0.02$, 0.03, and $0.05\ \mathrm{d}^{-1}$. Adapted from Batu (2010, p. 566, Figure 4).

Some segments of these curves corresponding to the increased x values are relatively closer to the third-type source fluxes profile, but they correspond to higher ν values. As the ν value increases, the mass balance violation rate increases significantly (Figure 6-8). Furthermore, for the first-type source solution, the advective–dispersive fluxes at $x = 0$ for $\nu = 0.02, 0.03$, and $0.05\ \mathrm{d}^{-1}$ are 64.0, 75.0, and 92.6 $\mathrm{mg/d/m^2}$, respectively, and the mass balance constraint at the inlet location is further violated.

6.7 Degradation Rate Estimation Method by Curve-Matching

6.7.1 Introduction

In this section, a degradation rate estimation method will be presented by curve-matching based on the third-type source solution given by Eq. (6-18) and first-type source solution given by Eq. (6-5). Based on the analyses in the previous sections, it has been proven that the first-type source solution is inappropriate due to the fact that it generates mass balance errors at the inlet. As a result, it generates significantly overestimated degradation rates (ν) depending on the values of dispersion coefficient (D_x) and retardation factor (R_d). This will be shown in an example below.

6.7.2 Curve-Matching Method for Degradation Rate Estimation

The curve-matching method for estimation of degradation rates of contaminant constituents can be used for the case depicted in Figure 6-9 in which the ground water flow has a uniform velocity U (the ratio between the Darcy flux q and effective porosity φ_e). If the measured concentrations represent equilibrium conditions (i.e., steady state) and the concentrations decrease from the source concentration C_M at $x = 0$ as a function of the x coordinate, the curve-matching method can be used. The observation wells $\mathrm{MW}-1, \mathrm{MW}-2, \ldots, \mathrm{MW}-n$ should be close enough to the x coordinate axis.

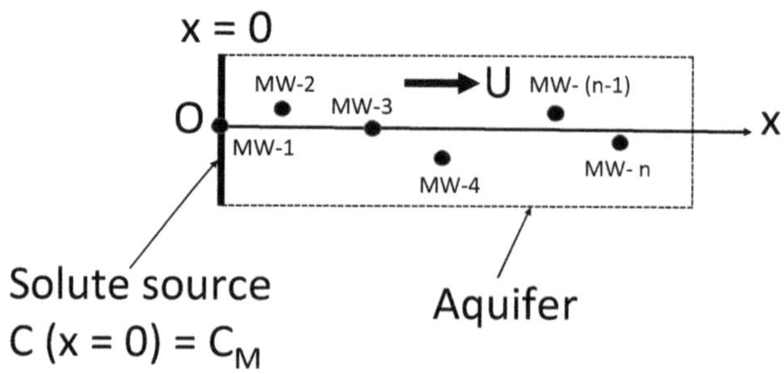

FIGURE 6-9
Schematic of one-dimensional solute transport in an aquifer under uniform flow conditions and observation wells around the x coordinate direction.

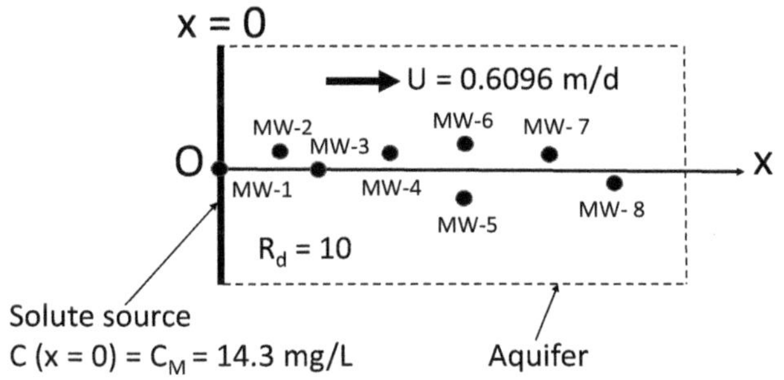

FIGURE 6-10
Schematic for Example 6-1.

TABLE 6-1

Measured Concentration Data for Example 6-1

Well ID	Distance from the Source (m)	Concentration (mg/L)
MW-1	0	14.30
MW-2	60.096	9.00
MW-3	12.192	6.00
MW-4	21.336	4.20
MW-5	30.480	1.60
MW-6	82.296	0.47
MW-7	85.344	0.19
MW-8	100.584	0.09

Example 6-1

Determine the degradation rate v for the dataset for a contaminant constituent having high sorbing characteristics. The schematic is shown in Figure 6-10. The measured concentration data are given in Table 6-1. The flow and solute transport parameters are given as follows: $K_h = 60.96$ m/d, $\varphi_e = 0.25$, $I = 0.0025$ m/m, $\alpha_L = 10$ m, $D^* = 2.0 \times 10^{-9}\ \text{m}^2/\text{s} = 1.728 \times 10^{-4}\ \text{m}^2/\text{d}$, and $R_d = 10$. Here, K_h is the horizontal hydraulic conductivity. Determine the degradation rate v value with the third-type source solution, compare it with the first-type source solution value, and discuss the results.

Solution

From Darcy's law, the ground water velocity is

$$U = \frac{K_h I}{\varphi_e} = \frac{\left(60.96\ \frac{\text{m}}{\text{d}}\right)\left(0.0025\ \frac{\text{m}}{\text{m}}\right)}{0.25} = 0.6096\ \frac{\text{m}}{\text{d}}$$

$$C_M = 14.3\ \text{mg/L}$$

From Eq. (6-21), the advective–dispersive flux at $x = 0$ is

$$F_{x_r}(x=0) = \varphi_e U C_M = (0.25)\left(0.6096\ \frac{\text{m}}{\text{d}}\right)\left(14.3\ \frac{\text{mg}}{\text{L}}\right) = 2.179\ \frac{\frac{\text{m}}{\text{d}}}{\text{m}^2}$$

Figure 6-11 presents curve-matching of Eq. (6-18) for the third-type source solution with the measured concentration data. By trial, it has been found that $v = 0.0035\ d^{-1}$ is established the best match. Figure 6-12 shows the curve-matching for the first-type source

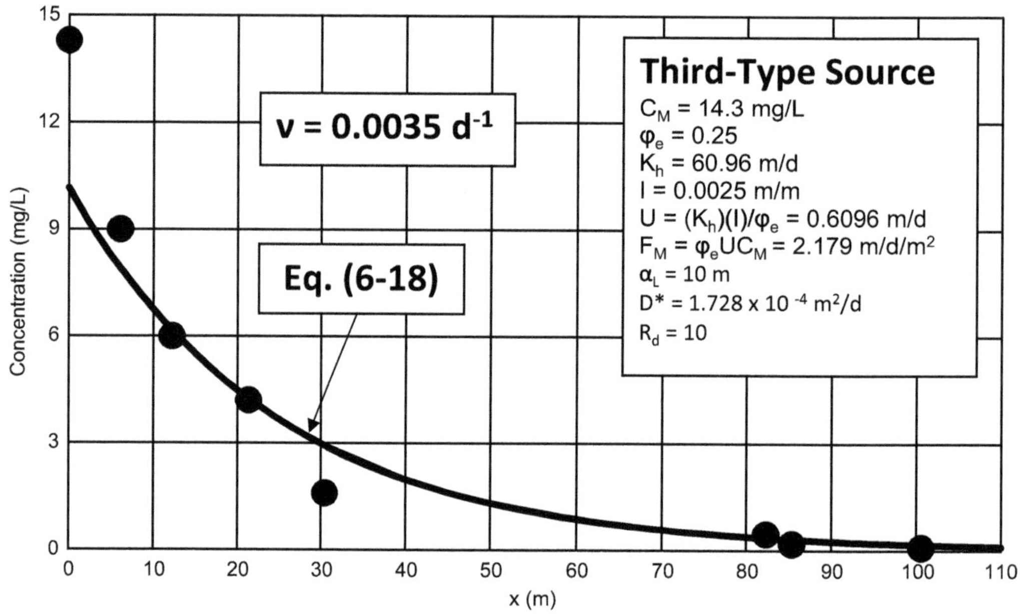

FIGURE 6-11
Determination of the degradation rate for the concentration data in Table 6-1 using the third-type source solution.

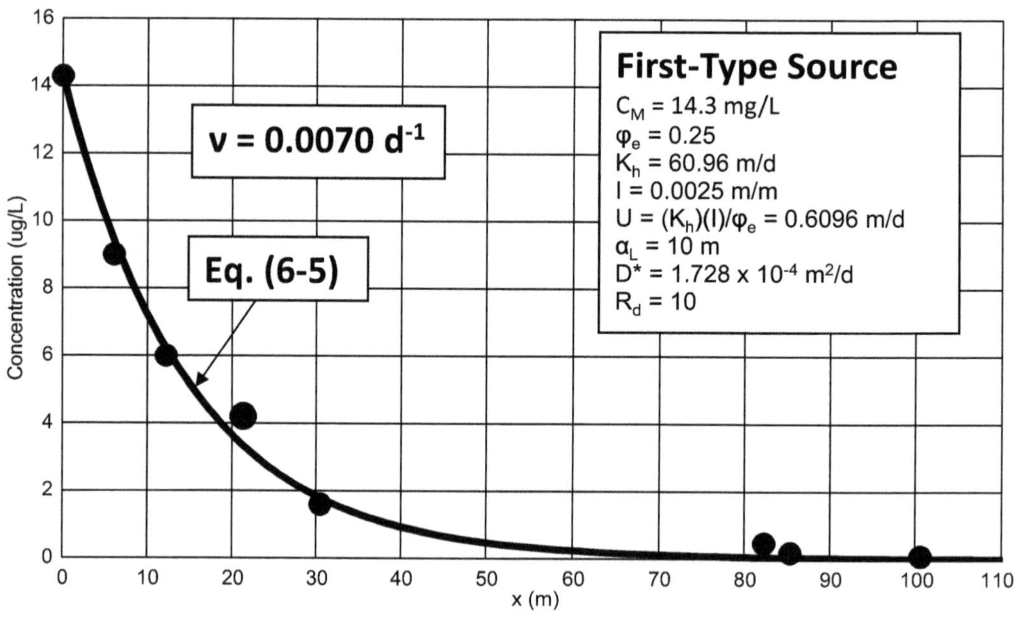

FIGURE 6-12
Determination of the degradation rate for the concentration data in Table 6-1 using the first-type source solution.

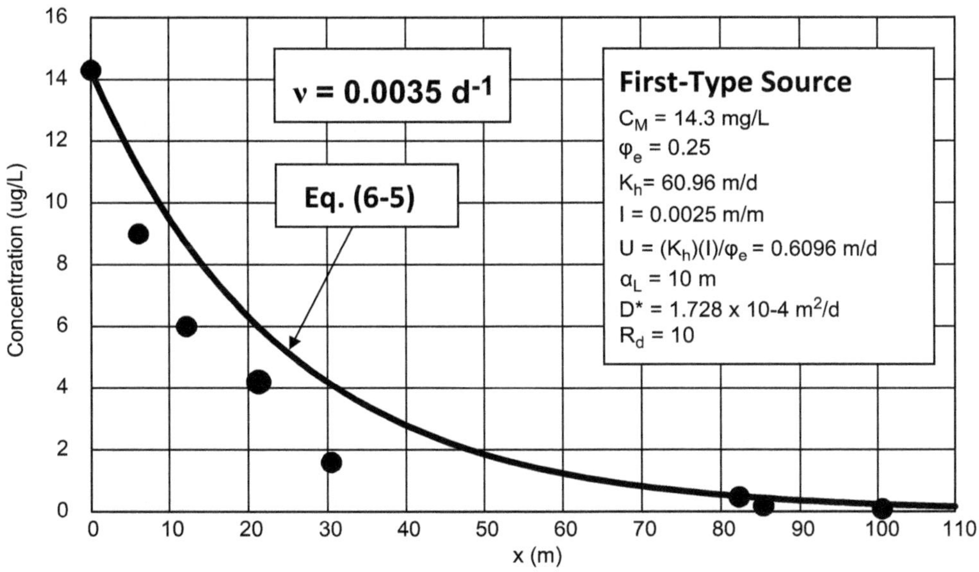

FIGURE 6-13
The concentration versus distance curve based on the first-type source solution for $\nu = 0.0035\,d^{-1}$ degradation rate.

solution given by Eq. (6-5) with the measured concentration data and the resulting degradation rate is $\nu = 0.0070\ d^{-1}$, which is twice that of the third-type source solution case. Figure 6-13 shows the concentration versus distance curve based on the first-type source solution for $\nu = 0.0035\ d^{-1}$ degradation rate, which clearly shows that there is no match. The above values clearly show that the mass balance violation of the first-type source solution at the inlet generates significantly overestimated degradation rate values depending on the dispersion coefficient (D_x) and retardation factor (R_d).

Problems

6.1 Describe the term "degradation" along with its terminology development process.

6.2 What are the main differences between the first-type and third-type source solutions? Answer the question by writing the source condition equations.

6.3 Evaluate the solute flux expression at the source for the first-type source case for mass balance point of view by writing the necessary equations. Plot $F_{x_f}(x=0)/F_{x_r}(x=0)$ as a function of $D_x R_d \nu / U^2$ and discuss the results.

7

Fluid Dynamics of Soil Venting and Its Associated Remediation Methods for Vadose Zones

7.1 Introduction

Besides aquifer contamination, volatile organic compound (VOC) contamination of unsaturated zones is also a serious widespread problem. The chemicals entering the ground due to leaking storage tanks, pipelines, and other means generate contaminant sources to the underlying aquifers. The conventional method for remediation of VOC-contaminated soils is excavation and disposal at approved off-site locations. This method is not only expensive but also problematic in transferring contaminant sources to other locations. Beyond this, at many contaminated sites, soil excavation may not be feasible because of the presence of ground structures, buried pipelines, utilities, and similar facilities.

The method of *soil venting* is to remove volatile and semivolatile contaminants from soil. A positive or negative air pressure is applied either passively or actively to soil to remove vapors that are appropriately treated. It is well known that petroleum products and organic solvents are serious contaminant constituents, especially in industrialized countries. Soil vapor extraction (SVE) and bioventing (BV) systems are being designed to remediate especially unsaturated zones.

The SVE method is one of the most effective and cost-efficient methods of removing volatile organic compounds (VOCs) from unsaturated soils. The BV method is a remediation technology that uses indigenous microorganisms to biodegrade organic constituents adsorbed to soils in the unsaturated zone. Soils in the capillary fringe and the saturated zone are not affected. When extraction wells are used for BV, the process is similar to SVE. However, while SVE removes constituents primarily through volatilization, BV systems promote biodegradation of constituents and minimize volatilization generally by using lower air flow rates than for SVE.

The dynamics of soil venting is the foundation of both SVE and BV. In this chapter, the mechanics of soil venting is presented in detail based on the scientific advancements during the past decades (e.g., Johnson et al., 1990; U.S. ACE, 2002).

7.2 Governing Equations of Soil Venting

7.2.1 Introduction

Since gas dynamics is involved in soil venting, the governing equations of soil venting are expressed with pressure instead of hydraulic head (e.g., Bear, 1979; Freeze and Cherry, 1979; Batu, 1998). Likewise, Darcy's law, which is given in Chapter 3, is the main governing

DOI: 10.1201/9781003739128-7

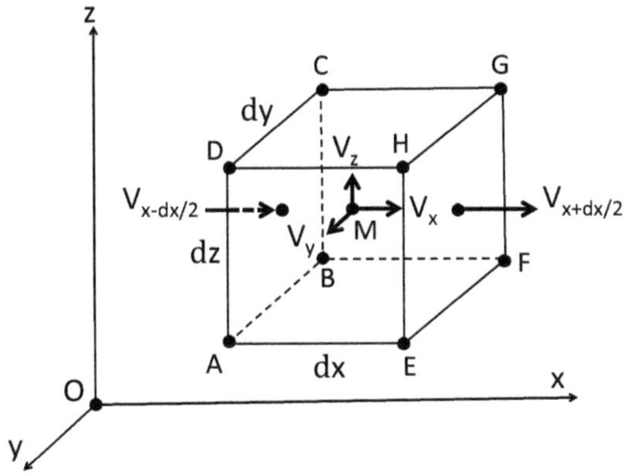

FIGURE 7-1
Schematic picture of a differential rectangular parallelepiped fluid volume at time t.

equation in deriving the partial differential equations of soil venting as well as their associated solutions.

In this section, the governing differential equations for soil venting are derived (e.g., Bear, 1972, 1979; Xue et al., 2020).

7.2.2 Equation of Continuity for Gases

7.2.2.1 The Mass Velocity and Darcy Velocity

Figure 7-1 is the schematic picture of a rectangular parallelepiped fluid volume at time t for the derivation of the continuity equation of fluid in the Cartesian coordinates. The mass velocity vector at the center of the rectangular parallelepiped is

$$\boldsymbol{V} = \rho \boldsymbol{q} \tag{7-1}$$

in which $\boldsymbol{V}$ is the mass flow velocity vector and $\boldsymbol{q}$ is the Darcy velocity vector. The mass velocity vector is

$$\boldsymbol{V} = V_x \boldsymbol{i} + V_y \boldsymbol{j} + V_z \boldsymbol{k} \tag{7-2}$$

in which V_x, V_y, and V_z are the mass flow velocity components along the x, y, and z coordinate directions, respectively, and $\boldsymbol{i}$, $\boldsymbol{j}$, and $\boldsymbol{k}$ are the unit vectors in the respective coordinate directions. Likewise,

$$\boldsymbol{q} = q_x \boldsymbol{i} + q_y \boldsymbol{j} + q_z \boldsymbol{k} \tag{7-3}$$

in which q_x, q_y, and q_z are the Darcy velocity components along the x, y, and z coordinate directions, respectively. From Eqs. (7-1), (7-2), and (7-3),

$$V_x \boldsymbol{i} + V_y \boldsymbol{j} + V_z \boldsymbol{k} = \rho q_x \boldsymbol{i} + q_y \boldsymbol{j} + q_z \boldsymbol{k} \tag{7-4}$$

Therefore, one can write

$$V_x = \rho q_x \quad V_y = \rho q_y \quad V_z = \rho q_z \tag{7-5}$$

7.2.2.2 Conservation of Mass Equation

The conservation of mass equation is

$$-\left[\frac{\partial(\rho q_x)}{\partial x}+\frac{\partial(\rho q_y)}{\partial y}+\frac{\partial(\rho q_z)}{\partial z}\right]=\frac{\partial(\rho n_a)}{\partial t} \tag{7-6}$$

in which n_a is the *air-filled porosity*. Eq. (7-6) is derived in Example 7-1.

In vector form.

$$-\nabla.(\rho \boldsymbol{q})=\frac{\partial(\rho n_a)}{\partial t} \tag{7-7}$$

in which ∇ is the vector differential *operator del* defined by

$$\nabla=\frac{\partial}{\partial x}\boldsymbol{i}+\frac{\partial}{\partial y}\boldsymbol{j}+\frac{\partial}{\partial z}\boldsymbol{k}\equiv \boldsymbol{i}\frac{\partial}{\partial x}+\boldsymbol{j}\frac{\partial}{\partial y}+\boldsymbol{k}\frac{\partial}{\partial z} \tag{7-8}$$

and $\boldsymbol{q}$ is given by Eq. (7-3). The dot in Eq. (7-7) is the *dot product* or *scalar product* of the ∇ and $\boldsymbol{q}$ vectors. Eq. (11-7) is exactly the same as in Xue et al. [2020, p. 60, Eq. (2-34)] with the exception of notation.

With $k \equiv k_a$, air permeability, for the flow mechanics in porous media, the equation of motion is generally expressed in the differential form of Darcy's flow as

$$\boldsymbol{q}=-\frac{k_a}{\mu}\nabla p \tag{7-9}$$

which is given equivalently in scalar form by Eq. (3-30). From Eqs. (7-7) and (7-9), one can write

$$\nabla.\left(\rho\frac{k_a}{\mu}\nabla p\right)=\frac{\partial(\rho n_a)}{\partial t} \tag{7-10}$$

which is exactly the same as U.S. Army Corps of Engineers [2002, pp. 2–13, Eq. (2-13)]. Klinkenberg's (1941) effects are assumed to be negligible with the form of Darcy's law given by Eq. (7-9).

Example 7-1

Derive the conservation of mass equation given by Eq. (7-6)

Solution

From Figure 7-1, the component of mass velocity along the x direction at the center of the ABCD area is

$$V_{x-\frac{dx}{2}}=\rho q_x-\frac{\partial(\rho q_x)}{\partial x}\frac{dx}{2} \tag{E7-1-1}$$

The component of mass velocity along the x direction at the center of the EFGH area is

$$V_{x+\frac{dx}{2}} = \rho q_x + \frac{\partial(\rho q_x)}{\partial x}\frac{dx}{2} \tag{E7-1-2}$$

Therefore, the mass inflow along the x direction through the ABCD area during dt time period is

$$\rho Q_{x-\frac{dx}{2}} = \left[\rho q_x - \frac{\partial(\rho q_x)}{\partial x}\frac{dx}{2}\right]dydzdt \tag{E7-1-3}$$

Likewise, the mass outflow along the x direction through the EFGH area during dt time period is

$$\rho Q_{x+\frac{dx}{2}} = \left[\rho q_x + \frac{\partial(\rho q_x)}{\partial x}\frac{dx}{2}\right]dydzdt \tag{E7-1-4}$$

As a result, from Eqs. (E7-1-3) and (E7-1-4), the mass inflow and outflow difference in the x direction is

$$\begin{aligned}\rho Q_{x-\frac{dx}{2}} - \rho Q_{x+\frac{dx}{2}} &= \left[\rho q_x - \frac{\partial(\rho q_x)}{\partial x}\frac{dx}{2}\right]dydzdt - \left[\rho q_x + \frac{\partial(\rho q_x)}{\partial x}\frac{dx}{2}\right]dydzdt \\ &= -\frac{\partial(\rho q_x)}{\partial x}dydzdt\end{aligned} \tag{E7-1-5}$$

Similarly, the following equations for the y and z directions can, respectively, be written as

$$\rho Q_{y-\frac{dy}{2}} - \rho Q_{y+\frac{dy}{2}} = -\frac{\partial(\rho q_y)}{\partial y}dydzdt \tag{E7-1-6}$$

$$\rho Q_{z-\frac{dz}{2}} - \rho Q_{z+\frac{dz}{2}} = -\frac{\partial(\rho q_z)}{\partial z}dydzdt \tag{E7-1-7}$$

By summing up Eqs. (E7-1-5)–(E7-1-7),

$$\begin{aligned}&\left[\rho Q_{x-\frac{dx}{2}} - \rho Q_{x+\frac{dx}{2}}\right] + \left[\rho Q_{y-\frac{dy}{2}} - \rho Q_{y+\frac{dy}{2}}\right] + \left[\rho Q_{z-\frac{dz}{2}} - \rho Q_{z+\frac{dz}{2}}\right] \\ &= -\left[\frac{\partial(\rho q_x)}{\partial x} + \frac{\partial(\rho q_y)}{\partial y} + \frac{\partial(\rho q_z)}{\partial z}\right]dxdydzdt\end{aligned} \tag{E7-1-8}$$

which is the mass difference between the inflow and outflow of the entire differential volume in Figure 7-1.

The difference in the mass inflow and outflow is due to the accumulation of air in the differential element volume and

$$\text{Total air mass in the } dxdydz \text{ volume} = n_a dxdydz \tag{E7-1-9}$$

in which n_a is the air-filled porosity. Therefore,

$$\text{The accumulation rate of air mass per unit time} = \frac{\partial(\rho n_a)}{\partial t} dxdydz \qquad \text{(E7-1-10)}$$

$$\text{Total accumulation of gass mass in } dt \text{ time} = \frac{\partial(\rho n_a)}{\partial t} dxdydzdt \qquad \text{(E7-1-11)}$$

Conservation of mass requires that Eqs. (E7-1-8) and (E7-1-11) must be equal to each other:

$$-\left[\frac{\partial(\rho q_x)}{\partial x} + \frac{\partial(\rho q_y)}{\partial y} + \frac{\partial(\rho q_z)}{\partial z}\right] dxdydzdt = \frac{\partial(\rho n_a)}{\partial t} dxdydzdt \qquad \text{(E7-1-12)}$$

Equation (E7-1-12) can be simplified as

$$-\left[\frac{\partial(\rho q_x)}{\partial x} + \frac{\partial(\rho q_y)}{\partial y} + \frac{\partial(\rho q_z)}{\partial z}\right] = \frac{\partial(\rho n_a)}{\partial t} \qquad \text{(E7-1-13)}$$

which is exactly the same as Eq. (7-6).

7.2.2.3 Conservation of Mass Equation in Scalar Form

Performing the dot product in Eq. (7-10), one gets

$$\frac{\partial}{\partial x}\left(\rho \frac{k_a}{\mu} \frac{\partial p}{\partial x}\right) + \frac{\partial}{\partial y}\left(\rho \frac{k_a}{\mu} \frac{\partial p}{\partial y}\right) + \frac{\partial}{\partial z}\left(\rho \frac{k_a}{\mu} \frac{\partial p}{\partial z}\right) = \frac{\partial(\rho n_a)}{\partial t} \qquad \text{(7-11)}$$

which is exactly the same as in Xue et al. [2020, p. 62, Eq. (2-52)] with the exception of notation.

7.2.3 Equation of Continuity for Vapor Flow

In case of vapor flow, a pressure–density relationship is also needed in addition to Eqs. (7-9) and (7-10) (e.g., Johnson et al., 1990; U.S. Army Corps of Engineers, 2002).

7.2.3.1 Pressure–Density Relationship for Vapor Flow

A reasonable assumption is that the vapor behaves like an ideal flow, so that the density expression is [e.g., Bear, 1972, p. 27, Eq. (2-1-2), first expression]

$$\rho(p) = \frac{p}{RT} \qquad \text{(7-12)}$$

For atmospheric pressure, Eq. (7-12) takes the form

$$\rho_{atm} = \frac{p_{atm}}{RT} \qquad \text{(7-13)}$$

in which ρ_{atm} is the vapor pressure at the reference pressure p_{atm}. From Eqs. (7-12) and (7-13), one can write [e.g., Johnson et al., 1990, p. 414, Eq. (3)]

$$\rho(p) = \rho_{atm}\left(\frac{p}{p_{atm}}\right) \tag{7-14}$$

7.2.3.2 Usage of Pressure–Density Relationship

If k_a and μ are constants, using Eq. (7-14) in Eq. (7-11), one can write

$$k_a \frac{\partial}{\partial x}\left(p\frac{\partial p}{\partial x}\right) + k_a \frac{\partial}{\partial y}\left(p\frac{\partial p}{\partial y}\right) + k_a \frac{\partial}{\partial z}\left(p\frac{\partial p}{\partial z}\right) = n_a \mu \frac{\partial p}{\partial t} \tag{7-15}$$

Equation (7-15) can also be written as

$$\frac{1}{2}k_a\frac{\partial p^2}{\partial x^2} + \frac{1}{2}k_a\frac{\partial p^2}{\partial y^2} + \frac{1}{2}k_a\frac{\partial p^2}{\partial z^2} = n_a\mu\frac{\partial p}{\partial t} \tag{7-16}$$

After some manipulations,

$$\frac{\partial p^2}{\partial x^2} + \frac{\partial p^2}{\partial y^2} + \frac{\partial p^2}{\partial z^2} = \frac{2n_a\mu}{k_a}\frac{\partial p}{\partial t} \tag{7-17}$$

With the ∇^2 *Laplacian operator,* it takes the form

$$\nabla^2 p^2 = \frac{2n_a\mu}{k_a}\frac{\partial p}{\partial t} \tag{7-18}$$

which is exactly the same as in the literature [e.g., Johnson et al., 1990, p. 414, Eq. (4); U.S. Army Corps of Engineers, 2002, pp. 2–13, Eq. (2-14)] with the exception of notation. (7-18) can also be written as

$$\nabla^2 p^2 = \frac{n_a\mu}{k_a}\frac{1}{p}\frac{\partial p^2}{\partial t} \tag{7-19}$$

which is exactly the same as in [e.g., Bear, 1972, p. 200, Eq. (6-2-7); Shan et al., 1992, p. 1106, Eq. (1); Falta, 1996, p. 751, Eq. (3)] with the exception of notation.

7.2.4 Linearized Equation of Continuity for Vapor Flow

7.2.4.1 Cartesian Coordinates Form

The pressure can be expressed in terms of the ambient pressure p_{atm} and deviation p' from this pressure. Namely,

$$p = p_{atm} + p' \tag{7-20}$$

The deviation pressure p' is equivalent to the vacuum that would be measured in the soil. Inserting Eq. (7-20) into Eq. (7-18),

$$\nabla^2\left(p_{atm} + p'\right)^2 = \frac{2n_a\mu}{k_a}\frac{\partial\left(p_{atm} + p'\right)}{\partial t} \tag{7-21}$$

From here,

$$\nabla^2\left[p_{atm}^2 + 2p_{atm}p' + (p')^2\right] = \frac{2n_a\mu}{k_a}\frac{\partial(p_{atm}+p')}{\partial t} \tag{7-22}$$

The derivatives of the first terms on the right-hand and left-hand sides are zero because p_{atm} is a constant. If $(p')^2$ is neglected, Eq. (7-22) becomes

$$\nabla^2 p' = \left(\frac{n_a\mu}{k_a p_{atm}}\right)\frac{\partial p'}{\partial t} \tag{7-23}$$

Massmann (1989) determined that the errors introduced by neglecting $(p')^2$ are negligible for vacuums less than 0.2 atmospheres, gauge. Accordingly, Massmann proposed that ground water flow models based on the linearized Boussinesq equation [e.g., Bear, 1972, p. 408, Eq. (8-4-1)] can be applied to vapor flow, with the substitution of pressure head, i.e., $p/(\rho g)$, for hydraulic head, and soil gas conductivity for hydraulic conductivity. Model simulations should be limited to vacuums less than 0.2 atmospheres, gauge, i.e., in accord with the assumption of incompressible flow (U.S. Army Corps of Engineers, 2002, pp. 2–19).

7.2.4.2 Cylindrical Coordinates Form

The two-dimensional form of Eq. (7-23) is

$$\frac{\partial p'}{\partial x^2} + \frac{\partial p'}{\partial y^2} = \left(\frac{n_a\mu}{k_a p_{atm}}\right)\frac{\partial p'}{\partial t} \tag{7-24}$$

With $k_a \equiv k_{ar}$, its cylindrical coordinates form is

$$\frac{\partial^2 p'}{\partial r^2} + \frac{1}{r}\frac{\partial p'}{\partial r} = \left(\frac{n_a\mu}{k_{ar} p_{atm}}\right)\frac{\partial p'}{\partial t} \tag{7-25}$$

for which the derivation is given in the literature [e.g., Batu, 2024, Eq. (29-176), derivational details are given in Example 29-17] Eq. (7-25) is exactly the same as Johnson et al. [1990, p. 414, Eq. (5)] and U.S. Army Corps of Engineers [2002, pp. 2–19, Eq. (2-30)] with the exception of notation.

7.3 Transient Analytical Solutions for Soil Gas Flow to Wells and Determination of Unsaturated Zone Properties – Fully-Penetrating Wells

7.3.1 Introduction

Determination of the unsaturated zone properties affecting the gas flow is an important part of the SVE system design process. Of these properties, the most important one is the gas permeability and soil gas pump tests need to be conducted to determine this quantity.

During the past decades, the well-known analytical solution in the aquifer hydraulics (Theis, 1935) has been adapted to gas flow in the unsaturated zones (e.g., Johnson et al., 1990). In this section, applications of this solution will be presented.

7.3.2 Fully-Penetrating Wells in Unsaturated Zones and Adaptation of the Theis Solution

7.3.2.1 Problem Description and Assumptions

To understand the derivation process for gas flow in the unsaturated zone, the foundation of the Theis (1935) solution needs to be understood (e.g., Batu, 2024, Section 29.15). The solution is based on the following assumptions (Figure 7-2): (1) the aquifer is homogeneous and isotropic; (2) the aquifer is horizontal and has a constant thickness b; (3) the well discharges water at a constant rate Q; (4) the aquifer is not leaky; (5) the aquifer has infinite horizontal extent; (6) the well is a line source; (7) the well fully penetrates the entire thickness of the aquifer; and (8) before water extraction, the piezometric head is the same at every point in the aquifer and is equal to h_0.

7.3.2.2 Governing Differential Equation

The governing differential equation is given by [e.g., Batu, 2024, Section 29.15, Eq. (29-377)]

$$\frac{\partial^2 h}{\partial r^2} + \frac{1}{r}\frac{\partial h}{\partial r} = \frac{S}{T}\frac{\partial h}{\partial t} \tag{7-26}$$

in which h is the hydraulic head, r is the radial distance, S is the storativity, T is the transmissivity; and t is the time. The *transmissivity* T and *storativity* S, respectively, are defined as

$$T = K_r b \quad S = S_s b \tag{7-27}$$

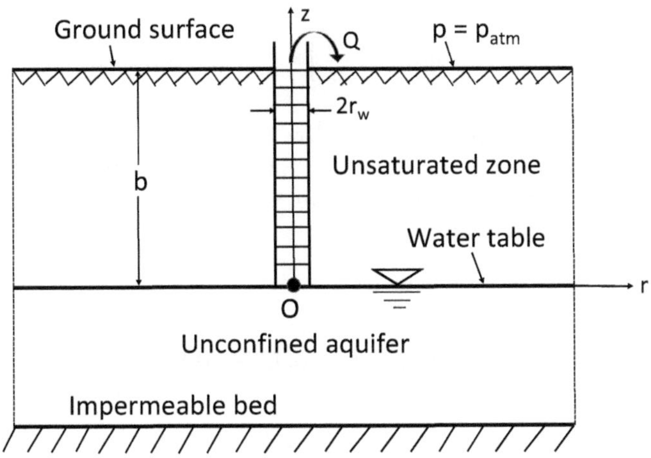

FIGURE 7-2
Schematic of gas flow to a fully penetrating well open to the atmosphere at ground surface in the unsaturated zone.

With the drawdown

$$s(r,t) = h_0 - h(r,t) = \Delta h \tag{7-28}$$

Equation (7-26) becomes

$$\frac{\partial^2 s}{\partial r^2} + \frac{1}{r}\frac{\partial s}{\partial r} = \frac{S}{T}\frac{\partial s}{\partial t} \tag{7-29}$$

7.3.2.3 Initial and Boundary Conditions

The initial condition for drawdown $s(r,t)$ is

$$s(r,0) = 0 \tag{7-30}$$

The boundary condition at infinity is

$$\lim_{r\to\infty} s(r,t) = 0 \tag{7-31}$$

The radial Darcy velocity is

$$q_r = -K_r\frac{\partial h}{\partial r} = -K_r\frac{\partial\,(h_0 - s\,)}{\partial r} = K_r\frac{\partial s}{\partial r} \tag{7-32}$$

The discharge Q at the well is

$$\lim_{r\to 0} A\,(r)\,q_r = \lim_{r\to 0}\,(2\pi rb)\left(K_r\frac{\partial s}{\partial r}\right) = (2\pi T)\lim_{r\to 0}\left(r\frac{\partial s}{\partial r}\right) = -Q \tag{7-33}$$

in which $A(r) = 2\pi rb$ is the area of the circular aquifer cylinder and $T = K_r b$. Eq. (7-33) can also be written as

$$\lim_{r\to 0}\left(K_r\frac{\partial s}{\partial r}\right) = -\frac{Q}{2\pi T} \tag{7-34}$$

Equation (7-30) states that initially the drawdown is zero everywhere in the aquifer. The boundary condition given by Eq. (7-31) states that the drawdown becomes zero when the radial distance r tends to infinity Eq. (7-33) states that near the extraction well the flow rate toward the well is equal to its discharge.

7.3.2.4 Theis Solution

The solution is [e.g., Batu, 2024, Section 29.15, Eq. (29-390)]

$$s(r,t) = \frac{Q}{4\pi T}\int_{\xi=u}^{\infty}\frac{e^{-\xi}}{\xi}d\xi \tag{7-35}$$

in which u is given by

$$u = \frac{r^2 S}{4Tt} \tag{7-36}$$

Equation (7-35) is known as the Theis solution in aquifer hydraulics and hydrogeology.

7.3.3 Adaptation of the Theis Solution to the Gas Flow in the Unsaturated Zone

7.3.3.1 Governing Differential Equation

The governing differential equation in the cylindrical coordinates is given by Eq. (7-25) as

$$\frac{\partial^2 p'}{\partial r^2} + \frac{1}{r}\frac{\partial p'}{\partial r} = \left(\frac{n_a \mu}{k_a p_{atm}}\right)\frac{\partial p'}{\partial t} \tag{7-37}$$

7.3.3.2 Initial and Boundary Conditions

The initial condition for drawdown $p'(r,t)$ is

$$p'(r,0) = 0 \tag{7-38}$$

The boundary condition at infinity is

$$\lim_{r\to\infty} p'(r,t) = 0 \tag{7-39}$$

From Eq. (7-9), the radial Darcy velocity is

$$q_r = -\frac{k_a}{\mu}\frac{\partial p}{\partial r} \tag{7-40}$$

Using Eq. (7-20),

$$q_r = -\frac{k_a}{\mu}\frac{\partial(p_{atm} + p')}{\partial r} = -\frac{k_a}{\mu}\frac{\partial p'}{\partial r} \tag{7-41}$$

The discharge Q at the well is

$$\lim_{r\to 0} A(r) q_r = \lim_{r\to 0}\,(2\pi rb)\left(\frac{k_a}{\mu}\frac{\partial p'}{\partial r}\right) = -Q \tag{7-42}$$

in which $A(r) = 2\pi rb$ is the area of the circular unsaturated zone cylinder Eq. (7-42) can also be written as

$$\lim_{r\to 0}\left(r\frac{\partial p'}{\partial r}\right) = \frac{Q}{2\pi b\dfrac{k_a}{\mu}} \tag{7-43}$$

Equation (7-38) states that initially the deviation pressure p' is zero everywhere in the unsaturated zone. The boundary condition given by Eq. (7-39) states that deviation pressure p' becomes zero when the radial distance r tends to infinity Eq. (7-42) states that near the extraction well the flow rate toward the well is equal to its discharge.

7.3.3.3 Adapted Theis Solution

Comparing Eqs. (7-29) and (7-37), it can be seen that s corresponds to p' and $\frac{S}{T}$ corresponds to $\frac{n_a \mu}{k_a p_{atm}}$. Also, comparing Eqs. (7-33) and (7-42), it can be seen that K_r corresponds to $\frac{k_a}{\mu}$. Under these conditions, from Eq. (7-35), one can write

$$p'(r,t) = p(r,t) - p_{atm} = \frac{Q}{4\pi b \frac{k_a}{\mu}} \int_{\xi=u}^{\infty} \frac{e^{-\xi}}{\xi} d\xi \tag{7-44}$$

Likewise, Eq. (7-36) takes the form

$$u = \frac{r^2}{4t}\left(\frac{S}{T}\right) = \frac{r^2}{4t}\left(\frac{n_a \mu}{k_a p_{atm}}\right) = \frac{r^2 n_a \mu}{4 k_a p_{atm} t} \tag{7.45}$$

Equation (7-44) is exactly the same as Johnson et al. [1990, p. 414, Eq. (7)] and U.S. Army Corps of Engineers [2002, p. 2-19, Eq. (2-31)] with the exception of notation.

7.3.3.4 Theis Well Function and Theis Type Curve

Equation (7-44) is usually expressed as

$$p'(r,t) = p(r,t) - p_{atm} = \frac{Q}{4\pi b \frac{k_a}{\mu}} W_T(u) \tag{7-46}$$

in which

$$W_T(u) = \int_{\xi=u}^{\infty} \frac{e^{-\xi}}{\xi} d\xi = Ei(u) \tag{7-47}$$

is the *Theis well function* and u is given by Eq. (7-45). In mathematics, $Ei(x)$ is called the *exponential integral* (e.g., Spiegel, 1968, p. 183). The series form of $W_T(u)$ is [e.g., Spiegel, 1968, p. 183, Eq. (35-8)]

$$W_T(u) = Ei(u) = -\gamma - \ln u + u - \frac{u^2}{2.2!} + \frac{u^3}{3.3!} - \frac{u^4}{4.4!} + \cdots \tag{7-48}$$

in which $\gamma = 0.577215664902\ldots$ is the *Euler constant*. Theis type curve is given in Figure 7-3 as $W_T(u)$ versus $1/u$ and Figure 7-4 as $W_T(u)$ versus u.

7.3.3.5 Cooper and Jacob Approximation

Cooper and Jacob (1946) noticed that in the Theis well function $W_T(u)$ given by Eq. (7-48) may be approximated by neglecting the series terms as

$$W_{CJ}(u) \cong -\gamma - \ln u = -0.577216 + \ln\left(\frac{1}{u}\right) \tag{7-49}$$

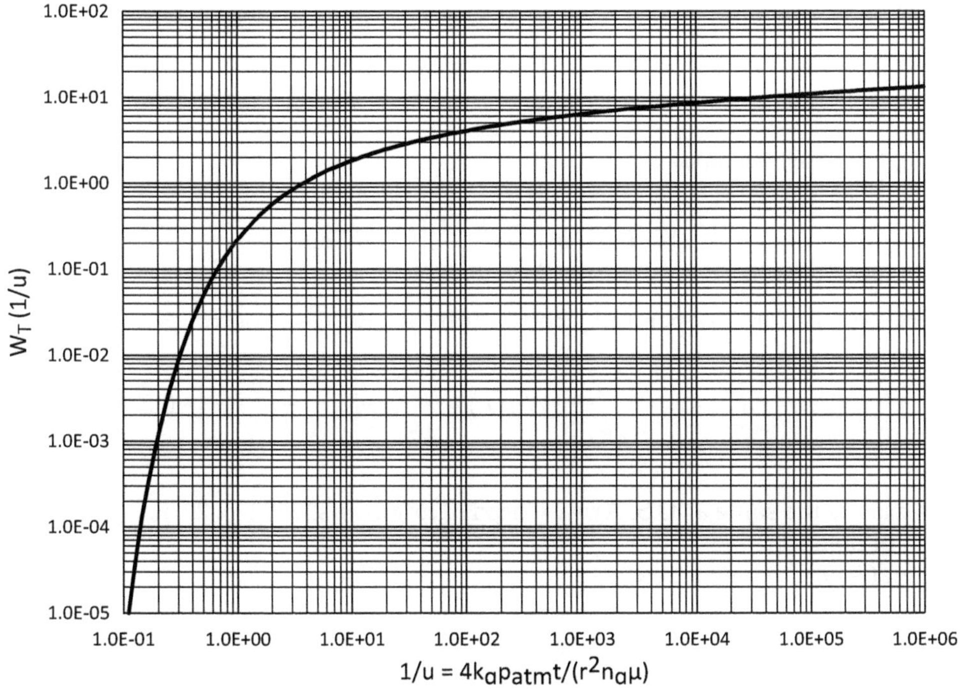

FIGURE 7-3
Adapted Theis type curve $W_T(u)$ versus $1/u$ for air flow in the unsaturated zone.

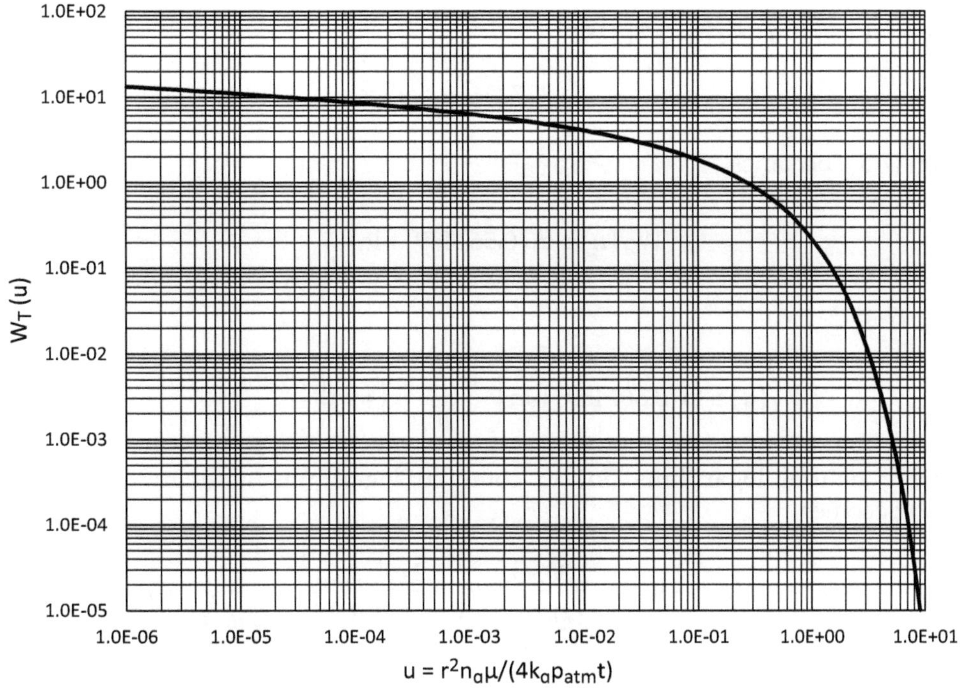

FIGURE 7-4
Adapted Theis type curve $W_T(u)$ versus u for air flow in the unsaturated zone.

From Eq. (7-45),

$$\frac{1}{u} = \frac{4k_a p_{atm} t}{r^2 n_a \mu} \quad (7\text{-}50)$$

Therefore, Eq. (7-49) becomes

$$W_{CJ}(u) = \ln\left(\frac{4k_a p_{atm} t}{r^2 n_a \mu}\right) - 0.577216 \quad (7\text{-}51)$$

in which *CJ* stands for "Cooper and Jacob." With this, Eq. (7-46) becomes

$$p'(r,t) \cong p(r,t) - p_{atm} = \frac{Q}{4\pi b \frac{k_a}{\mu}} W_{CJ}(u) = \frac{Q}{4\pi b \frac{k_a}{\mu}}\left[\ln\left(\frac{4k_a p_{atm} t}{r^2 n_a \mu}\right) - 0.577216\right] \quad (7\text{-}52)$$

which is exactly the same as U.S. Army Corps of Engineers [2002, p. 20, Eq. (2-33)] with the exception of notation.

Equation (7-49) can be written as

$$W_{CJ}(u) \cong -0.577215 - \ln u = \ln x + \ln\left(\frac{1}{u}\right) \quad (7\text{-}53)$$

Therefore,

$$-0.577216 = \ln x \quad (7\text{-}54)$$

Solving it for x and using $e = 2.718282$,

$$x = e^{-0.577216} = 0.561459 \quad (7\text{-}55)$$

Equation (7-53) becomes

$$W_{CJ}(u) \cong \ln(0.561459) + \ln\left(\frac{1}{u}\right) \quad (7\text{-}56)$$

With Eq. (7-45),

$$W_{CJ}(u) \cong \ln(0.561459) + \ln\left(\frac{4k_a p_{atm} t}{r^2 n_a \mu}\right) = \ln\left(\frac{2.246 k_a p_{atm} t}{r^2 n_a \mu}\right) \quad (7\text{-}57)$$

Converting natural logarithms (ln) to common logarithms with $\ln(u) = 2.303\log(u)$, Eq. (7-52) becomes

$$p'(r,t) \cong p(r,t) - p_{atm} = \frac{2.303Q}{4\pi b \frac{k_a}{\mu}} \log\left(\frac{2.246 k_a p_{atm} t}{r^2 n_a \mu}\right) = \frac{Q}{4\pi b \frac{k_a}{\mu}} W_{CJ}(u) \quad (7\text{-}58)$$

in which u from Eq. (7-50) can be written as

$$u = \frac{r^2 n_a \mu}{4 k_a p_{atm} t} \tag{7-59}$$

Equation (7-58) is known as the *Cooper and Jacob equation* in aquifer hydraulics and hydrogeology. The approximate $W_{CJ}(u)$ given by Eq. (7-49) is valid for the condition $1/u > 100$ or $u < 0.01$ given in many papers and books (e.g., Lohman, 1972, p. 19; Hantush, 1964, p. 410; Bear, 1979, p. 321).

Note that the pressure drawdown $p'(r,t) = p(r,t) - p_{atm}$ given by Eq. (7-52) or (7-58) varies linearly with $\ln(t)$. These equations are commonly used for transient air permeability test analysis (e.g., U.S. Army Corps of Engineers, 2002, pp. 2–20).

7.3.3.6 Comparison of the $W_T(u)$ and $W_{CJ}(u)$ Well Functions

The Theis well function $W_T(u)$ and Cooper and Jacob approximate well function $W_{CJ}(u)$ are compared in Figure 7-5 as a function of $1/u$. Figure 7-6 presents the same well functions with u being the abscissa. As can be seen from Figures 7-5 and 7-6, for the values of $1/u > 10$ or $u < 0.1$, the values of $W_{CJ}(u)$ become closer and closer to the values of $W(u)$, which can be observed explicitly from Table 7-1. The following values can be noted: $u = 0.126(1/u = 7.937)$, $W_T(u) = 1.616$, $W_{CJ}(u) = 1.494$; $u = 0.090(1/u = 11.112)$, $W_T(u) = 1.919$, $W_{CJ}(u) = 1.831$; $u = 0.064(1/u = 15.557)$, $W_T(u) = 2.231$, $W_{CJ}(u) = 2.167$; $u = 0.046$ $(1/u = 21.780)$, $W_T(u) = 2.549$, $W_{CJ}(u) = 2.504$; $u = 0.033(1/u = 30.491)$, $W_T(u) = 2.873$, $W_{CJ}(u) = 2.840$; $u = 0.023(1/u = 42.688)$, $W_T(u) = 3.200$, $W_{CJ}(u) = 3.177$; $u = 0.017(1/u = 59.763)$,

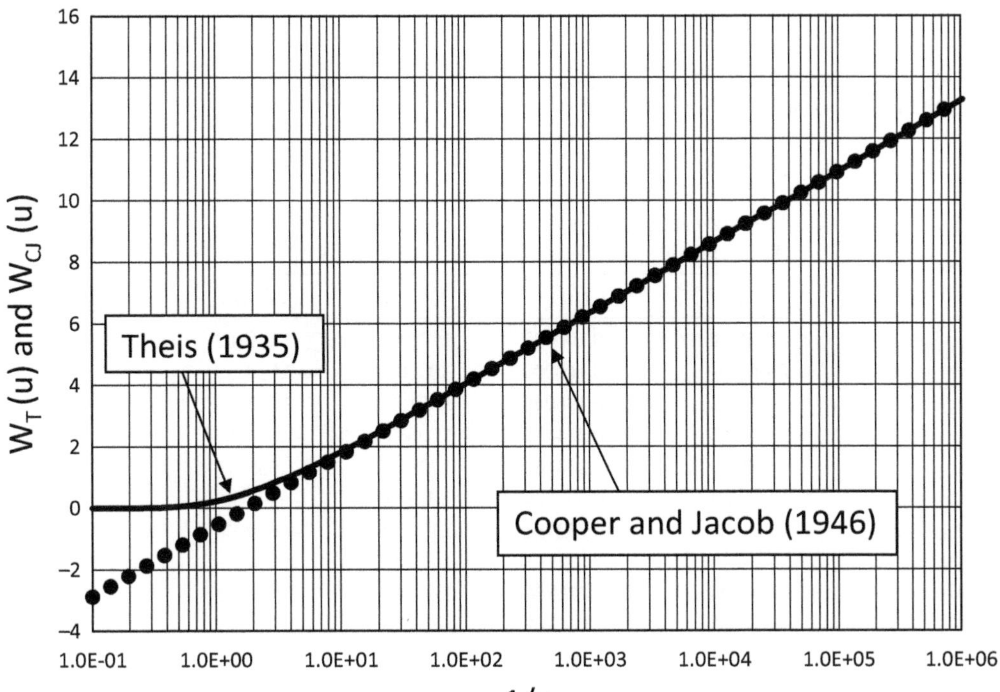

FIGURE 7-5
Comparison of the Theis (1946) well function $W_T(u)$ with the Cooper and Jacob (1946) approximate well function $W_{CJ}(u)$ with the abscissa $1/u$ for a confined nonleaky aquifer.

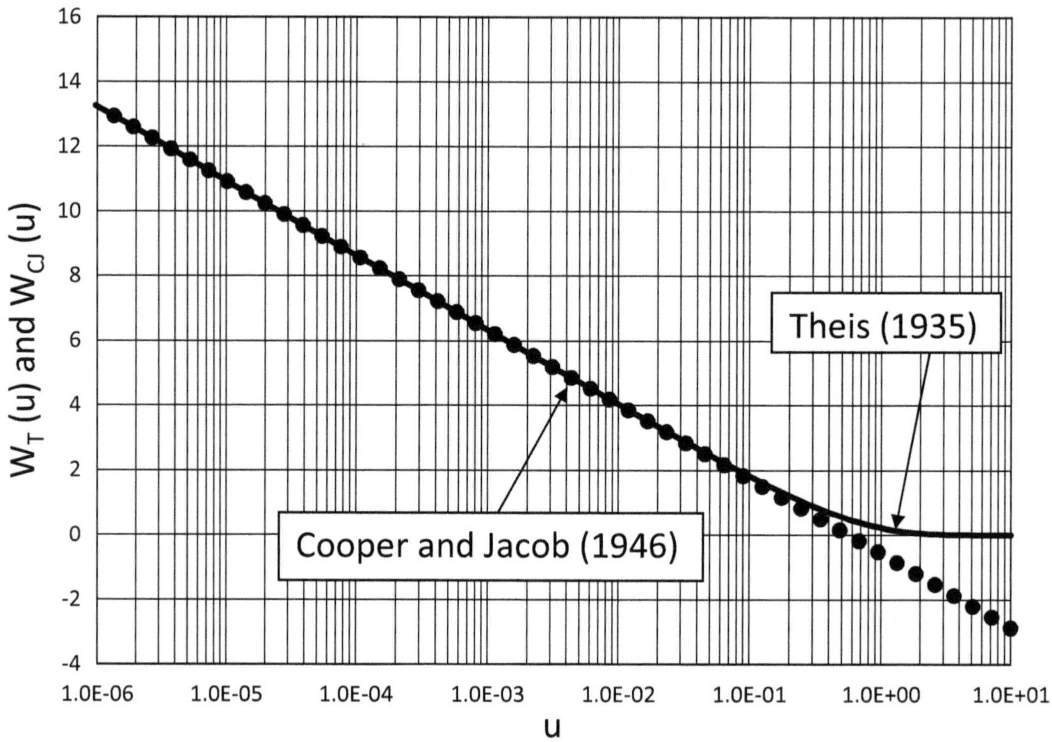

FIGURE 7-6
Comparison of the Theis (1946) well function $W_T(u)$ with the Cooper and Jacob (1946) approximate well function $W_{CJ}(u)$ with the abscissa u for a confined nonleaky aquifer.

TABLE 7-1
Theis Well Function $W_T(u)$ and Cooper and Jacob Approximate Well Function $W_{CJ}(u)$ Data Comparison

u	$1/u$	$W_T(u)$	$W_{CJ}(u)$
10.000000	0.100000	0.000004	−2.879802
7.142857	0.140000	0.000098	−2.543329
5.102041	0.196000	0.001019	−2.206857
3.644315	0.274400	0.005833	−1.870385
2.603082	0.384160	0.021762	−1.533913
1.859344	0.537824	0.059506	−1.197440
1.328103	0.752954	0.129703	−0.860968
0.948645	1.054135	0.239290	−0.524496
0.677604	1.475789	0.390100	−0.188024
0.484003	2.066105	0.579656	0.148449
0.345716	2.892547	0.802912	0.484921
0.246940	4.049565	1.053888	0.821393
0.176386	5.669391	1.326769	1.157865
0.125990	7.937148	1.616468	1.494337

(Continued)

TABLE 7-1 (*Continued*)

Theis Well Function $W_T(u)$ and Cooper and Jacob Approximate Well Function $W_{CJ}(u)$ Data Comparison

u	$1/u$	$W_T(u)$	$W_{CJ}(u)$
0.089993	11.112007	1.918818	1.830810
0.064281	15.556810	2.230545	2.167282
0.045915	21.779533	2.549148	2.503754
0.032796	30.491347	2.872756	2.840226
0.023426	42.687885	3.199989	3.176699
0.016733	59.763040	3.529834	3.513171
0.011952	83.668255	3.861560	3.849643
0.008537	117.135558	4.194635	4.186115
0.006098	163.989781	4.528677	4.522588
0.004356	229.585693	4.863411	4.859060
0.003111	321.419970	5.198641	5.195532
0.002222	449.987958	5.534226	5.532004
0.001587	629.983141	5.870064	5.868477
0.001134	881.976398	6.206083	6.204949
0.000810	1,34.766957	6.542231	6.541421
0.000578	1,28.673740	6.878472	6.877893
0.000413	2,20.143236	7.214779	7.214365
0.000295	3,88.200530	7.551133	7.550838
0.000211	4,43.480742	7.887521	7.887310
0.000151	6,40.873038	8.223933	8.223782

$W_T(u) = 3.530$, $W_{CJ}(u) = 3.513$. The approximate $W_{CJ}(u)$ values indicate that the condition $1/u > 100$ or $u < 0.01$ given in many papers and books (e.g., Lohman, 1972, p. 19; Hantush, 1964, p. 410; Bear, 1979, p. 321) is overly conservative.

7.3.3.7 Can Steady-State Drawdown Solution Be Determined from the Theis Solution?

As mentioned in Section 7.3.2.1, the aquifer has infinite horizontal extent. Regarding the answer to the heading, Jacob (1950, p. 366) makes the following statements:

> When a well begins discharging at a constant rate from an effectively infinite aquifer, the influence of the discharge progresses radially outward from the well in such a manner that the product of the rate of decline of head and the storage coefficient integrated over the area of influence equals the discharge. Inasmuch as there are no lateral boundaries to support the discharge of the well, it must necessarily come from reduction of storage within the aquifer (i.e., from the expansion of the water and the concomitant compression or compaction of the sand). Accordingly, there can be no steady state of flow in such an aquifer; the head will continue to decline as long as the aquifer is effectively infinite. ...

Therefore, the key conclusion is that the solution given by Eq. (11-46) does not approach a steady-state solution for $p'(r,t) = p(r,t) - p_{atm}$, namely, $p'(r,t)$ continuously increases as shown in Figure 7-3.

7.3.4 Determination of Air Permeability and Air-Filled Porosity

Using field-measured data, the k_a air permeability and n_a air-filled porosity can be determined using the solutions given above.

7.3.4.1 Theis Type-Curve Method

The Theis-type curve method can be used in determining the k_a and n_a values (e.g., U.S. Army Corps of Engineers, 2002, Appendix D, pp. D-5–D-7). The method is similar to the Theis type-curve method used in aquifer hydraulics (e.g., Batu, 1998, pp. 158–163).

Equation (7-46) can be solved for k_a air permeability as

$$k_a = \frac{Q\mu W_T(u)}{4\pi b\left[p(r,t) - p_{atm}\right]} \tag{7-60}$$

which is exactly the same as U.S. Army Corps of Engineers [2002, Appendix D, p. D-7, Eq. (D-13)] with the exception of notation. Eq. (7-59) can be solved for n_a air-filled porosity as

$$n_a = \frac{4k_a p_{atm} t}{r^2 \mu} u \tag{7-61}$$

which is exactly the same as U.S. Army Corps of Engineers [2002, Appendix D, pp. D-5 – D-7, Eq. (D-14)] with the exception of notation.

By fitting a log–log plot of $p - p_{atm}$ versus time t to the Theis-type curve (Figure 7-3), a point along the type curve can be selected where values of $p - p_{atm}$ and t correspond to a particular point $1/u$ and $W(1/u)$. These values can be substituted into Eqs. (7-60) and (7-61) to determine values of k_a air permeability and n_a air-filled porosity.

7.3.4.2 Cooper and Jacob Method

The Cooper and Jacob method can be used in determining the k_a and n_a values (e.g., U.S. Army Corps of Engineers, 2002, Appendix D, pp. D-7–D-8). The method is similar to the Cooper and Jacob method used in aquifer hydraulics (e.g., Batu, 1998, pp. 163–171). This method does not require type-curve matching.

From Eq. (7-52),

$$p'(r,t) \cong p(r,t) - p_{atm} = \frac{Q\mu}{4\pi b k_a}\left[\ln\left(\frac{4k_a p_{atm} t}{r^2 n_a \mu}\right) - 0.577216\right] \tag{7-62}$$

which for $1/u > 100$ or $u < 0.01$ should show a straight line with slope

$$m = \frac{Q\mu}{4\pi b k_a} \tag{7-63}$$

in which m is the change in pressure over one log cycle. Eq. (7-62) is exactly the same as Eq. (D.15) of U.S. Army Corps of Engineers [2002, p. D-7, Eq. (D-15)] with the exception of notation.

The time period intercept when $p - p_{atm} = 0$ should occur is

$$\ln\left(\frac{4k_a p_{atm} t_0}{r^2 n_a \mu}\right) = 0.577216 \tag{7-64}$$

in which t_0 is the time intercept when $p - p_{atm} = 0$.

From Eq. (7-63), the k_a air permeability is

$$k_a = \frac{Q\mu}{4\pi bm} \tag{7-65}$$

From Eq.(7-64),

$$\frac{4k_a p_{atm} t_0}{r^2 n_a \mu} = e^{0.577216} \tag{7-66}$$

With $e = 2.718282$ and solving it for n_a, one gets

$$n_a = 2.246 \frac{k_a p_{atm} t_0}{r^2 \mu} \tag{7-67}$$

which is exactly the same as of U.S. Army Corps of Engineers [2002, Appendix D, p. D-8, Eq. (D-19)] with the exception of notation.

With

$$\ln(u) = 2.303 \log(u) \tag{7-68}$$

Equation (7-58) becomes

$$p'(r,t) \cong p(r,t) - p_{atm} = \frac{2.303Q}{4\pi b \dfrac{k_a}{\mu}} \log(u) = \frac{2.303Q}{4\pi b \dfrac{k_a}{\mu}} \log\left(\frac{2.246 k_a p_{atm} t}{r^2 n_a \mu}\right) \tag{7-69}$$

Equation (7-69) corresponds to the commonly used *Cooper and Jacob equation* in aquifer hydraulics and hydrogeology [e.g., Batu, 2024, p. 1138, Eq. (29-400)].

Equation (7-69) can also be written as

$$p'(r,t) = \frac{2.303Q}{4\pi b \dfrac{k_a}{\mu}} \log\left(\frac{2.246 k_a p_{atm}}{r^2 n_a \mu}\right) + \frac{2.303Q}{4\pi b \dfrac{k_a}{\mu}} \log(t) \tag{7-70}$$

Equation (7-70) is an asymptotic expression that predicts $p'(r,t) = p(r,t) - p_{atm}$ very closely under the condition that a sufficient time has elapsed since the SVE started. Eq. (7-70) is the equation of a straight line on a semilogarithmic coordinate system $p'(r,t)$ versus $\log(t)$ shown in Figure 7-7. The slope of the straight line is $\dfrac{2.303Q}{4\pi b \dfrac{k_a}{\mu}}$, and this line intercepts the time axis where $p'(r,t) = 0$. Consequently, the interception point has the coordinate $p'(r,t) = 0$ and $t = t_0$. Introducing these in Eq. (7-69) gives

$$0 = \frac{2.303Q}{4\pi b \dfrac{k_a}{\mu}} \log\left(\frac{2.246 k_a p_{atm} t_0}{r^2 n_a \mu}\right) \tag{7-71}$$

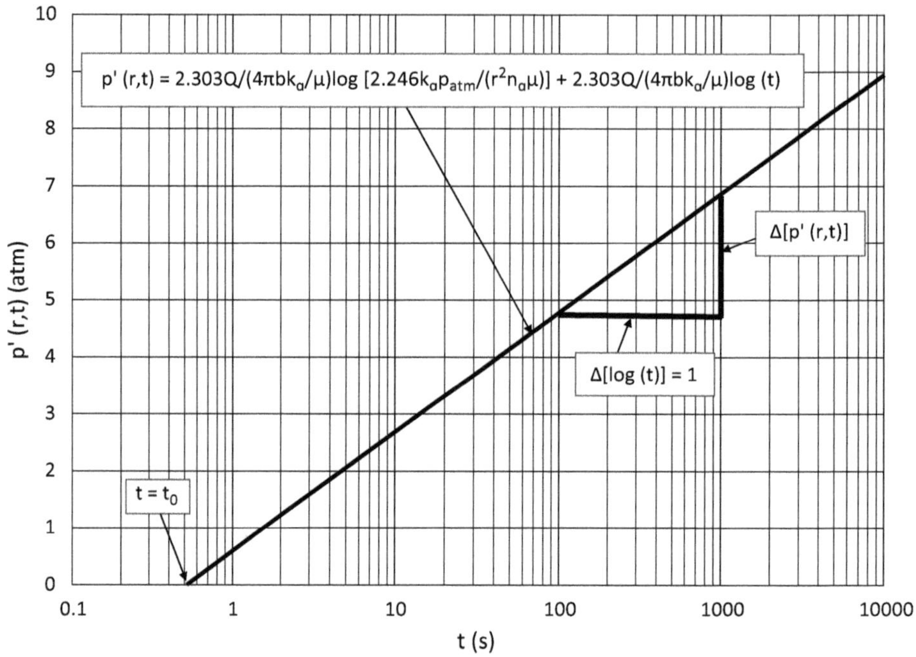

FIGURE 7-7
Schematic figure for the application of the Cooper and Jacob method to soil vapor extraction to determine k_a and n_a values.

Because $\frac{2.303Q}{4\pi b \frac{k_a}{\mu}}$ cannot be zero, it follows that

$$\frac{2.246k_a p_{atm} t_0}{r^2 n_a \mu} = 1 \quad (7\text{-}72)$$

And solving for air-filed porosity n_a,

$$n_a = \frac{2.246k_a p_{atm} t_0}{r^2 \mu} \quad (7\text{-}73)$$

which is exactly the same as Eq. (7-67). Because the derivative of a constant is zero, the derivation of Eq. (7-70) with respect to $\log(t)$ is

$$d\left[p'(r,t)\right] = \frac{2.303Q}{4\pi b \frac{k_a}{\mu}} d\left[\log(t)\right] \quad (7\text{-}74)$$

And writing it in finite difference form

$$\Delta\left[p'(r,t)\right] = \frac{2.303Q}{4\pi b \frac{k_a}{\mu}} \Delta\left[\log(t)\right] \quad (7\text{-}75)$$

and solving it for k_a gives

$$k_a = \frac{2.303Q\mu}{4\pi b\Delta\left[p'(r,t)\right]}\Delta\left[\log(t)\right] \tag{7-76}$$

To apply the Cooper and Jacob method, the following conditions should be satisfied (Ferris et al, 1962): (1) the assumptions of the Theis (1935) solution (e.g., Batu, 1998, pp. 145–146) and (2) small values of u (Generally less than 0.01). In other words, r is small and t is large.

Example 7-2

For a SVE test, the following values are given: $Q = 47.5\ \text{L/s}$; $\mu_{air} = 1.81\times10^{-4}\ \text{g/(cm.s)}$; $b = 3\ \text{m}$; and $r = 9.6\ \text{m}$. After plotting the measured data points, the following values are determined in accordance with the Cooper and Jacob method: $\Delta\left[p'(r,t)\right] = 4\ \text{atm}$ for $\Delta\left[\log(t)\right] = 1$. Also, it is found out $t_0 = 4{,}320$ s. Determine the values of k_a and n_a.

Solution

GIVEN DATA

The given data are

$$Q = 47.5\ \frac{\text{L}}{\text{s}} = 47{,}500\ \frac{\text{cm}^3}{\text{s}}$$

$$\mu_{air} = 1.81\times10^{-4}\ \frac{\text{g}}{\text{cm.s}}$$

$$b = 3\ \text{m} = 300\ \text{cm}$$

$$r = 9.6\ \text{m} = 960\ \text{cm}$$

SEMILOGARITHMIC STRAIGHT LINE VALUES

Equation (7-70) is the equation of a straight line on a semilogarithmic coordinate system $p'(r,t)$ versus $\log(t)$ shown in Figure 7-7. From the $p'(r,t)$ versus t graph, the following value is determined per log cycle, i.e., $\Delta\left[\log(t)\right] = 1$:

$$\Delta\left[p'(r,t)\right] = 4\ \text{atm}$$

$$p_{atm} = 101{,}325\ \text{Pa}$$

$$1\ \text{Pa} = 1\ \frac{\text{N}}{\text{m}^2}$$

$$1\ \text{N} = 1\ \frac{\text{kg.m}}{\text{s}^2}$$

Therefore,

$$p_{atm} = 101{,}325\ \text{Pa} = 101{,}325\ \frac{\text{N}}{\text{m}^2} = 101{,}325\ \frac{\text{kg}}{\text{m}^2\text{s}^2} = 1.013\times10^6\ \frac{\text{g}}{\text{cm.s}^2}$$

Using the values in Eq. (7-76),

$$k_a = \frac{2.303Q\mu}{4\pi b\Delta\left[p'(r,t)\right]}\Delta\left[\log(t)\right] = \frac{2.303\left(47{,}500\,\frac{\text{cm}^3}{\text{s}}\right)\left(1.81\times10^{-4}\,\frac{\text{g}}{\text{cm.s}}\right)}{4\pi(300\text{ cm})\left(4\times1.013\times10^{6}\,\frac{\text{g}}{\text{cm.s}^2}\right)}(1)$$

$$= 12.96\times10^{-10}\text{ cm}^2$$

Since

$$1\text{ darcy} = 9.87\times10^{-9}\text{ cm}^2$$

$$k_a = 12.96\times10^{-10}\text{ cm}^2 = 12.96\times10^{-10}\left(\frac{1\text{ darcy}}{9.87\times10^{-9}\text{ cm}^2}\right) = 0.131\text{ darcy}$$

From Eq. (7-73),

$$n_a = \frac{2.246k_a p_{atm} t_0}{r^2\mu} = \frac{2.246\left(12.96\times10^{-10}\text{ cm}^2\right)\left(1.013\times10^{6}\,\frac{\text{g}}{\text{cm.s}^2}\right)(4{,}320\text{ s})}{(960\text{ cm})^2\left(1.81\times10^{-4}\,\frac{\text{g}}{\text{cm.s}}\right)} = 0.076$$

7.4 Transient Analytical Solutions for Soil Gas Flow for Partially Penetrating Wells to Determine Unsaturated Zone Properties

7.4.1 Introduction

As mentioned in Section 7.3.1, determination of the unsaturated zone properties affecting the gas flow is an important part of the SVE system design process. Of these properties, the most important one is the gas permeability and soil gas pump tests need to be conducted to determine this quantity. During the past decades, the well-known analytical solution under partially penetrating well conditions in the aquifer hydraulics (Hantush, 1964) has been adapted to gas flow in the unsaturated zones (e.g., Baehr and Hult, 1989, 1991; Shan et al., 1992; Falta, 1993, 1996). In this section, some of the solutions will be presented.

7.4.2 Partially Penetrating Well in an Unsaturated Zone Below the Ground Surface without a Leaky Confining Layer

7.4.2.1 Governing Differential Equation

Equation (7-19) can also be written in a generalized form as

$$k_{ax}\frac{\partial^2 p^2}{\partial x^2} + k_{ay}\frac{\partial^2 p^2}{\partial y^2} + k_{az}\frac{\partial^2 p^2}{\partial z^2} = \frac{n_a\mu}{p}\frac{\partial p^2}{\partial t} \quad (7\text{-}77)$$

in which k_{ax}, k_{ay}, and k_{az} are the air permeabilities in the x, y, and z coordinate directions, respectively. With $k_{ax} = k_{ax} = k_{ar}$, in circular cylindrical coordinates, Eq. (7-77) can be expressed as [e.g., Batu, 2024, pp. 1052–1056, Eq. (29-175)]

$$k_{ar}\left(\frac{\partial^2 p^2}{\partial r^2}+\frac{1}{r}\frac{\partial p^2}{\partial r}\right)+k_{az}\frac{\partial^2 p^2}{\partial z^2}=\frac{n_a\mu}{p}\frac{\partial p^2}{\partial t} \tag{7-78}$$

which is exactly the same as Baehr and Hult [1989, p. 58, Eq. (4)] with the exception of notation.

Equation (7-78) is a nonlinear differential equation and may be linearized by replacing p in the denominator of the right-hand side with the ambient gas pressure, p_{atm}:

$$k_{ar}\left(\frac{\partial^2 p^2}{\partial r^2}+\frac{1}{r}\frac{\partial p^2}{\partial r}\right)+k_{az}\frac{\partial^2 p^2}{\partial z^2}=\frac{n_a\mu}{p_{atm}}\frac{\partial p^2}{\partial t} \tag{7-79}$$

which is exactly the same as Falta [1996, p. 751, Eq. (4)] with the exception of notation.

With this assumption, the gas is assumed to be compressible with a constant compressibility factor $1/p_{atm}$. The accuracy of this linearization depends on the magnitude of the pressure variation in the system, and the linearization is less accurate when pressure variation is large. For a vacuum of 0.5 atm or less, this linearization gives results with a few percent of exact solutions of Eq. (7-79) based on Massmann (1989).

7.4.2.2 Problem Description and Initial and Boundary Conditions

The schematic diagram of the problem is shown in Figure 7-8, which shows that there is no leaky confining layer below the ground surface. The initial condition is

$$t = 0 \quad p^2 = p_{atm}^2 \tag{7-80}$$

The boundary conditions at the ground surface and water table, respectively, are

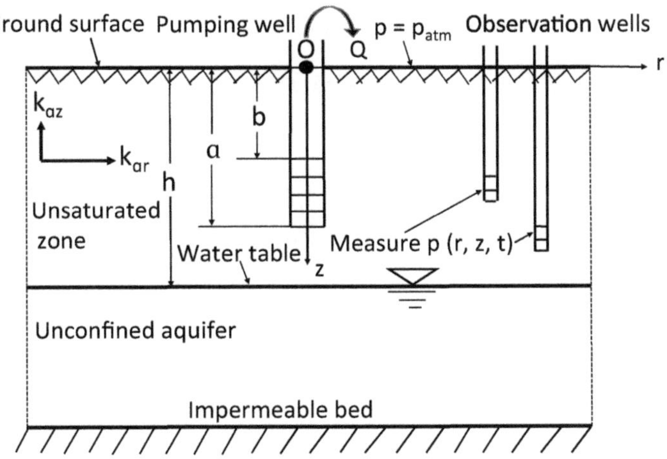

FIGURE 7-8
Schematic diagram of a soil gas pump test in a formation open to the atmosphere at the ground surface. Adapted from Falta (1996, p. 751, Figure 1).

$$z = 0 \quad p^2 = p_{atm}^2 \tag{7-81}$$

$$z = h \quad \frac{\partial p^2}{\partial z} = 0 \tag{7-82}$$

At infinity, the boundary condition is

$$r \to \infty \quad p^2 = p_{atm}^2 \tag{7-83}$$

The line source boundary conditions at the well are

$$\lim_{r \to 0}\left(r\frac{\partial p^2}{\partial r}\right) = 0 \quad 0 < z < b \tag{7-84}$$

$$\lim_{r \to 0}\left(r\frac{\partial p^2}{\partial r}\right) = \frac{Qp^*\mu}{\pi k_{ar}(a-b)} \quad b < z < a \tag{7-85}$$

$$\lim_{r \to 0}\left(r\frac{\partial p^2}{\partial r}\right) = 0 \quad a < z < h \tag{7-86}$$

Derivation of Eq. (7-85) is given in Example 7-3.

Example 7-3

Derive Eq. (7-85).

Solution

From Eq. (7-9), the radial Darcy velocity is

$$q_r = -\frac{k_{ar}}{\mu}\frac{\partial p}{\partial r} = -\frac{k_{ar}}{\mu}\left(\frac{1}{2p}\frac{\partial p^2}{\partial r}\right) \tag{E7-3-1}$$

The air flow rate at the distance r from the well is

$$Q(r) = -2\pi r(a-b)q_r \tag{E7-3-2}$$

Using Eq. (E7-3-1) in Eq. (E7-3-2),

$$Q(r) = -2\pi r(a-b)\left[-\frac{k_{ar}}{\mu}\left(\frac{1}{2p}\frac{\partial p^2}{\partial r}\right)\right] = \pi(a-b)\frac{k_{ar}}{\mu}r\frac{1}{p}\frac{\partial p^2}{\partial r} \tag{E7-3-3}$$

From here,

$$r\frac{\partial p^2}{\partial r} = \frac{Qp\mu}{\pi k_{ar}(a-b)} \tag{E7-3-4}$$

For $r \to 0$ and $p = p^*$, one can write

$$\lim_{r \to 0}\left(r\frac{\partial p^2}{\partial r}\right) = \frac{Qp^*\mu}{\pi k_{ar}(a-b)} \tag{E7-3-5}$$

which is exactly the same as Eq.7-85).

Equation (7-85) is the line source boundary condition which is written on the assumption that the volumetric gas flow rate of Q (negative for gas removal) measured at a pressure of p^*, flowing from a well screened from $z = b$ to $z = a$ as shown in Figure 7-8. The gas pressure factor p^* is used to correct Q for compression effects, and this formulation is exactly equivalent to assuming a constant gas mass flow rate from the well (Falta, 1996, p. 752).

7.4.2.3 Solution

The solution to Eqs. (7-79) through (7-86) was determined using successive integral transforms by Falta (1993) as

$$p^2 = p_{atm}^2 - \frac{Q\,p^*\mu}{\pi^2 k_{ar}(a-b)} \sum_{n=1}^{\infty} \left\{ \frac{1}{m} \left[\cos\left(\frac{m\pi b}{h}\right) - \cos\left(\frac{m\pi a}{h}\right) \right] \sin\left(\frac{m\pi z}{h}\right) W(u,\beta) \right\} \quad (7\text{-}87)$$

with the exception of notation. In Eq. (7-87),

$$m = n - \frac{1}{2} \quad (7\text{-}88)$$

$$u = \frac{r^2 \left(\dfrac{n_a}{p_{atm}}\right)}{4\left(\dfrac{k_{ar}}{\mu}\right)t} \quad (7\text{-}89)$$

$$\beta = r\left[\frac{k_{az}}{k_{ar}}\left(\frac{m\pi}{h}\right)^2\right]^{\frac{1}{2}} \quad (7\text{-}90)$$

and $W(u,\beta)$ is the *Hantush and Jacob well function for leaky aquifers* defined as [Hantush and Jacob, 1955, p. 98, Eq. (14); Hantush, 1964, p. 321]

$$W(u,\beta) = \int_u^{\infty} \frac{1}{y} \exp\left(-y - \frac{\beta^2}{4y}\right) dy \quad (7\text{-}91)$$

Equation (7-87) is also exactly the same as Falta [1996, p. 752, Eq. (9)] with the exception of notation.

7.4.3 Partially Penetrating Well in an Unsaturated Zone Below the Ground Surface with a Leaky Confining Layer

7.4.3.1 Governing Differential Equation

The modified form of Eq. (7-79) with the inclusion of the leakage term is given in Baehr and Hult (1989) based on the leaky aquifer drawdown differential equation for a leaky confined aquifer by Hantush [1964, p. 349, Eq. (72)]. With the leakage term, Eq. (7-79) takes the form

$$k_{ar}\left(\frac{\partial^2 p^2}{\partial r^2}+\frac{1}{r}\frac{\partial p^2}{\partial r}\right)+k_{az}\frac{\partial^2 p^2}{\partial z^2}+\frac{k_a'}{hh'}\left(p_{atm}^2-p^2\right)=\frac{n_a\mu}{p_{atm}}\frac{\partial p^2}{\partial t} \tag{7-92}$$

in which h' is the thickness of the leaky confining layer and k_a' is the permeability of the leaky confining layer Eq. (7-92) is exactly the same as Baehr and Hult [1989, p. 58, Eq. (5a)] and Falta [1996, p. 752, Eq. (13)] with the exception of notation.

7.4.3.2 Problem Description and Initial and Boundary Conditions

The schematic diagram of the problem is shown in Figure 7-9, which shows that there is a leaky confining layer below the ground surface. The initial condition is

$$t=0 \quad p^2=p_{atm}^2 \tag{7-93}$$

The boundary conditions at the ground surface and water table, respectively, are

$$z=0 \quad \frac{\partial p^2}{\partial z}=0 \tag{7-94}$$

$$z=h \quad \frac{\partial p^2}{\partial z}=0 \tag{7-95}$$

At infinity, the boundary condition is

$$r\to\infty \quad p^2=p_{atm}^2 \tag{7-96}$$

The line source boundary conditions at the well are

$$\lim_{r\to 0}\left(r\frac{\partial p^2}{\partial r}\right)=0 \quad 0<z<b \tag{7-07}$$

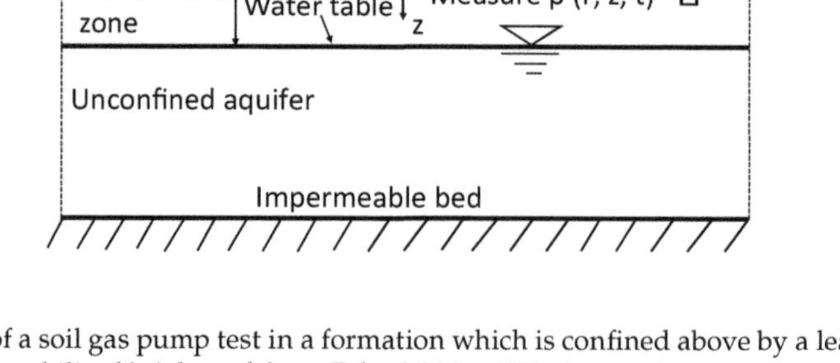

FIGURE 7-9
Schematic diagram of a soil gas pump test in a formation which is confined above by a leaky confining layer of thickness h' and permeability k_a'. Adapted from Falta (1996, p. 752, Figure 2).

$$\lim_{r\to 0}\left(r\frac{\partial p^2}{\partial r}\right)=\frac{Qp^*\mu}{\pi k_{ar}(a-b)}\quad b<z<a \tag{7-98}$$

$$\lim_{r\to 0}\left(r\frac{\partial p^2}{\partial r}\right)=0\quad a<z<h \tag{7-99}$$

7.4.3.3 Solution

The solution to Eqs. (7-92) through (7-99) was determined by Baehr and Hult [1989, p. 58, Eq. 6] based on the solution of Hantush [1964, p. 350, Eq. (73)] as

$$p^2=p_{atm}^2-\frac{Qp^*\mu}{2\pi k_{ar}h}W(u,\beta_1)-\frac{Qp^*\mu}{\pi^2 k_{ar}(a-b)}\sum_{n=1}^{\infty}\left\{\frac{1}{n}\left[\sin\left(\frac{n\pi a}{h}\right)-\sin\left(\frac{n\pi b}{h}\right)\right]\right.$$
$$\left.\cos\left(\frac{n\pi z}{h}\right)W(u,\beta_2)\right\} \tag{7-100}$$

In Eq. (7-100), $W(u,\beta)$ is the *Hantush and Jacob well function for leaky aquifers* defined by Eq. (7-91) with u defined by Eq. (7-89), and

$$\beta_1=r\left(\frac{k_a'}{k_{ar}hh'}\right)^{\frac{1}{2}} \tag{7-101}$$

and

$$\beta_2=r\left[\frac{k_{az}}{k_{ar}}\left(\frac{m\pi}{h}\right)^2+\frac{k_a'}{k_{ar}hh'}\right]^{\frac{1}{2}} \tag{7-102}$$

Equation (7-100) is exactly the same as Baehr and Hult [1989, p. 58, Eq. (6)] and Falta [1996, p. 752, Eq. (15)] with the exception of notation.

7.4.4 Special Solutions for Steady-State Cases

Steady-state solutions can be determined from Eqs. (7-87) and (7-100) with $t=\infty$. With this, Eq. (7-89) becomes $u=0$. Then, Eq. (7-91) becomes (Hantush, 1964, p. 321)

$$W(0,\beta)=2K_0(\beta) \tag{7-103}$$

in which $K_0(\beta)$ is the modified Bessel function of zero order Eq. (7-103) is also used by Baehr and Hult [1989, p. 59, Eq. (7)] and Falta [1996, p. 753, Eq. (18)].

7.4.4.1 Steady-State Solution without Leaky Confining Layer

With Eq. (7-103), Eq. (7-87) becomes

$$p^2=p_{atm}^2-\frac{2Qp^*\mu}{\pi^2 k_{ar}(a-b)}\sum_{n=1}^{\infty}\left\{\frac{1}{m}\left[\cos\left(\frac{m\pi b}{h}\right)-\cos\left(\frac{m\pi a}{h}\right)\right]\sin\left(\frac{m\pi z}{h}\right)K_0(\beta)\right\} \tag{7-104}$$

in which m and β are given by Eqs. (7-88) and (7-90), respectively.

Fully penetrating well case is analogous to the Theis solution and does not have a steady-state solution (see Section 7.3.3.7). Therefore, Eq. (7-104) is not valid for fully penetrating well cases.

7.4.4.2 Steady-State Solution with Leaky Confining Layer

With Eq. (7-103), Eq. (7-100) becomes

$$p^2 = p_{atm}^2 - \frac{Qp^*\mu}{\pi k_{ar} h} K_0(\beta_1) - \frac{2Qp^*\mu}{\pi^2 k_{ar}(a-b)} \sum_{n=1}^{\infty} \left\{ \frac{1}{n} \left[\sin\left(\frac{n\pi a}{h}\right) - \sin\left(\frac{n\pi b}{h}\right) \right] \cos\left(\frac{n\pi z}{h}\right) K_0(\beta_2) \right\} \quad (7\text{-}105)$$

in which β_1 and β_2 are given by Eqs. (7-101) and (7-102), respectively.

Fully penetrating well cases are accounted for by evaluating Eq. (7-105) for $a = h$ and $b = 0$. Substituting these values into Eq. (7-105), the first and second sine terms become $\sin(n\pi) = 0$ and $\sin(0) = 0$, which indicate that Eq. (7-105) is not valid for fully penetrating well cases.

7.4.5 Evaluation of the Solutions and Air Permeability Determination from Field Data

7.4.5.1 Evaluation of the Solutions with Field Data

Falta (1996, p. 754, Figure 3) compared the results of partially penetrating well solutions with the field-measured values of Massmann (1989). The author states that the analytical solutions were found to provide very good matches with the experimental data.

7.4.5.2 Air Permeability Determination from Field Data

Falta (1996) presents a software called GASSSOLV, which analyzes soil gas pump test data to determine air permeability. The FORTRAN program is interactive and the user is prompted for the various pump test operating conditions (Falta, 1996, p. 755).

7.5 Steady-State Analytical Solutions for Soil Gas Flow to Wells and Determination of the Unsaturated Zone Properties

7.5.1 Introduction

As mentioned in Section 7.1, SVE and BV systems are being designed to remediate especially unsaturated zones. Since most SVE/BV systems are designed for long-term operation, steady-state flow models are appropriate in designing SVE/BV systems. Steady-state solutions can be used in determining k_a from air permeability tests in an unsaturated zone under the conditions that sufficient time is elapsed for flow to stabilize.

In this section, solutions will be presented by adapting some aquifer/well hydraulics solutions for water flow in aquifers to air flow from wells in unsaturated zones.

7.5.2 Estimation of Steady-State Reaching Time in a Typical Sand Formation

As mentioned in Section 7.3.3.4, Theis solution given by Eq. (7-47) or (7-48) does not approach steady-state solution for $p'(r,t) = p(r,t) - p_{atm}$, namely, $p'(r,t)$ continuously increases with u decreases as shown in Figure 7-3. However, as can be seen from Figure 7-3, the tangential slope to the curve tends to decrease for $u < u_{min} = 0.001, u < u_{min} = 0.0001, u < u_{min} = 0.00001, \ldots$ From here (Johnson et al., 1990, p. 414),

$$u = \frac{r^2 n_a \mu}{4 k_a p_{atm} t} < u_{min} \tag{7-106}$$

Johnson et al. (1990, p. 414) used $u_{min} = 0.001$ and from Eq. (7-106), the authors estimated the time is required for u to become smaller than $u_{min} = 0.001$ using the following set of parameters is typical for a sand formation: $n_a = 0.30$; $\mu = 1.8 \times 10^{-4}$ g/(cm.s); $k_a = 10^{-7} - 10^{-6}$ cm^2 $= 10 - 100$ darcy [1 darcy $\cong 1.0 \times 10^{-8}$ cm^2, see Eq. (3-24) in Chapter 3]; and $p_{atm} = 101{,}325$ Pa $= 1.013 \times 10^6$ g/(cm.s^2). The dimensions conversion of p_{atm} is shown in Example 7-4. In the following sections, $u_{min} = 0.001$ and $u_{min} = 0.0001$ are used with the above given values.

Example 7-4

Write $p_{atm} = 101{,}325$ Pa with units of g/(cm.s^2).

Solution

$$p_{atm} = 101{,}325 \text{ Pa} = 101{,}325 \frac{\text{N}}{\text{m}^2}$$

$$1\,\text{N} = 1 \frac{\text{kg.m}}{\text{s}^2}$$

$$p_{atm} = 101{,}325 \frac{\text{N}}{\text{m}^2} = 101{,}325 \frac{\frac{\text{kg.m}}{\text{s}^2}}{\text{m}^2} = 101{,}325 \frac{\text{kg}}{\text{m.s}^2}$$

$$= 101{,}325 \frac{1{,}000 \text{ g}}{100 \text{ cm.s}^2} = 101{,}325 \times 10 \frac{\text{g}}{\text{cm.s}^2}$$

$$p_{atm} = 1.013 \times 10^6 \frac{\text{g}}{\text{cm.s}^2}$$

7.5.2.1 Estimation of Steady-State Reaching Time for $u_{min} = 0.001$

For $u_{min} = 0.001$, using Eq. (7-106) with $n_a = 0.30$; $\mu = 1.8 \times 10^{-4}$ g/(cm.s); $k_a = 10^{-7} - 10^{-6}$ cm^2 $=$ $10 - 100$ darcy [1 darcy $\cong 1.0 \times 10^{-8}$ cm^2, see Eq. (3-26) in Chapter 3]; and $p_{atm} = 101{,}325$ Pa $=$ 1.013×10^6 g/(cm.s^2), time (r) versus radial distance (r) are shown in Figure 7-10 for $k_a = 10$ darcy and $k_a = 100$ darcy in the SI system. Figure 7-11 shows the curves in the English FSS system.

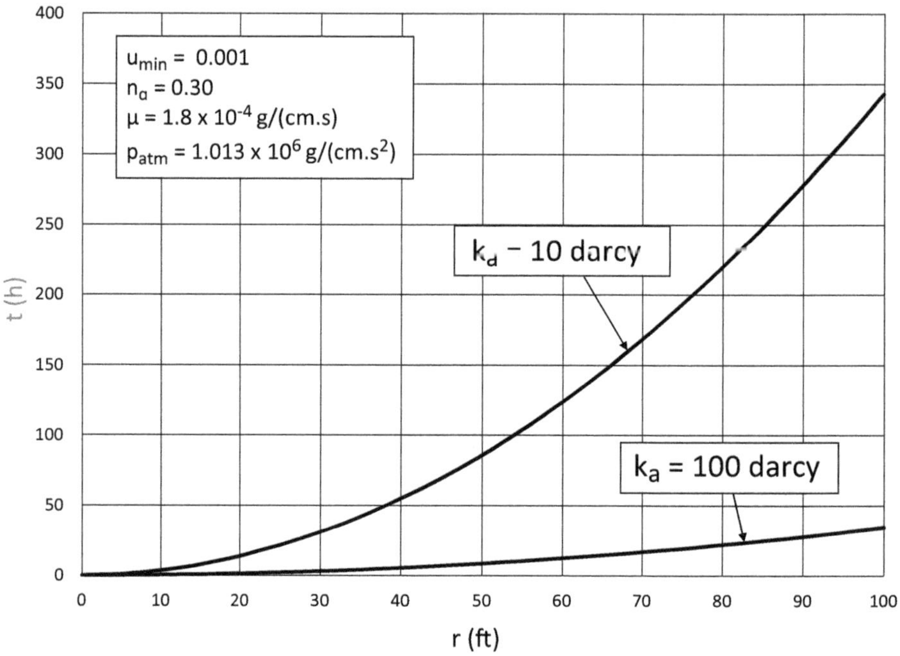

FIGURE 7-10
Estimated times for vapor flow to reach steady state for $u_{min} = 0.001$ in the English FSS units. Adapted from Johnson et al. (1990, p. 415, Figure 2).

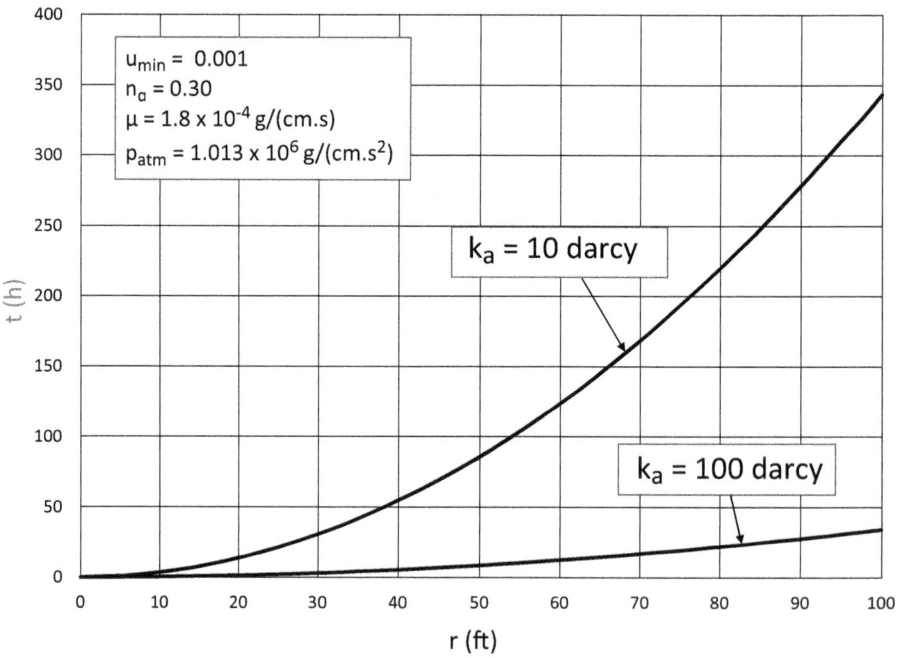

FIGURE 7-11
Estimated times for vapor flow to reach steady state for $u_{min} = 0.001$ in the SI units. Adapted from Johnson et al. (1990, p. 415, Figure 2).

For a typical sandy soil ($10 < k_a < 100$ darcy), generally, it may take as little as 1 day, or as long as a week to reach steady state. Figures 7-10 and 7-11 are directly applicable to field vapor permeability tests, which are often performed before designing or installing a venting system. For example, at $r = 20\ \text{m}(66\ \text{ft})$ distance, from Figures 7-10 and 7-11 for a soil having $k_a = 10$ darcy permeability, one can find that it requires 150 h to reach steady state. For $k_a = 100$ darcy permeability, at the same r distance, it requires 15 h to reach steady state.

7.5.2.2 Estimation of Steady-State Reaching Time for u_{min} = 0.00001

Likewise, for $u_{min} = 0.0001$, using Eq. (7-106) with $n_a = 0.30$; $\mu = 1.8\times10^{-4}\ \text{g}/(\text{cm.s})$; $k_a = 10^{-7} - 10^{-6}\ \text{cm}^2 = 10 - 100$ darcy [$1\ \text{darcy} \cong 1.0\times10^{-8}\ \text{cm}^2$, see Eq. (3-24) in Chapter 3]; and $p_{atm} = 101,325\ \text{Pa} = 1.013\times10^6\ \text{g}/(\text{cm.s}^2)$, time ($r$) versus radial distance (r) are shown in Figure 7-12 for $k_a = 10$ darcy and $k_a = 100$ darcy in the SI system. Figure 7-13 shows the curves in the English FSS system.

For a typical sandy soil ($10 < k_a < 100$ darcy), generally, it may take as little as 1 day, or as long as a week to reach steady state. Figures 7-12 and 7-13 are directly applicable to field vapor permeability tests, which are often performed before designing or installing a venting system. For example, at $r = 20\ \text{m}(66\ \text{ft})$ distance, from Figures 7-12 and 7-13 for a soil having $k_a = 10$ darcy permeability one can find that it requires 1,500 h to reach steady state. For $k_a = 100$ darcy permeability, at the same r distance it requires 150 h to reach steady state.

As expected, reaching steady state for $u_{min} = 0.0001$ is 10 times increased due to the fact that the $W_T(u)$ versus $r^2 n_a \mu/(4k_a p_{atm} t)$ curve (Figure 7-4) does not converge to an exact steady-state equation. For $u = 0.001$ and 0.0001, $W_T(u) = 6.54$ and 8.56, respectively.

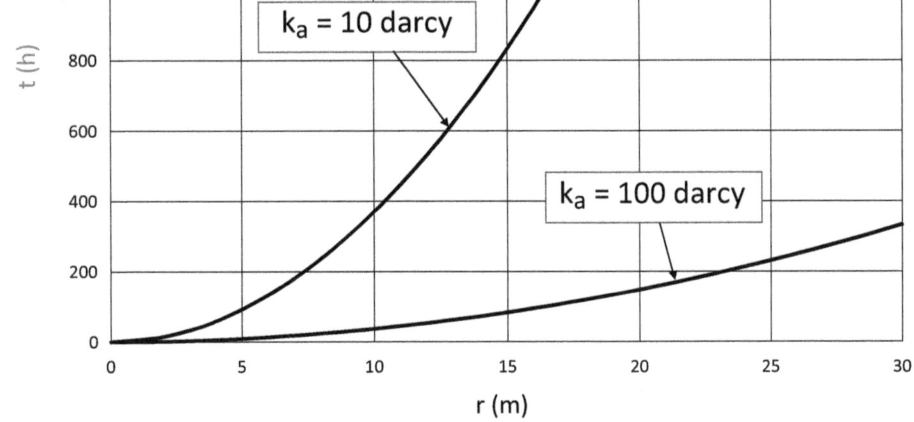

FIGURE 7-12
Estimated times for vapor flow to reach steady state for $u_{\text{min}} = 0.0001$ in the English FSS units.

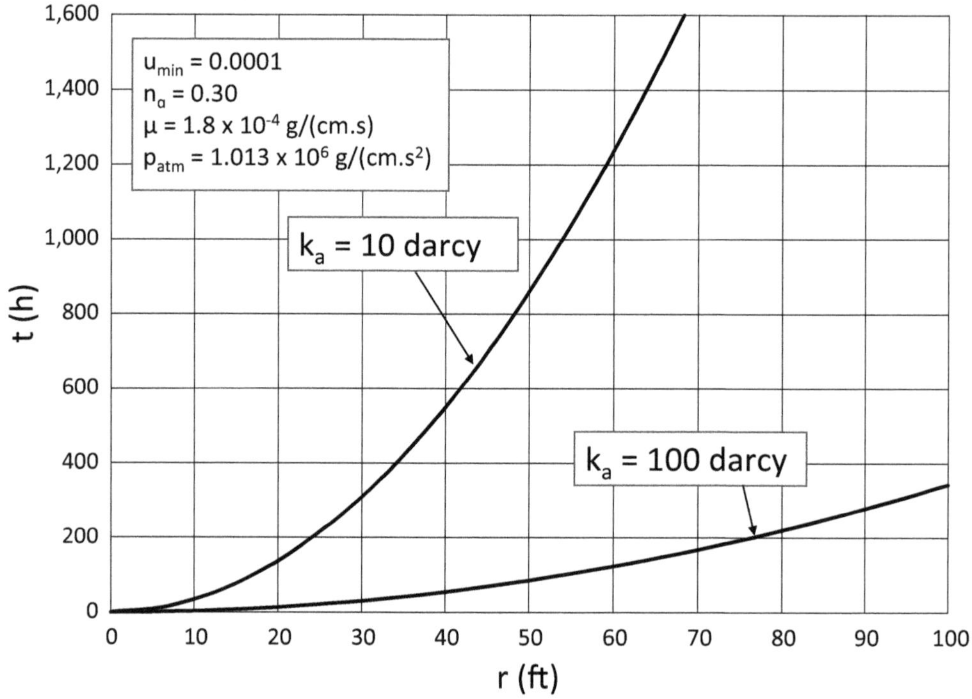

FIGURE 7-13
Estimated times for vapor flow to reach steady state for $u_{\min} = 0.0001$ in the SI units.

Therefore, the ratio $(8.56 - 6.54)/8.56 = 0.24$ indicates the error for the $W_T(u)$ values, and this error is not significant for all practical purposes.

7.5.3 Steady-State Soil Gas Flow to a Fully-Penetrating Well in the Unsaturated Zone

As mentioned in Section 7.1, SVE and BV systems are being designed to remediate especially unsaturated zones. Because SVE and BV systems are designed for long-term operation, steady-state gas flow models are appropriate for remediation systems design for contaminated unsaturated zones. Steady-state gas flow solutions can be used for air permeability tests under the conditions that sufficient time is elapsed for the stabilization of air flow. To analyze for air flow data, steady-state well hydraulics solutions developed since the invention of Darcy's law in 1856 have been adapted to gas flow. In this section, the adapted solutions will be presented with examples.

7.5.3.1 Governing Equations and Boundary Conditions for Steady-State Gas Flow in Unsaturated Zones

Under steady-state conditions, the pressure p becomes time-independent. Therefore, the three-dimensional partial differential equation for steady-state gas flow in the Cartesian coordinates given by Eq. (7-17) becomes

$$\frac{\partial p^2}{\partial x^2} + \frac{\partial p^2}{\partial y^2} + \frac{\partial p^2}{\partial z^2} = 0 \tag{7-107}$$

which is equivalent to the Laplace differential equation in p^2.

For two-dimensional gas flow, Eq. (7-107) takes the form

$$\frac{\partial p^2}{\partial x^2} + \frac{\partial p^2}{\partial y^2} = 0 \tag{7-108}$$

Its form in circular cylindrical coordinates is

$$\frac{d^2 p^2}{dr^2} + \frac{1}{r}\frac{dp^2}{dr} = 0 \tag{7-109}$$

which can be derived from Eq. (7-108) using several methods (e.g., Batu, 2024, pp. 1052–1056).

From Eq. (7-9), for two-dimensional flows, radial Darcy's velocity is

$$q_r = -\frac{k_a}{\mu}\frac{dp}{dr} \qquad q_z = 0 \tag{7-110}$$

Figure 7-2 shows the schematic of gas flow to a fully penetrating well open to the atmosphere at ground surface in the unsaturated zone. The boundary conditions are

$$p = p_w \qquad r = r_w \tag{7-111}$$

$$p = p \qquad r = r \tag{7-112}$$

$$p = p_I = p_{atm} \qquad r = R_I \tag{7-113}$$

in which p_I is the pressure at the R_I *radius of the influence of the well* and p_{atm} is the atmospheric pressure. These boundary conditions are the *first-type boundary condition* or *Dirichlet boundary condition.*

The boundary condition given by Eq. (7-111) states that at the r_w radius of the well, the pressure is equal to p_w. The boundary condition given by Eq. (7-112) states that at r distance, the pressure is equal to the pressure p. Equation (7-113) states that at the zone of distance R_I, the pressure is equal to the atmospheric pressure.

7.5.3.2 Solution for Steady-State Gas Flow Boundary-Value Problem from r_w to r

The solution of the boundary-value problem defined in Section 7.5.3.1 is

$$p^2 - p_w^2 = \frac{p\mu Q_v}{\pi k_a b}\ln\left(\frac{r}{r_w}\right) \tag{7-114}$$

which is derived in Example 7-5.

With $p = p^*$, which is absolute pressure at the point of flow measurement (Falta, 1996, p. 752; U.S. ACE, 2002, p. 2.15), Eq. (7-114) becomes

$$p^2 - p_w^2 = \frac{p^*\mu Q_v}{\pi k_a b}\ln\left(\frac{r}{r_w}\right) \tag{7-115}$$

Example 7-5

Derive Eq. (7-114).

Solution

Let

$$\frac{d^2p^2}{dr^2} \equiv p_{rr}^2 \quad \frac{dp^2}{dr} \equiv p_r^2 \tag{E7-5-1}$$

Then, Eq. (7-109) becomes

$$\frac{p_{rr}^2}{p_r^2} + \frac{1}{r} = 0 \tag{E7-5-2}$$

which can be written as

$$\frac{d}{dr}\left[\ln\left(p_r^2\right) + \ln(r)\right] = 0 \tag{E7-5-3}$$

From here, one can write

$$\ln\left(p_r^2 r\right) = \ln(C_1) \tag{E7-5-4}$$

$$\frac{dp^2}{dr} = C_1 \frac{1}{r} \tag{E7-5-5}$$

in which C_1 is a constant. The continuity equation is

$$-Q_v = 2\pi r b q_r \tag{E7-5-6}$$

in which Q_v is the volumetric flow rate. In Eq. (E7-5-6), $-Q_v$ corresponds to withdrawal of gas and $+Q_v$ corresponds to injection of gas. q_r is the radial flow given by Eq. (7-110), which can be written as

$$q_r = -\frac{k_a}{\mu}\frac{dp}{dr} = -\frac{k_a}{\mu}\frac{1}{2p}\frac{dp^2}{dr} \tag{E7-5-7}$$

Using Eq. (E7-5-5) in Eq. (E7-57),

$$\frac{dp^2}{dr} = C_1 \frac{1}{r} = -\frac{2p\mu q_r}{k_a} \tag{E7-5-8}$$

Solving q_r from Eq. (E7-5-6) and introducing it in Eq. (E7-5-8) gives

$$C_1 = \frac{p\mu Q_v}{\pi k_a b} \tag{E7-5-9}$$

With this, Eq. (E7-5-8) becomes

$$\frac{dp^2}{dr} = \frac{1}{r}\frac{p\mu Q_v}{\pi k_a b} \tag{E7-5-10}$$

$$dp^2 = \frac{1}{r}\frac{p\mu Q_v}{\pi k_a b}dr \tag{E7-5-11}$$

Based on the boundary conditions given by Eqs. (7-111) and (7-112), integrating the left-hand side of Eq. (E7-5-11) from p_w to p and the right-hand side from r_w to r,

$$\int_{p_w}^{p} dp^2 = \int_{r_w}^{r} \frac{p\mu Q_v}{\pi k_a b}\frac{dr}{r} \tag{E7-5-12}$$

$$\left|p^2\right|_{p_w}^{p} = p^2 - p_w^2 = \frac{p\mu Q_v}{\pi k_a b}\left|\ln r\right|_{r_w}^{r} = \frac{p\mu Q_v}{\pi k_a b}[\ln(r) - \ln(r_w)] \tag{E7-5-13}$$

$$p^2 - p_w^2 = \frac{p\mu Q_v}{\pi k_a b}\ln\left(\frac{r}{r_w}\right) \tag{E7-5-14}$$

which is exactly the same as Eq. (7-114).

Also, Eq. (2.20a) of US ACE (2002, p. 2.15) is derived in Example 7-6.

Example 7-6

With $p = p^*$, from Eq. (E7-5-10) of Example 7-5, derive Eq. (2-20a) of US ACE (2002, p. 2.15)

Solution

With $p = p^*$, Eq. (E7-5-10) of Example 7-5 becomes

$$\frac{dp^2}{dr} = \frac{1}{r}\frac{p^*\mu Q_v}{\pi k_a b} \tag{E7-6-1}$$

Using the boundary conditions given by Eqs. (7-112) and (7-113) with $p = p_{atm} = p_I$,

$$\int_{p}^{p_I} dp^2 = \frac{p^*\mu Q_v}{\pi k_a b}\int_{r}^{R_I}\frac{dr}{r} \tag{E7-6-2}$$

Integrating it,

$$p_I^2 - p^2 = \frac{p^*\mu Q_v}{\pi k_a b}\left[\ln(R_I) - \ln(r)\right] \tag{E7-6-3}$$

$$p_I^2 - p^2 = \frac{p^*\mu Q_v}{\pi k_a b}\ln\left(\frac{R_I}{r}\right) \tag{E7-6-4}$$

which is exactly the same as Eq. (2-20a) of US ACE (2002, p. 2.15) with the exception of notation as well as the sign convention of Q_v.

7.5.3.3 Solution for Steady-State Gas Flow Boundary-Value Problem from r_w to R_I

From Eq. (7-115), it follows that $p = p\ (r)$ will increase indefinitely as r increases. This is impossible. As a result, Eq. (7-115) is valid only around the close proximity of the well at which steady-state flow has been reached (e.g., Bear, 1979, p. 305). With the values in Eqs. (7-111) and (7-113), Eq. (E7-6-4) of Example 7-6 becomes

$$p_{atm}^2 - p_w^2 = \frac{p^* \mu Q_v}{\pi k_a b} \ln\left(\frac{R_I}{r_w}\right) \tag{7-116}$$

7.5.3.4 Gas Pressure Distribution Equation

Dividing Eq. (7-115) to Eq. (7-116),

$$\frac{p^2 - p_w^2}{p_{atm}^2 - p_w^2} = \frac{\ln\left(\dfrac{r}{r_w}\right)}{\ln\left(\dfrac{R_I}{r_w}\right)} \tag{7-117}$$

which can be written as

$$p^2 - p_w^2 = \left(p_{atm}^2 - p_w^2\right)\frac{\ln\left(\dfrac{r}{r_w}\right)}{\ln\left(\dfrac{R_I}{r_w}\right)} \tag{7-118}$$

Equation (7-114) is exactly the same as Johnson et al. [1990, p. 415, Eq. (9)] with the exception of notation. In Eq. (7-118), p_w is the pressure at the vacuum well having r_w and R_I is the radius of influence, where the pressure, by definition, is equal to the ambient atmospheric pressure p_{atm}.

One should notice that Eq. (7-118) appears to predict that the pressure distribution is independent of soil properties. In Eq. (7-118), the radius of influence R_I is the only quantity that depends on the soil properties. Johnson et al. (1990, p. 415) state that the radial pressure distribution is fairly insensitive to significant changes to R_I. This is illustrated in Figures 7-14 and 7-15 of Example 7-7, in the English FSS units and SI units, respectively.

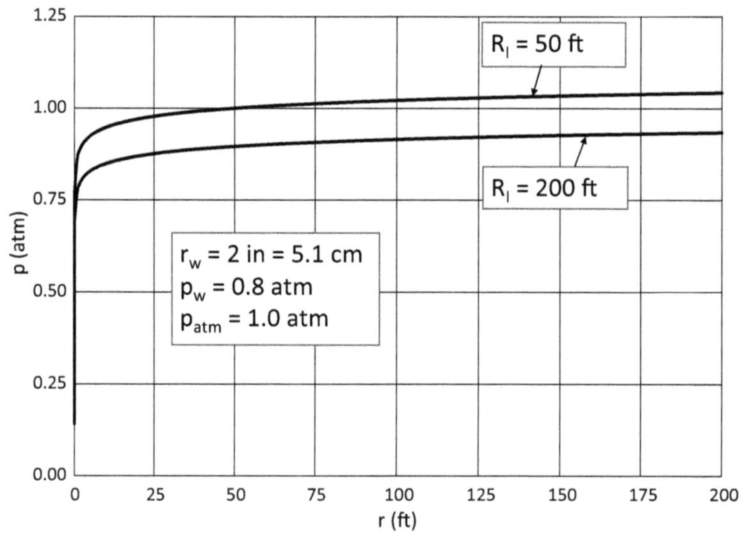

FIGURE 7-14
Variation of pressure p with the radial distance in the English FSS units for $R_I = 50\ \text{ft}\ (15.24\ \text{m})$ and $R_I = 200\ \text{ft}\ (60.96\ \text{m})$. Adapted from Johnson et al. (1990, p. 415, Figure 3).

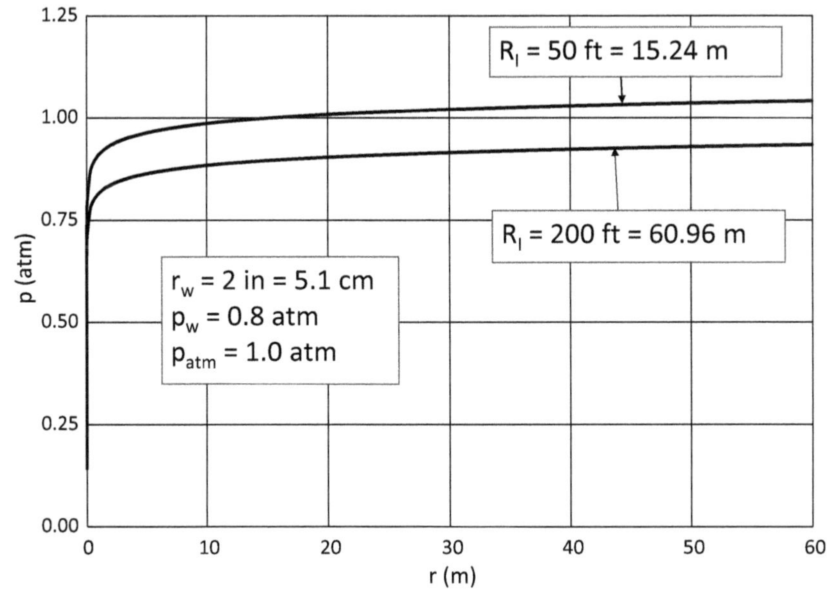

FIGURE 7-15
Variation of pressure p with the radial distance in the SI units for $R_I = 50\text{ ft}(15.24\text{ m})$ and $R_I = 200\text{ ft}(60.96\text{ m})$. Adapted from Johnson et al. (1990, p. 415, Figure 3).

Example 7-7

This example is adapted from Johnson et al. (1990, p. 415, Figure 3). In Figure 7-2, the following values are given: $r_w = 2\text{ in} = 5.1\text{ cm}$; $p_w = 0.8\text{ atm}$; and $p_{atm} = 1\text{ atm}$. Determine the variation of p as a function of r for $R_I = 50\text{ ft}(15.24\text{ m})$ and $R_I = 200\text{ ft}(60.96\text{ m})$ and discuss the results.

Solution

Using Eq. (7-118), the pressure p versus r distance in the English FSS units system is shown in Figure 7-14 and its corresponding form in the SI units system is shown in Figure 7-15. These figures show that the pressure distribution is not too sensitive with the radius of influence R_I. For example, Figures 7-14 and 7-15 show that at $r = 50\text{ ft}(15.24\text{ m})$, $p = 1.00\text{ atm}$ and 0.90 atm for $R_I = 50\text{ ft}(15.24\text{ m})$ and $R_I = 200\text{ ft}(60.96\text{ m})$, respectively. Likewise, at $r = 200\text{ ft}(60.96\text{ m})$, $p = 1.04\text{ atm}$ and 0.93 atm for $R_I = 50\text{ ft}(15.24\text{ m})$ and $R_I = 200\text{ ft}(60.96\text{ m})$, respectively.

7.5.3.5 Determination of the Radius of Influence (R_I)

By extrapolating Eq. (7-116) to the radius at $p = p_{atm}$, the radius of influence (R_I) can be determined. In a practical sense, R_I is the limit of measurable pressure influence under extraction conditions from a well. The radius of influence (R_I) may be determined by fitting measured data from multiple observation wells to Eq. (7-116) or it can be determined by preparing on a single logarithmic paper, distance (logarithmic) versus steady-state drawdown of each well (e.g., Batu, 1998, pp. 128–131). However, the radius of influence (R_I) is both somewhat problematic and misunderstood (US ACE, 2002, pp. 2–16).

Mass balance requires that for continuous withdrawal of air from a formation with impermeable upper and lower boundaries, R_I must increase with time. This conclusion is resulted from analysis of transient radial air flow, which indicates that R_I increases in proportion to the square root of time (McWhorter and Sunada, 1977; McWhorter, 1990). However, the widespread acceptance of a fixed R_I reflects the common field observation that the limit of radial pressure influence often shows little change with time. This phenomenon may be explained by leakage of air through the upper and lower boundaries (US ACE, 2002, pp. 2–16).

7.5.3.6 Vapor Flow Velocity Equation for a Given Applied Vacuum Pressure at the Well

In this section, first, the radial Darcy velocity is determined from Eqs. (7-110) and (7-118). From Eq. (7-118), the radial Darcy velocity $q_r(r)$ can be derived as

$$q_r(r) = -\left(\frac{k_a}{2\mu}\right)\frac{\left[\dfrac{p_w}{r\ln\left(\dfrac{r_w}{R_I}\right)}\right]\left(1-\dfrac{p_{atm}^2}{p_w^2}\right)}{\left\{1+\left(1-\dfrac{p_{atm}^2}{p_w^2}\right)\left[\dfrac{\ln\left(\dfrac{r}{r_w}\right)}{\ln\left(\dfrac{r_w}{R_I}\right)}\right]\right\}^{\frac{1}{2}}} \tag{7-119}$$

which is exactly the same as Johnson et al. [1990, p. 415, Eq. (10)] with the exception of notation Eq. (7-119) is derived in Example 7-8. In Eq. (7-119), the negative sign is due to the reason that the flow is directed toward the well.

Example 7-8

Derive Eq. (7-119).

Solution

Equation (7-118) can be written as

$$p^2 = p_w^2 + A\left[\ln(r) - \ln(r_w)\right] \tag{E7-8-1}$$

in which

$$A = \frac{p_{atm}^2 - p_w^2}{\ln\left(\dfrac{R_I}{r_w}\right)} \tag{E7-8-2}$$

From Eq. (E7-8-2),

$$p = \left\{p_w^2 + A\left[\ln(r) - \ln(r_w)\right]\right\}^{\frac{1}{2}} \tag{E7-8-3}$$

Taking its derivative,

$$\frac{dp}{dr} = \frac{A}{2r\left[p_w^2 + A\ln(r) - A\ln(r_w)\right]^{\frac{1}{2}}} \tag{E7-8-4}$$

Using this in Eq. (7-100),

$$q_r = -\left(\frac{k_a}{2\mu}\right)\frac{A}{r\left[p_w^2 + A\ln(r) - A\ln(r_w)\right]^{\frac{1}{2}}} \tag{E7-8-5}$$

Using the A expression from Eq. (E7-8-2),

$$q_r = -\left(\frac{k_a}{2\mu}\right)\frac{\dfrac{p_{atm}^2 - p_w^2}{r\ln\left(\dfrac{R_I}{r_w}\right)}}{\left[p_w^2 + \left(\dfrac{p_{atm}^2 - p_w^2}{\ln\left(\dfrac{R_I}{r_w}\right)}\right)\ln\left(\dfrac{r}{r_w}\right)\right]^{\frac{1}{2}}}$$

$$= -\left(\frac{k_a}{2\mu}\right)\frac{-\left[\dfrac{p_w^2}{r\ln\left(\dfrac{R_I}{r_w}\right)}\right]\left(1 - \dfrac{p_{atm}^2}{p_w^2}\right)}{\left\{p_w^2 + \left[\dfrac{p_{atm}^2 - p_w^2}{\ln\left(\dfrac{R_I}{r_w}\right)}\right]\ln\left(\dfrac{r}{r_w}\right)\right\}^{\frac{1}{2}}}$$

$$= -\left(\frac{k_a}{2\mu}\right)\frac{-\left[\dfrac{p_w^2}{r\ln\left(\dfrac{R_I}{r_w}\right)}\right]\left(1 - \dfrac{p_{atm}^2}{p_w^2}\right)}{p_w\left\{1 + \left(\dfrac{p_{atm}^2}{p_w^2} - 1\right)\left[\dfrac{\ln\left(\dfrac{r}{r_w}\right)}{\ln\left(\dfrac{R_I}{r_w}\right)}\right]\right\}^{\frac{1}{2}}}$$

$$= -\left(\frac{k_a}{2\mu}\right)\frac{-\left[\dfrac{p_w}{r\ln\left(\dfrac{R_I}{r_w}\right)}\right]\left(1 - \dfrac{p_{atm}^2}{p_w^2}\right)}{\left\{1 + \left(\dfrac{p_{atm}^2}{p_w^2} - 1\right)\left[\dfrac{\ln\left(\dfrac{r}{r_w}\right)}{\ln\left(\dfrac{R_I}{r_w}\right)}\right]\right\}^{\frac{1}{2}}}$$

$$= -\left(\frac{k_a}{2\mu}\right)\frac{-\left[\dfrac{p_w}{-r\ln\left(\dfrac{r_w}{R_I}\right)}\right]\left(1-\dfrac{p_{atm}^2}{p_w^2}\right)}{\left\{1-\left(-\dfrac{p_{atm}^2}{p_w^2}+1\right)\left[\dfrac{\ln\left(\dfrac{r}{r_w}\right)}{-\ln\left(\dfrac{r_w}{R_I}\right)}\right]\right\}^{\frac{1}{2}}} \tag{E7-8-6}$$

And finally one gets

$$q_r(r) = -\left(\frac{k_a}{2\mu}\right)\frac{\left[\dfrac{p_w}{r\ln\left(\dfrac{r_w}{R_I}\right)}\right]\left(1-\dfrac{p_{atm}^2}{p_w^2}\right)}{\left\{1+\left(1-\dfrac{p_{atm}^2}{p_w^2}\right)\left[\dfrac{\ln\left(\dfrac{r}{r_w}\right)}{\ln\left(\dfrac{r_w}{R_I}\right)}\right]\right\}^{\frac{1}{2}}} \tag{E7-8-7}$$

which is exactly the same as Johnson et al. [1990, p. 415, Eq. (10)] with the exception of notation.

Example 7-9

This example is adapted from Johnson et al. (1990, p. 416, Figure 6). In Figure 7-2, the following values are given: $r_w = 2$ in = 5.1 cm; $p_w = 0.8$ atm; and $p_{atm} = 1$ atm. Determine the variation of p as a function of r for $R_I = 75$ ft(22.86 m) and discuss the results.

Solution

Using Eq. (7-119), the velocity $q(r)/n_e$ versus r distance in the English FSS units system is shown in Figure 7-16, and its corresponding form in the SI units system is shown in Figure 7-17. These figures do not show the negative sign in Eq. (7-119). Comparison of Figure 7-16 and Figure 6 of Johnson et al. (1990, p. 415) indicates that the results are the same.

7.5.3.7 Volumetric Air Flow Rate Equation for a Given Applied Vacuum Pressure at the Well

From Figure 7-2, the volumetric flow rate can be expressed as

$$Q = 2\pi r_w q_w b \tag{7-120}$$

in which q_w is the gas velocity at $r = r_w$. With $r = r_w$, Eq. (7-119) becomes

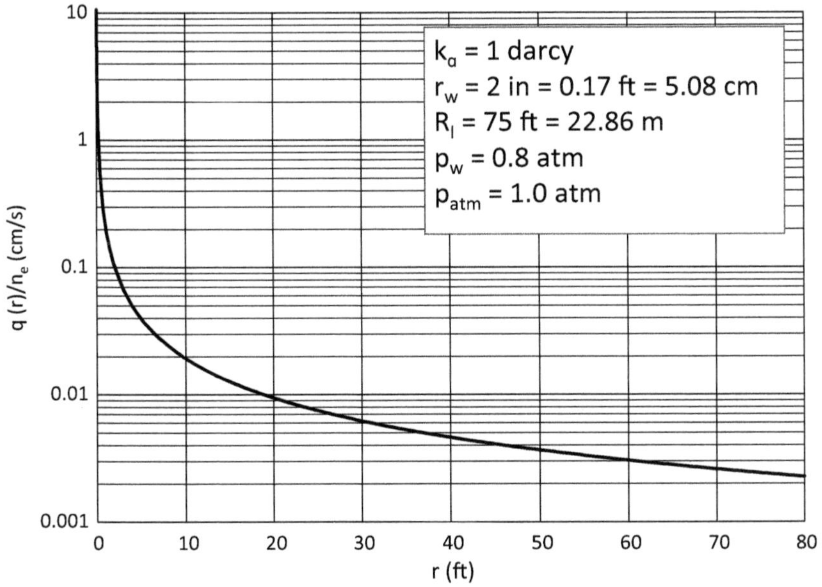

FIGURE 7-16
Variation of pressure $q(r)/n_e$ with the radial distance in the English FSS units for $R_I = 75\ \text{ft}(22.86\ \text{m})$. Adapted from Johnson et al. (1990, p. 416, Figure 6).

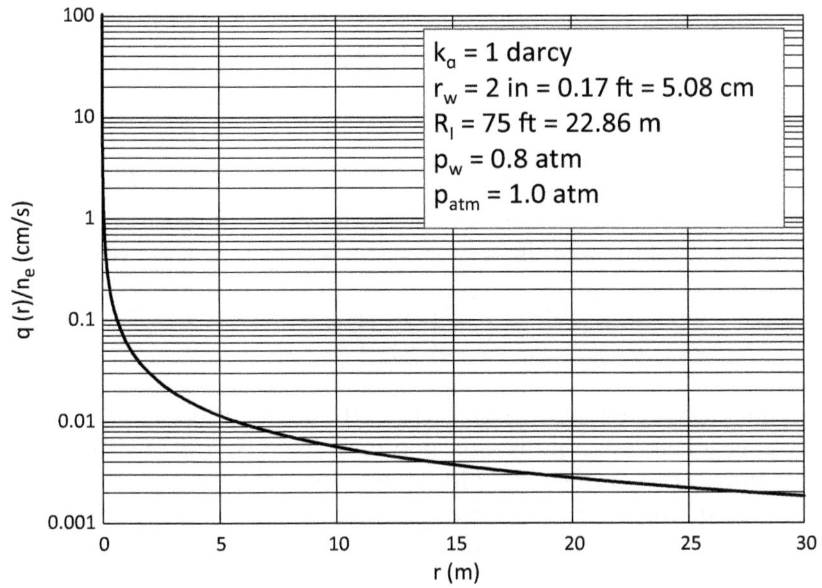

FIGURE 7-17
Variation of pressure $q(r)/n_e$ with the radial distance in the SI units for $R_I = 75\ \text{ft}(22.86\ \text{m})$. Adapted from Johnson et al. (1990, p. 415, Figure 6).

$$q_w = -\left(\frac{k_a}{2\mu}\right)\frac{\left[\dfrac{p_w}{r_w \ln\left(\dfrac{r_w}{R_I}\right)}\right]\left(1-\dfrac{p_{atm}^2}{p_w^2}\right)}{\left\{1+\left(1-\dfrac{p_{atm}^2}{p_w^2}\right)\left[\dfrac{\ln\left(\dfrac{r_w}{r_w}\right)}{\ln\left(\dfrac{r_w}{R_I}\right)}\right]\right\}^{\frac{1}{2}}} \tag{7-121}$$

Since $\ln\left(\frac{r_w}{r_w}\right) = \ln(1) = 0$, omitting the negative sign, Eq. (7-121) becomes

$$q_w = \left(\frac{k_a}{2\mu}\right)\frac{p_w}{r_w \ln\left(\dfrac{r_w}{R_I}\right)}\left(1-\frac{p_{atm}^2}{p_w^2}\right) \tag{7-122}$$

Using this in Eq. (7-120),

$$Q = b\left(\frac{\pi k_a}{\mu}\right)\frac{p_w\left(1-\dfrac{p_{atm}^2}{p_w^2}\right)}{\ln\left(\dfrac{r_w}{R_I}\right)} \tag{7-123}$$

which is exactly the same as Johnson et al. [1990, p. 415, Eq. (11)] with the exception of notation.

7.5.3.8 Converting Actual Vapor Flowrates to Equivalent Standard Vapor Flowrates

To convert actual vapor flow rates to equivalent standard vapor rates, Q^* $(p = 1\text{ atm})$, the following correction factor is used [e.g., Jonson et al., 1990, p. 415, Eq. (13)]:

$$Q^* = \left(\frac{p_w}{p_{atm}}\right)Q \tag{7-124}$$

in which p_w is the absolute pressure expressed in the same units as p_{atm}.

When searching for an air compressor, one may see a couple of different terms used to measure air flow rate, specifically CFM (stands for Cubic Feet per Minute) and SCFM (stands for Standard Cubic Feet per Minute). Regarding these, one useful information Internet site is as follows: https://www.aavsales.com/blog/cfm-vs-scfm-whats-the-difference%EF%BF%BC/.

CFM is the standard flow rate in the local conditions. This is important because it can be significantly different from SCFM based on the differences in static pressure, temperature, and humidity mentioned above, depending on where the unit is tested versus the location. CFM is a measurement of inlet cubic feet per minute, or how much air density in the air compressor can take per minute, which is calculated using the local temperature and actual pressure. As an example of why this is important, 1 CFM at standard conditions used at 100 psi will equal 0.947 CFM at 3,000 ft; 0.912 CFM at 5,000 ft; 0.868 CFM at 7,500 ft; and 0.824 CFM at 10,000 ft of elevation, making a big difference in how much air volume in the system is actually being taken in use.

7.5.3.9 Prediction of the Vapor Flow Rate

Equations (7-123) and (7-124) can be used to predict the vapor flow rate achieved for a given applied vacuum, the vapor velocity as a function of position, and the time it takes vapor to travel from a distance r to the well, after steady state has been reached.

To show the applications of Eqs. (7-123) and (7-124), the following input data taken from Johnson et al. (1990, p. 416) are used: $R_I = 75$ ft $= 22.96$ m; $r_w = 2$ in $= 0.17$ ft $= 5.08$ cm; $p_{atm} = 1$ atm; and $\mu = 1.8 \times 10^{-4}$ g/(cm.s). The results are presented below.

Figure 7-18 shows the variation of pressure Q^*/b with p_w in the English FSS units for $R_I = 75\text{ ft}(22.86\text{ m})$ for different k_a values (adapted from Johnson et al., 1990, p. 416, Figure 4). Likewise, Figure 7-19 shows the variation of pressure Q^*/b with p_w in the SI units for $R_I = 75\text{ ft}(22.86\text{ m})$ for different k_a values.

Figure 7-20 shows the variation of pressure Q^*/b with k in the English FSS units for $R_I = 75\text{ ft}(22.86\text{ m})$ for different p_w values (adapted from Johnson et al., 1990, p. 416, Figure 5). Likewise, Figure 7-21 shows the variation of pressure Q^*/b with k in the SI units for $R_I = 75\text{ ft}(22.86\text{ m})$ for different p_w values.

7.5.3.10 Water Table Upwelling

During a soil venting operation, the vacuum or reduced pressure in the soil potentially may cause a rise in the water table and this phenomenon is called *water table upwelling.*

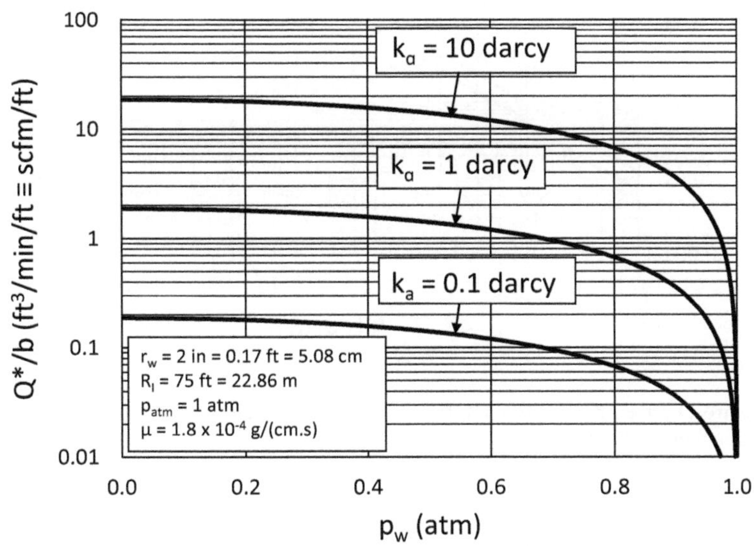

FIGURE 7-18
Variation of pressure Q^*/b with p_w in the English FSS units for $R_I = 75\text{ ft}(22.86\text{ m})$ for different k_a values. Adapted from Johnson et al. (1990, p. 416, Figure 4).

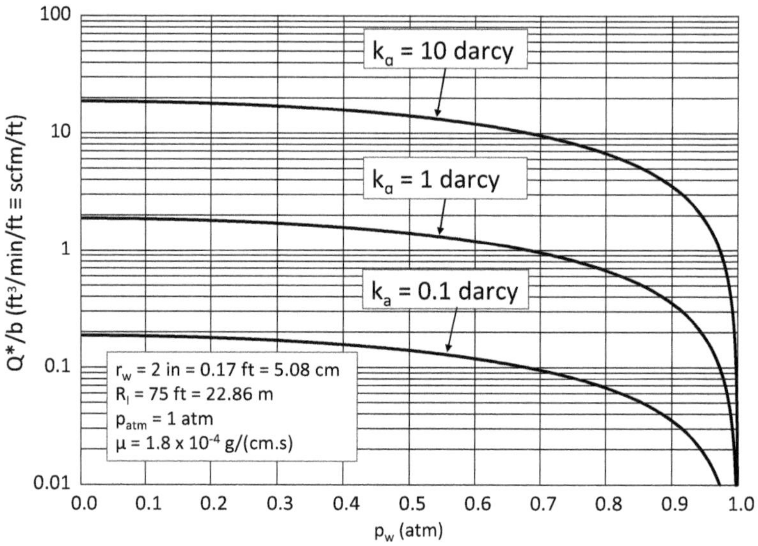

FIGURE 7-19
Variation of pressure Q^*/b with p_w in the SI units for $R_I = 75\,\text{ft}\,(22.86\,\text{m})$ for different k_a values. Adapted from Johnson et al. (1990, p. 416, Figure 4).

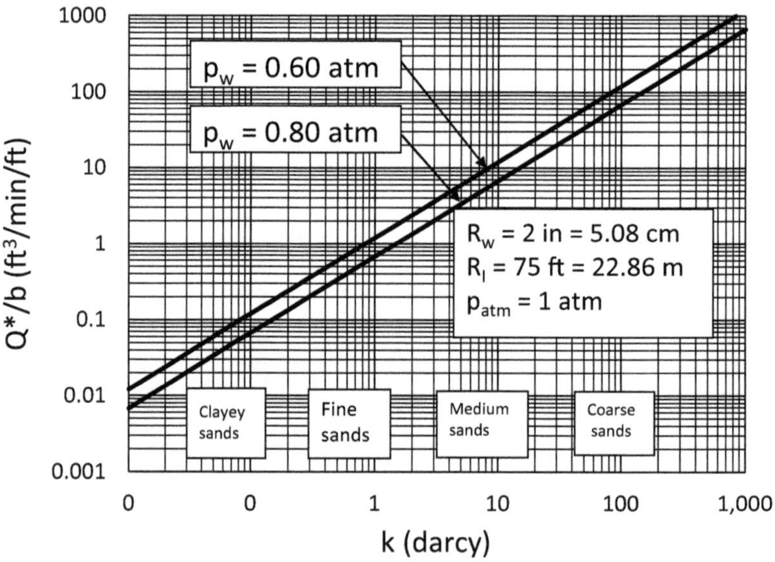

FIGURE 7-20
Variation of pressure Q^*/b with k in the English FSS units for $R_I = 75\,\text{ft}\,(22.86\,\text{m})$ for different p_w values. Adapted from Johnson et al. (1990, p. 416, Figure 5).

The change in the water table height can easily be determined from the predicted radial pressure distribution, because the rise at any point, H_{rise} (measured in cm of H_2O), will be equivalent to the change in pressure at that point, expressed as equivalent cm of H_2O is [e.g., Johnson et al., 1990, p. 417, Eq. (15)].

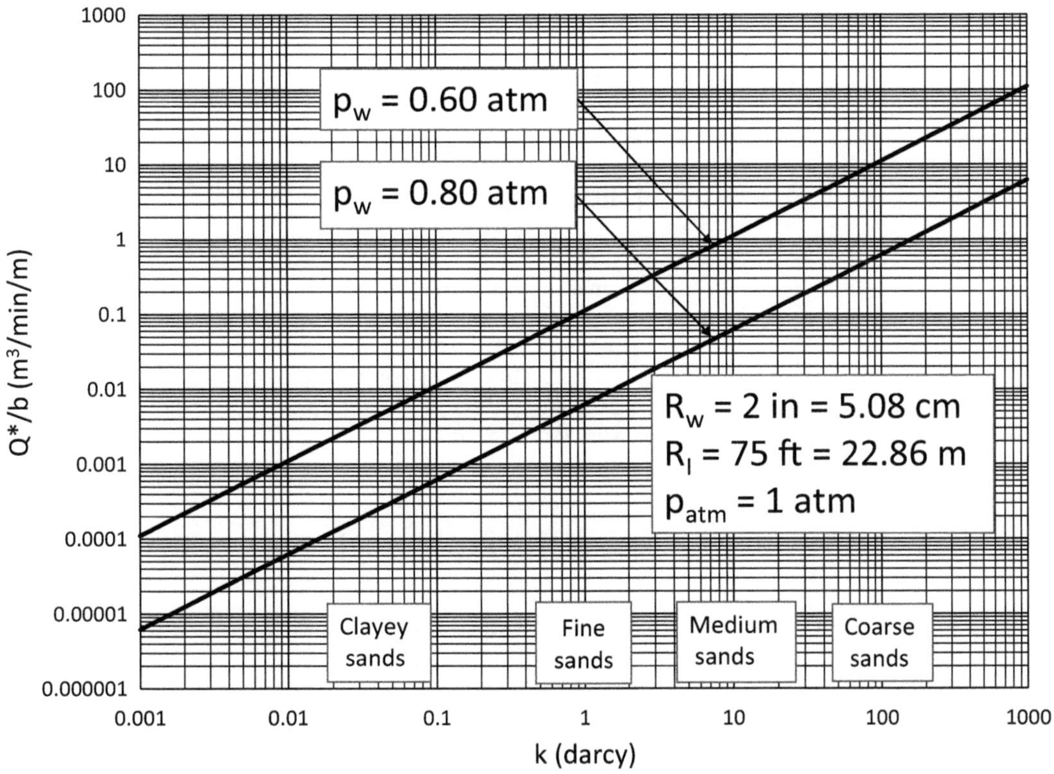

FIGURE 7-21
Variation of pressure Q^*/b with k in the SI units for $R_I = 75\,\text{ft}\,(22.86\,\text{m})$ for different p_w values. Adapted from Johnson et al. (1990, p. 416, Figure 5).

$$H_{rise} = (1{,}033\ \text{cm})\left[1 - p(r)\right] \tag{7-125}$$

in which $p\ (r)$ is the pressure in atm at the r distance Eq. (7-125) is derived in Example 7-10.

Example 7-10

Derive Eq. (7-125).

Solution

Some key input parameter values are given as

$$\rho = 1\,\frac{\text{g}}{\text{cm}^3}$$

$$g = 981\,\frac{\text{cm}}{\text{s}^2}$$

$$p_{atm} = 101{,}325\ \text{Pa}$$

$$1\ \text{Pa} = 1\,\frac{\text{N}}{\text{m}^2}\quad 1\ \text{N} = 1\,\frac{\text{kg.m}}{\text{s}^2}$$

Therefore,

$$p_{atm} = 1\,\text{atm} = 101,325\,\frac{\text{N}}{\text{m}^2} = 101,325\left(\frac{1}{\text{m}^2}\right)\left(\frac{\text{kg.m}}{\text{s}^2}\right)$$

$$= 101,325\left[\frac{1}{(100\,\text{cm})^2}\right]\left[\frac{(1,000\,\text{g}).(100\,\text{cm})}{s^2}\right]$$

$$= 1,013,250\,\frac{\text{g}}{\text{cm.s}^2}$$

$$h_{atm} = \frac{p_{atm}}{\rho g} = \frac{1\,\text{atm}}{\rho g} = \frac{1,013,250\,\frac{\text{g}}{\text{cm.s}^2}}{\left(1\,\frac{\text{g}}{\text{cm}^3}\right)\left(981\,\frac{\text{cm}}{\text{s}^2}\right)} = 1,032.87\,\text{cm} \cong 1,033\,\text{cm}$$

The H_{rise} water rise is

$$H_{rise} = h_{atm} - h_{atm}p\,(r) = h_{atm}\left[1 - p(r)\right]$$

$$H_{rise} = (1,033\,\text{cm})\left[1 - p(r)\right]$$

which is exactly the same as Eq. (7-125).

Figure 7-22 shows the variation H_{rise} with r in the English FSS units for $R_I = 75\,\text{ft}(22.86\,\text{m})$ for different p_w values (adapted from Johnson et al., 1990,

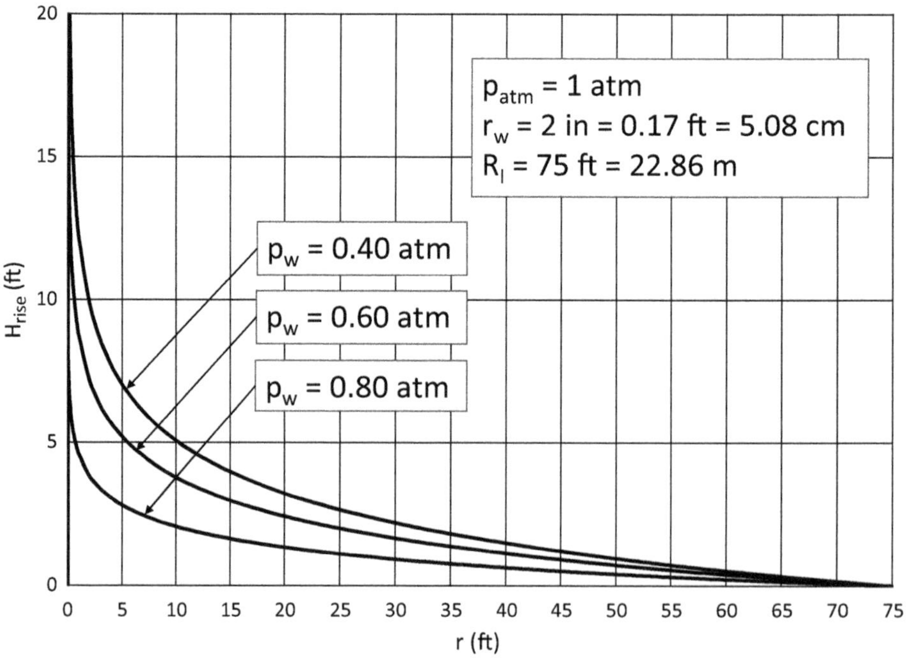

FIGURE 7-22
Variation of pressure H_{rise} with r in the English FSS units for $R_I = 75\,\text{ft}(22.86\,\text{m})$ for different p_w values. Adapted from Johnson et al. (1990, p. 417, Figure 8).

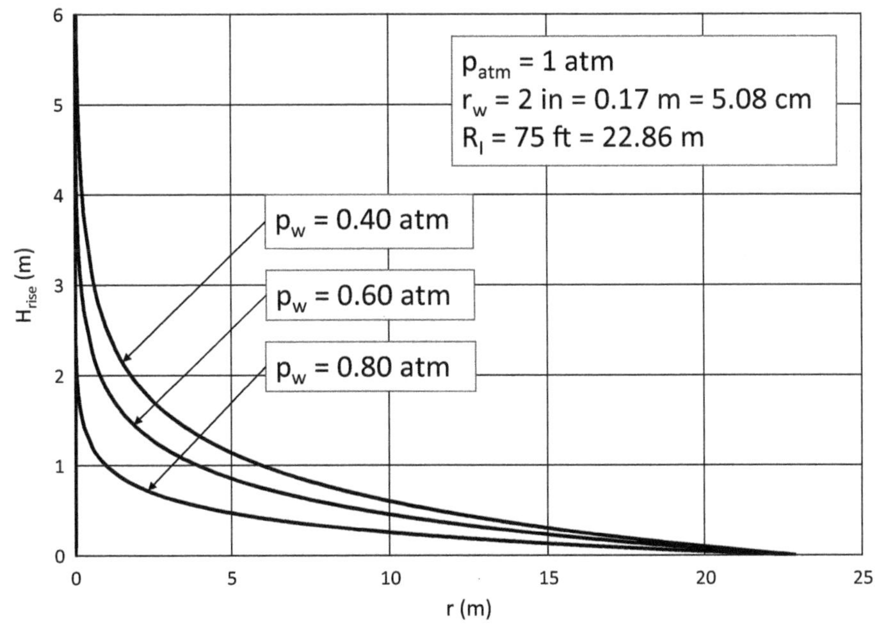

FIGURE 7-23
Variation of pressure H_{rise} with r in the SI units for $R_I = 75\ \text{ft}\left(22.86\ \text{m}\right)$ for different p_w values. Adapted from Johnson et al. (1990, p. 417, Figure 8).

p. 417, Figure 8). Likewise, Figure 7-23 shows the variation of pressure H_{rise} with r in the SI units for $R_I = 75\ \text{ft}\left(22.86\ \text{m}\right)$ for different p_w values. These figures show that the water table upwelling can be significant depending on the values of p_w. If the predicted water table rise occurs and the contaminated zone lies above the water table, which is often the case, then the ground water will become contaminated. As a result, the ground water will become contaminated, and the contaminated zone may become saturated with water, which will cause reduced mass removal. These results suggest that most soil venting operations should include a ground water extraction well to keep the ground water level below the zone of contamination (Johnson et al., 1990, p. 417).

Problems

7.1 Describe the method of *soil venting* including soil vapor extraction (SVE) and bio-venting (BV).

7.2 For a soil vapor extraction test, the following values are given: $Q = 56.4\ \text{L/s}$; $\mu_{air} = 1.81 \times 10^{-4}\ \text{g/(cm.s)}$; $b = 5\ \text{m}$; and $r = 12.5\ \text{m}$. After plotting the measured data points, the following values are determined in accordance with the Cooper and Jacob method: $\Delta\left[p'(r,t)\right] = 5.5\ \text{atm}$ for $\Delta\left[\log(t)\right] = 1$. Also, it is found out $t_0 = 5{,}400\ \text{s}$. Determine the values of k_a and n_a.

7.3 Describe the physical meaning of the boundary conditions in deriving the gas pressure distribution given by Eq. (7-114).

7.4 Write the steady-state gas pressure distribution equation and describe the meaning of its terms and evaluate them critically.

8

Pump-and-Treat Method

8.1 Introduction

Containment and cleanup of contaminated aquifers are among the primary objectives of environmental programs of many countries in the world. For example, in the United States, CERCLA (Comprehensive Environmental Response, Compensation, and Liability Act; also known as Superfund) and RCRA (Resource Conservation and Recovery Act) remediation programs are among them. A common, frequently used practice is to extract contaminated ground water from an aquifer and then clean it at the ground surface to discharge it back to the same aquifer. This is called conventional pump-and-treat (P&T) remediation.

Conventional P&T remediation is widely used in many remedial systems. However, P&T remediation systems may not be feasible to clean up in portions of many sites due to hydrogeologic and contaminant-related limitations such as those presented by significant accumulations of dense nonaqueous phase liquids (DNAPLs) trapped below the water table. Such limitations will directly affect the effectiveness of P&T at many sites and the selection of remedial actions.

As mentioned above, P&T is one of the most widely used ground water remediation technologies for dissolved contaminants in ground water. However, under complex site conditions, it may be necessary to combine several remedial actions for a given site. In a P&T system, contaminated ground water is extracted to the ground surface and after the cleaning process is completed, it is injected back into the aquifer. Selection of a remedial technology is a function of the physical characteristics, reactivity and mobility of the dissolved contaminant, and characteristics of the aquifer. If the aquifer is very tight, extraction of water at a reasonable rate may cause the well to go dry, creating a narrow and deep capture zone. For such conditions, an intercept drain may be more appropriate.

There may be cases for which the P&T technology will not be effective in removing contaminants. An example is dense nonaqueous phase liquids (DNAPLs) at residual saturation. This is a very difficult problem for which other remedial options may not be effective either. If the residual DNAPLs are shallow, then excavation may be a reasonable option. If they are too deep to excavate, then P&T technology is a possible remedial action to hydraulically contain any dissolved contamination.

8.2 General Description of a Pump-and-Treat System

In this section, the general description of a P&T system is provided (Mercer et al., 1990; Cohen et al., 1997).

DOI: 10.1201/9781003739128-8

8.2.1 Sources of Ground water Contamination

Potentially, ground water contaminant sources are leaky tanks, landfills, spills, and agricultural practices. Chemicals in the soil dissolving from nonaqueous phase liquids (NAPLs) or chemicals desorbing from the soil matrix are also among them.

8.2.2 Options for Ground water Cleanup

8.2.2.1 Source Removal

Several options can be used to contain and/or clean up contaminated ground water. Source removal is also part of the ground water cleanup activities. Typically, source removal is to excavate and remove wastes and/or contaminated soil. Vacuum extraction is also source removal.

8.2.2.2 Ground water Source Containment

Ground water containment can be achieved by constructing barriers with low hydraulic conductivity around the contaminant source area. The other option is hydraulic containment using extraction wells and intercept trenches or drains as appropriate. In some cases, several methods may be combined to effect complete cleanup.

8.2.3 A Typical Pump-and-Treat System

In Figure 8-1, the schematic picture of a typical P&T system for a contaminated aquifer is shown. In this system, contaminated ground water is extracted and sent to a nearby treatment facility at the ground surface. After treatment, it is injected into the aquifer through

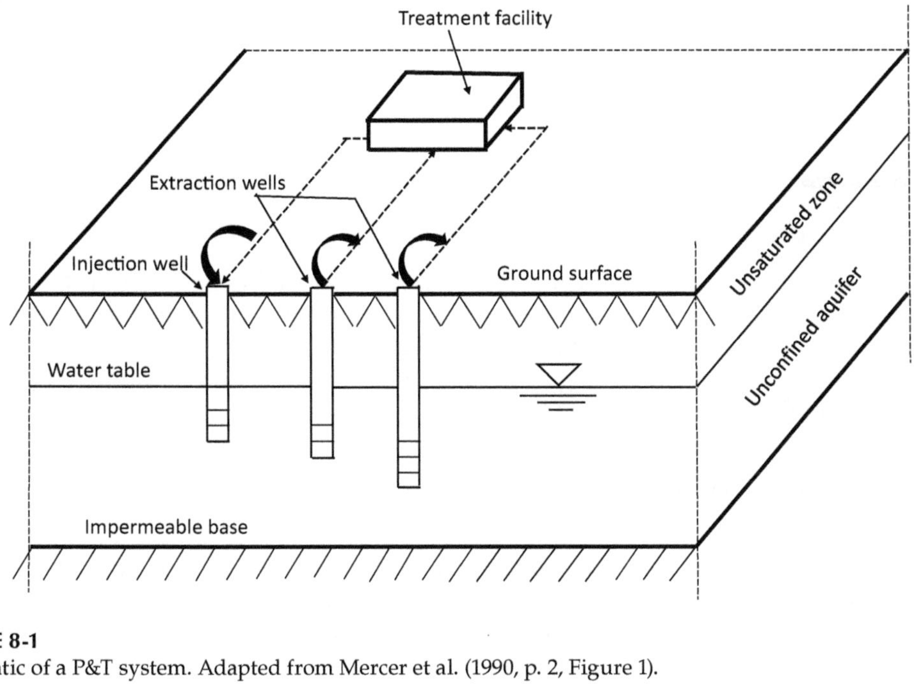

FIGURE 8-1
Schematic of a P&T system. Adapted from Mercer et al. (1990, p. 2, Figure 1).

an injection well near the extraction wells. The injection well(s) increase the hydraulic gradient toward the extraction wells. In this way, the efficiency of the extraction wells increases so that the time required to reach a cleanup goal is reduced.

The technique of P&T can also be used as a *hydraulic barrier* to prevent off-site migration of contaminant plumes from contaminant sources (e.g., landfills) or residual NAPLs. The basic principle of a hydraulic barrier system is to lower the hydraulic heads, which results in the ground water flow toward the extraction wells.

8.2.4 Components of a Typical P&T System

The main objectives of a P&T system are to reduce contaminant concentrations to the required levels and to protect the ground water from further contamination. The components of a P&T system are as follows: (1) a set of goals or objectives; (2) constructed facilities such as extraction wells, injection wells, treatment facilities, and pumps; (3) operational and monitoring guidelines; and (4) termination criteria.

Each of the above-listed components must be addressed in the design and evaluation of the P&T system.

8.2.5 Main Limitations of P&T Systems

One of the main requirements is that the subsurface flow system must have certain transmissive capabilities such that both the horizontal (K_h or K_r) and vertical (K_v or K_z) hydraulic conductivities must be high enough. Under these conditions, the contaminants can be transported to the cones of depression of the extraction wells.

In the following cases, contaminants potentially can have limited transport capabilities to the cones of depression of the extraction wells: (1) heterogeneous and anisotropic with low K_h and K_v values; (2) sorbed or precipitated contaminants on the aquifer materials; (3) slowly desorb or dissolve back into the ground water as chemical equilibrium changes in the response to the ground water extraction activities; and (4) prolonged dissolution of the immobile or NAPLs.

8.3 Data Requirements of a P&T System

First, using the site's hydrogeological model, a *conceptual model* of the ground water contamination problem needs to be generated before an appropriate remedial action can be selected. A conceptual model provides an understanding of the general physical characteristics related to regional hydrology, geology, hydrogeology, and aquitards. These processes and related data are discussed below.

8.3.1 Hydrogeologic Data

One of the key datasets affecting the P&T system design is the characterization of the subsurface flow system and they are as follows: (1) horizontal and vertical hydraulic conductivities of the aquifers and aquitards; (2) storage coefficient and specific yield; (3) total porosity and effective porosity; (4) aquifer and aquitard thicknesses; and (5) precipitation recharge rate and net precipitation recharge rate. For long-term extraction conditions, the

TABLE 8-1

The Information Typically Used to Identify and Quantify the Important Aspects of a Ground water System

Geologic Aspects	
1	Type of water-bearing unit or aquifer (overburden, bedrock)
2	Thickness, areal extent of water-bearing units and aquifers
3	Type of porosity (primary, such as intergranular pore space, or secondary, such as bedrock discontinuities, e.g., fracture or solution cavities)
4	Presence or absence of impermeable units or confining layers
5	Depths to water table; thickness of vadose zone
Hydraulic Aspects	
1	Hydraulic properties of water-bearing unit or aquifer (horizontal and vertical hydraulic conductivities, transmissivity, storativity, specific yield, effective porosity)
2	Type of aquifers (confined, unconfined, leaky confined)
3	Ground water flow directions (horizontal and vertical hydraulic gradients), volumes (specific discharge), rate (average linear velocity)
4	Recharge and discharge areas
5	Ground water or surface water interactions; areas of ground water discharge to surface water
6	Seasonal variations of ground water conditions
Ground water Use Aspects	
1	Existing or potential underground sources of drinking water
2	Existing or near-site use of ground water

Source: Adapted from USEPA: Mercer et al. (1990, p. 4, Table 1).

effects of both storage coefficient and specific yield become insignificant. Table 8-1 presents the information typically used to identify and quantify the important characteristics of a ground water system.

8.3.2 Contaminant Data

Contaminant source characterization and spatial and temporal concentration distribution in the ground water flow field are some of the required key information components for a P&T system. Source characterization consists of the released chemical volume and the time duration of releases. Generally, the contaminant release information is difficult to determine. The required contaminant data are described below.

8.3.2.1 Chemical Data

To conduct a quantitative characterization of the subsurface chemistry, sampling in the vadose and saturated zones needs to be made to determine the concentration distributions in ground water and soil. A network of monitoring wells, which are also necessary for the hydrogeologic data, needs to be installed to collect depth-discrete ground water samples. Wells should be located in areas to provide information on ground water as well as plume chemistry.

After laboratory analysis of the samples, isoconcentration maps in the vadose zone as well as in the aquifers should be generated in order to properly delineate the extent of the contaminant plumes.

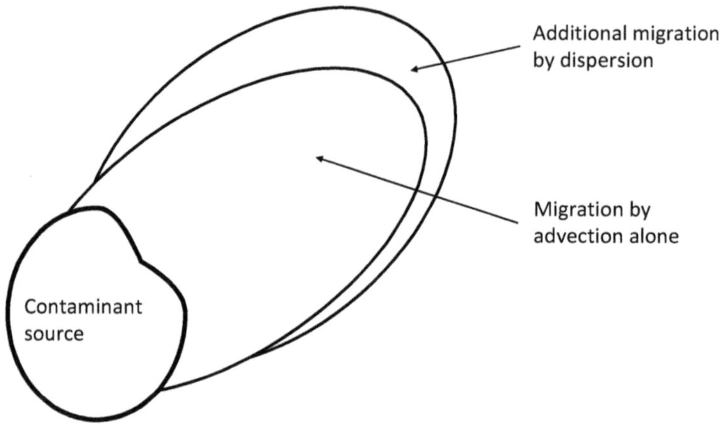

FIGURE 8-2
Plan view of a contaminant plume caused by advection and dispersion. Adapted from Keely (1989).

8.3.2.2 Solute Transport Data

Plume migration of nonreactive dissolved contaminants in saturated porous media is mainly controlled by advection and hydrodynamic dispersion (Figure 8-2). Advection is a function of the horizontal hydraulic conductivity (K_h), hydraulic gradient (I), and effective porosity (n_e). Hydrodynamic dispersion is the combined effect of mechanical mixing (or mechanical dispersion) and molecular diffusion. Dispersion causes the volume of contaminated ground water to occupy a greater volume than it would under advection only. Advection causes a plume to move in the direction and velocity of ground water flow. Hydrodynamic dispersion causes an increase in the plume volume. In coarse porous materials such as sands and gravels, the effects of molecular diffusion are significantly smaller than mechanical dispersion. However, in fine-textured materials such as clays, the molecular diffusion effects may become greater than mechanical dispersion.

Additional processes such as sorption, desorption, and chemical reactions affect the migration of contaminants in subsurface flow. Some of these processes cause to retard in the migration of contaminants. Because of these effects, the plume of a reactive contaminant migrates more slowly and the concentration becomes less than that of an equivalent nonreactive contaminant. This retarding effect increases the cleanup time of a P&T system. The data requirements for contamination characterization are included in Table 8-2.

For a P&T system, chemical properties of the plume are required to characterize the migration of the chemicals and to evaluate the feasibility. The following properties affecting the mobility of dissolved contaminants in ground water need to be considered for plume migration and cleanup.

8.3.2.3 Aqueous Solubility

Solubility indicates the maximum possible concentrations. Solubility also determines the degree to which the chemical will dissolve in water. High solubility indicates low sorption tendencies, e.g., methylene chloride.

TABLE 8-2

Data Pertinent to Ground water Contamination Characterization

General Category	Specific Data
Site physical framework	Estimates of hydrodynamic dispersion parameters Effective porosity distribution Natural (background) aquifer constituent concentration
Distributions	Fluid density and relationship to concentrations
System stresses	Pollution source locations Pollutant releases
Chemical/biological framework	Mineralogy Organic content Ground water temperature Solute properties Major ion chemistry Minor ion chemistry Eh–pH environment
Observable responses	Areal and temporal distributions of water, solid, and vapor phase contaminants Stream flow quality distributions over space and time

Source: Adapted from USEPA: Mercer et al. (1990, p. 6, Table 2).

8.3.2.4 Henry's Law Constant

High values of this constant may signify volatilization from the aqueous phase as an important migration process, e.g., dichlorodifluoromethane. Henry's law constant is used in conjunction with vapor pressure.

8.3.2.5 Density

For high concentrations, the density of the contaminated fluid may be greater than the density of pure water, e.g., the density of trichloroethylene (TCE) is 1.46 g/cm^3. This causes the downward vertical movement of contaminants.

8.3.2.6 Octanol–Water Partition Coefficient

The octanol–water partition coefficient indicates the tendency of a chemical to partition between the ground water and the soil. A large octanol–water partition coefficient corresponds to a highly hydrophobic compound, which indicates strong sorption. Typical hydrophobic materials include normal oils, fats, and greasy substances.

8.3.2.7 Organic Carbon Partition Coefficient

Organic carbon partition coefficient is another indicator of the chemical tendency to partition between ground water and the soil. For certain chemicals, the organic carbon partition coefficient (K_{oc}) is directly related to the distribution coefficient (K_d) with the fraction of organic carbon content (f_{oc}).

8.3.2.8 Degradation

Degradation or biodegradation may influence the behavior of contaminants in ground water. A broad range of organic compounds such as pesticides, halogenated hydrocarbons, aromatic hydrocarbons, amines, and alcohols have shown degradation.

8.3.2.9 NAPLs Data

NAPLs are hydrocarbons and exist as a separate, immiscible phase when in contact with water and/or air. Differences in the physical and chemical properties of water and NAPLs affect the interaction such that the mixing of the two fluids is prevented. NAPLs are typically classified as either light nonaqueous phase liquids (LNAPLs), which have densities less than that of water, or DNAPLs, which have densities greater than that of water.

Additional data requirements for NAPLs include (1) fluid specific gravity (density); (2) fluid viscosity; (3) residual saturation; (4) the relationships between relative permeability, saturation, and capillary pressure relationships, and (5) NAPL thickness and distribution. After a spill or release, LNAPLs tend to spread over the water table. DNAPLs tend to move below the water table until reaching a low-permeability barrier such as a confining layer. The chemicals most commonly found as DNAPLs at contaminated sites are chlorobenzenes, chloromethanes, polychlorinated biphenyls (PCBs) (electrical transformer oils), tetra and trichloroethane (solvents), tetra and trichloroethene (solvents, degreasers), creosote (wood treaters), and coal tar. LNAPLs are slightly soluble organic compounds that float on water, gasoline, diesel fuel, home heating oil, and related petroleum hydrocarbons are the most common examples.

8.3.3 Data Collection

8.3.3.1 Sources of General Information

First, a site background data search needs to be conducted before attempting to generate site-specific data. This effort reduces the amount of information that will have to be collected in the field. As shown in Table 8-3, various sources of general information on specific sites are available.

TABLE 8-3

Potential Sources of Information (Knox et al., 1986)

Problem-specific information	Federal or state geological surveys, university libraries, geology and engineering departments, state health departments, property owner, county records, well drillers
Site-specific information	Weather bureaus, state water resource boards, census bureaus, soil and water conservation districts, employment commissions, corporation commissions, Department of Agriculture, Forest Service
Other information	Medical libraries, state or federal environmental protection agencies, state attorney general's office

TABLE 8-4

Data Collection Methods

Category	Commonly Used Methods	Advantages/Disadvantages
Geophysics (indirect data method)	Electromagnetics	Good for delineation of high conductivity plumes
	Resistivity	Useful in locating fractures
	Seismic	Limited use in shallow studies
	Ground penetrating radar	Useful in very shallow soil studies
Drilling	Augering	Poor stratigraphic data
	Augering with split-spoon sampling	Good soil sampling
	Air/water rotary	Rock sample information
	Mud rotary	Fills fractures—needs intensive developments
	Coring	Complete details on bedrock
	Jetting/driving	No subsurface data
Ground water sampling	Bailer	Allows escape of volatiles (operator dependent)
	Centrifugal pump	Can produce turbid samples increasing chance of misrepresenting contamination
	Peristaltic/bladder pumps	Gives more representative samples
Soil sampling	Soil boring	Restricted to shallow depths
Aquifer tests	Pump test	Samples from a large aquifer section
	Slug test	Does not require liquid disposal

Source: Adapted from USEPA: Mercer et al. (1990, p. 9, Table 4).

8.3.3.2 Common Data Collection Methods

Table 8-4 presents common data collection methods. Selection of appropriate methods depends on the overall scope of the project and site conditions. Site conceptualization and contamination problem should be described properly and updated as data become available. During all phases of the project, it is necessary to document all well construction details, sampling episodes, etc. in order to develop an accurate evaluation of the entire site. Understanding of the site geology and hydrogeology and the extent of contamination are important in generating a successful field study. Formulating adequate design plans ensures that wells are developed to a proper depth and stratigraphic layer. In this way, the extent of contamination is not exacerbated by cross-contamination.

8.3.3.3 Methods for Determining Hydraulic Characteristics

Methods for determining hydraulic characteristics of aquifers and aquitards are primarily conducted by pump tests, slug tests, and constant-head injection tests.

8.3.3.3.1 Pump Tests

In a pump test, water is extracted from a well and drawdowns are measured in nearby observation wells. Since the invention of Darcy's law in 1856, many solutions have been developed for well hydraulics for unconfined and confined aquifers under steady and transient conditions. Based on these solutions, several methods have been developed (e.g., Hantush, 1964; Freeze and Cherry, 1979; Batu, 1998) to determine horizontal and vertical hydraulic conductivities, storage coefficient, and specific yield.

8.3.3.3.2 Slug Tests

Slug tests may be categorized as *rising head tests* and *falling head tests*. Both tests are conducted by causing a sudden change in the water level in a well or piezometer by removing or introducing a known volume of water. Likewise, many solutions have been developed for slug tests since the 1950s, and based on these solutions, methods were developed for determining only the horizontal hydraulic conductivity and storage coefficient (e.g., Batu 1998; Butler, 2019). The advantage of a slug test is that little or no contaminated water will be generated. Unfortunately, with slug tests, the horizontal hydraulic conductivity can only be measured around the test well, whereas with pump tests, the aquifer characteristics can be measured in a much larger volume.

8.3.3.3.3 Flow Directions and Hydraulic Gradients in Porous Media

To determine flow directions and vertical and horizontal hydraulic gradients, water levels must be measured and converted to elevations relative to a specified datum. The datum is usually selected as the mean sea level. Water-level measurements may be taken by several means including (1) chalk and tape, (2) an electrical water-level probe, and (3) a pressure transducer. These techniques are discussed in Acker (1974) and Streltsova (1988). Horizontal hydraulic gradients are determined using water-level data from wells whose screens are open to the same hydrologic unit and/or at the same elevation but separated areally. Vertical hydraulic gradients are determined using water-level data from wells in the same location but open to different elevations. The hydraulic gradient is the difference in water levels divided by the distance between the measurement locations. Since water levels often yield a complex three-dimensional surface, one must be careful in computing the hydraulic gradients. The hydraulic gradient determines the direction of flow. Ground water velocity is determined by multiplying the hydraulic gradient by the horizontal or vertical hydraulic conductivity and dividing by effective porosity.

8.3.3.3.4 Site Characterization of Fractured Media and Karst Formations

Site characterization and remediation designs for fractured media and karst formations are even more difficult. Techniques such as fracture trace analysis (Lattman and Parizek, 1964) and the use of geophysical instrumentation may be useful for locating the more permeable zones where contaminants are most likely to be located and, thus, where extraction wells should be placed. Other characterization techniques include continuous coring, aquifer tests, and tracer tests (IAHS, 1988). A more detailed discussion on flow in the special heterogeneous conditions of fractured media can be found in Streltsova (1988); for karst formations, in Bögli (1980), IAHS (1988), and Quinlan and Ewers (1985).

8.3.3.3.5 Quality Assurance (QA) and Quality Control (QC)

During the ground water sampling process, strict protocols must be followed in the field for QA and QC. The pH, temperature, and specific conductance of a sample should be measured. Ideally, before a sample is taken, water should be extracted from the well until these parameters have stabilized. This practice will help ensure that the sample is from the formation. Proper sample storage and shipment to a qualified laboratory are also important. A sampling plan should address issues such as sampling frequency, locations, and statistical relevance of samples. More details on sampling guidance can be found in Cartwright and Shafer (1987), Barcelona et al. (1983), and Barcelona et al. (1985).

8.3.4 Data Interpretation

8.3.4.1 Uncertainties for Contaminated Sites

Addressing uncertainties for contaminated sites is an indispensable part of designing a remediation system. Uncertainties for contaminant sites are (1) characterization of the contaminant source and (2) extrapolation and interpretation of subsurface geologic and hydrogeologic data. There are, however, limits to the ability of science, and a mathematical model will always be a simplification of reality, and the predictions it makes will always be uncertain.

Interpretation of point ground water level data for an unconfined aquifer and potentiometric head data begins by plotting the data spatially and temporally. Generation of hydrographs and hydraulic head contours is part of this practice. Hydrograph is a graphic displaying water-level fluctuations over a period of time in unconfined and confined aquifers. Water-level data for specific times is contoured to form potentiometric head maps that are interpreted with respect to geologic sections and information on the horizontal and vertical hydraulic conductivities. For a steady flow in an aquifer, a region of higher hydraulic gradient on the head contour maps should correspond to a region of lower horizontal hydraulic conductivity. Likewise, the generation of an isoconcentration contours map for a contaminant constituent is also a part of data interpretation.

8.3.4.2 Data Interpretation Using Mathematical Models

Data interpretation can be made using mathematical models varying from simple Darcy's equation-based models up to three-dimensional analytical and numerical models. Darcy's equation is one of the simplest mathematical models interpreting flow and contaminant data. It requires horizontal (or vertical) hydraulic conductivity, hydraulic gradient, and effective porosity in predicting advective (or convective) transport. If the advective transport is not comparable to the mean plume movement, it is likely that there is uncertainty in the source release or location or sorption effects are also dominant factors. Inconsistencies among data need to be explained. Resolving inconsistencies with data increases the understanding of the site and reduces uncertainty.

There are numerous mathematical approaches in interpreting ground water hydraulic head and contaminant concentration data. The most frequently used ones are briefly described below.

8.3.4.2.1 Geochemical Analysis

Common geochemical techniques applied to deciphering ground water–surface water exchange include mass balance geochemical mixing modeling, application of stable and radioactive isotope data to differentiate water histories and sources, and use of natural and introduced environmental tracers. During the past decades, geochemical models have been developed to assess flow in an aquifer by comparing ion activity and solubility products in a state of saturation and ground water mineralization. Among these models are MINEQL (Morel and Morgan, 1972), WATEQ2 (Ball et al., 1979), EQ3 (Wolery, 1979), and MINTEQA1 (U.S. EPA, 1987).

8.3.4.2.2 Geostatistical Analysis

Geostatistical analysis is based on statistical models including autocorrelation, i.e., statistical relationships among the measured points. Kriging techniques assume that at least some of the spatial variation observed in natural phenomena can be modeled by random processes with spatial autocorrelation, which requires that the spatial autocorrelation be explicitly modeled. Kriging techniques can be used to describe and model spatial patterns, predict values at unmeasured locations, and assess the uncertainty associated with a predicted value at the unmeasured locations. Kriging methods can also be used to quantify spatial variability inherent especially in the horizontal hydraulic conductivity field of an aquifer (e.g., Englund and Sparks, 1988; Journel and Huijbregts, 2003).

8.3.4.2.3 Mathematical Modeling

Mathematical flow and solute transport models can be used to simulate flow patterns and changes resulting from the operation of a P&T system. To address uncertainty, one may use discrete sensitivity analysis where a parameter is varied and its impact on the concentration can be determined. There are a number of analytical and numerical ground water flow and solute transport models, which are available publicly.

8.3.4.3 Parameter Uncertainties

Uncertainties regarding a flow or transport parameter result from the estimation procedure and spatial and temporal variability of the parameter. Numerous techniques are available to assess the effects of parameter uncertainty in ground water flow. These techniques are mainly full distribution analyses and second moment analyses (Dettinger and Wilson, 1981). Full distribution analyses require a complete specification of the *probability density functions* of the random variables or parameters. These probability density functions are either known or assumed. The most common full distribution techniques are the method of derived distributions (Benjamin and Cornell, 1970), the Monte Carlo method (Kalos and Whitlock, 1986), and the Latin hypercube method (Iman and Shortencarier, 1984).

8.4 Conceptual Design

8.4.1 Aquifer Contamination Cleanup

Figure 8-3 presents the flow diagram for the decision regarding ground water contamination (USEPA: Mercer et al., 1990). As shown in Figure 8-3, the first decision is related to whether a remedial action is necessary. If a risk assessment results in a need for a remedial action, the potential options are containment (G4), in situ treatment (G5), and pump and treat (G6). If G5 is selected, then other decisions are necessary but not discussed here. If G4 is selected, then the containment can be either physical containment (G7) or hydraulic containment (G8). The physical containment option may not work well due to the migration by molecular diffusion against flow (Chapter 5). Hydraulic containment can be achieved by P&T technologies (G11).

If the P&T (G6) option is selected, the next decision is whether to use wells (G9) or drains (G10). For low-permeable materials, drains may be required. After wells are selected, a decision is required regarding extraction wells (G12), injection wells (G13), or a combination of them. Injection wells potentially can reduce the cleanup time by flushing contaminants toward the extraction wells. For the extraction wells, decisions need to be made regarding continuous extraction (G16), pulsed extraction (G17), and/or extraction combined with containment. Continuous extraction can generate inward hydraulic gradients,

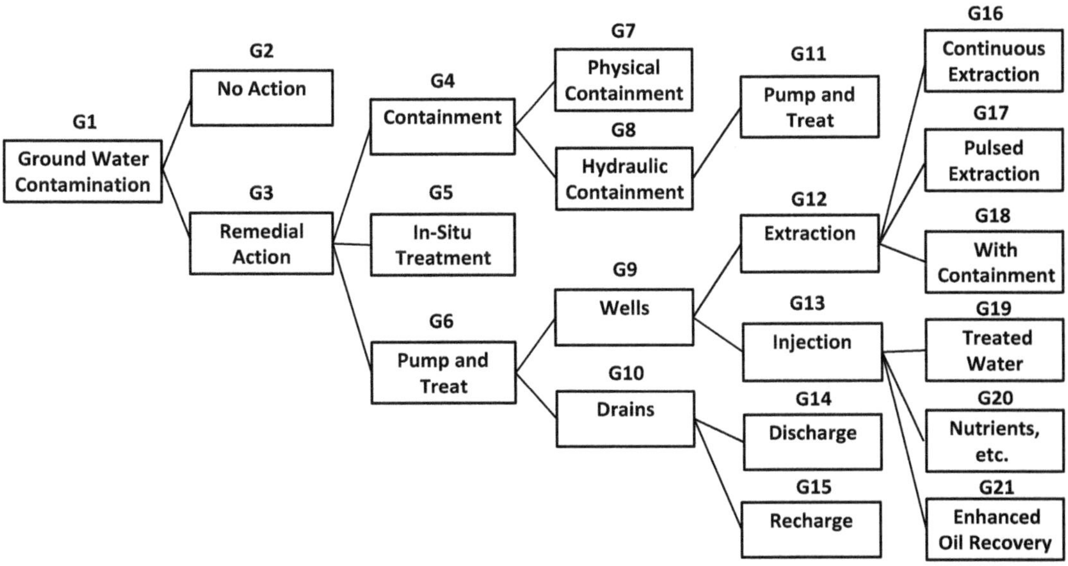

FIGURE 8-3
Decision-flow diagram for ground water contamination. Adapted from Mercer et al. (1990, p. 12, Figure 6a; EPA-600/8-90/003).

pulsed extraction allows maximum concentrations to be extracted efficiently, and containment can be used to limit the inflow of clean water. The injected water can be treated water (G19); for degradable contaminants, it can contain nutrients and/or electron acceptors (G20) to enhance in situ degradation, or for NAPLs, it can consist of enhanced oil recovery materials (G21).

8.4.2 Soil Contamination Cleanup

For soil contamination, a similar decision process is shown in Figure 8-4 (USEPA: Mercer et al., 1990). As shown in Figure 8-4, the first decision is either no action (S2) or remedial action (S3). For a remedial action, the alternatives are excavation (S4), in situ treatment (S5), and cap/cover (S6). For in situ soil treatment, the alternatives are fixation (S7), vacuum extraction (S8), thermal (S9), and bioremediation (S10). Vacuum extraction is possible for volatile contaminants. Most contamination cases will affect both soil and ground water. For such cases, a combination of alternatives in Figures 8-3 and 8-4 may be required.

8.4.3 Calculating the Estimated Cleanup Time for Nonreactive Solutes

8.4.3.1 Extraction Rate Estimation

Consider an aquifer having horizontal area A, saturated thickness b, and total porosity n. The volume of water in the aquifer is

$$V_w = Abn \tag{8-1}$$

If this volume is extracted within t_{ext} amount of time, the extraction flow rate will be

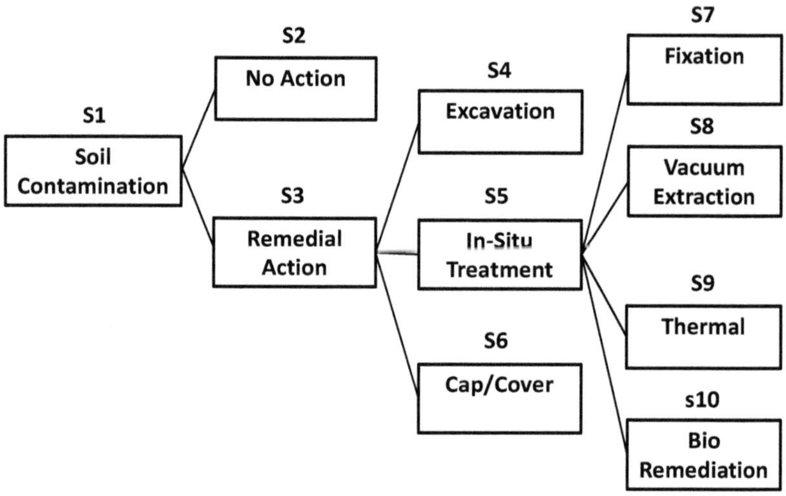

FIGURE 8-4
Decision-flow diagram for ground water contamination. Adapted from Mercer et al. (1990, p. 13, Figure 6a; EPA-600/8-90/003).

$$Q_{ext} = \frac{V_w}{t_{ext}} = \frac{Abn}{t_{ext}} \tag{8-2}$$

Example 8-1

For an aquifer, the following values are given: $A = 40,470\ \text{m}^2$, $b = 16.75\ \text{m}$, $n = 0.38$, and $t_{ext} = 1\ \text{yr}$. Determine the average extraction rate Q_{ext}.

Solution

From Eq. (8-1),

$$V_w = Abn = (40,470\ \text{m}^2)(16.75\ \text{m})(0.38) = 257,591.55\ \text{m}^3$$

Using this value in Eq. (8-2),

$$Q_{ext} = \frac{V_w}{t_{ext}} = \frac{257,591.55\ \text{m}^3}{365\ \text{d}} = 705.73\ \frac{\text{m}^3}{\text{d}} = 8.17\ \frac{\text{L}}{\text{s}}$$

Example 8-2

For an aquifer, the following values are given: $A = 12$ acres, $b = 60$ ft, $n = 0.40$, and $t_{ext} = 1\ \text{yr}$. Determine the average extraction rate Q_{ext}.

Solution

From Eq. (8-1),

$$V_w = Abn = (12)(43,560\ \text{ft}^2)(60\ \text{ft})(0.40) = 12,545,280\ \text{ft}^3$$

$$1\ \text{ft}^3 = 7.48\ \text{gal}$$

$$V_w = \left(12{,}545{,}280\ \text{ft}^3\right)\left(7.48\ \frac{\text{gal}}{\text{ft}^3}\right) = 93{,}838{,}694.40\ \text{gal}$$

From Eq. (8-2),

$$Q_{ext} = \frac{V_w}{t_{ext}} = \frac{93{,}838{,}694.40\ \text{gal}}{365\ \text{day}} = \left(270{,}092.3134\ \frac{\text{gal}}{\text{day}}\right)\left[\frac{\text{day}}{24(60\ \text{min})}\right] = 178.54\ \frac{\text{gal}}{\text{min}}$$

$$= 178.54\ \text{gpm}$$

8.4.4 Calculating the Estimated Cleanup Time for Reactive Solutes

For a reactive solute, the time required to remove the contaminant by extraction is increased, and this will be shown by some examples below.

Example 8-3

For a square aquifer, the following values are given: $A = (200\ \text{m})(200\ \text{m}) = 40{,}000\ \text{m}^2$, $b = 16.75\ \text{m}$, $n = 0.40$, $n_e = 0.30$, $\rho_b = 1{,}600\ \text{kg/m}^3$, $\rho_w = 1{,}000\ \text{kg/m}^3$, $K_d = 0.75\ \text{mL/g}$, and $t_{ext} = 1\ \text{yr}$. The flow in the aquifer is from one side of the contaminant zone to the other side with a volume discharged $Q = 6\ \text{L/s}$. Determine the ground water velocity under sorbing conditions and compare it with the nonsorbing conditions case.

Solution

Since $Q = 6\ \text{L/s} = 0.006\ \text{m}^3/\text{s}$, the ground water flux q (Darcy velocity) in the aquifer is

$$q = \frac{Q}{A} = \frac{0.006\ \frac{\text{m}^3}{\text{s}}}{(200\ \text{m})(16.75\ \text{m})} = 17.91\times10^{-7}\ \frac{\text{m}}{\text{s}} = 56.482\ \frac{\text{m}}{\text{yr}}$$

From Eq. (3-6), the ground water velocity is

$$V = \frac{q}{n_e} = \frac{56.482\ \frac{\text{m}}{\text{yr}}}{0.30} = 188.273\ \frac{\text{m}}{\text{yr}}$$

From Eq. (4-68), the retardation factor is

$$R_d = 1 + \frac{\rho_b}{n} K_d = 1 + \left(\frac{1{,}600\ \frac{\text{kg}}{\text{m}^3}}{0.40}\right)\left(0.00075\ \frac{\text{m}^3}{\text{kg}}\right) = 4$$

Notice that $K_d = 0.75\ \text{mL/g} = 0.00075\ \text{m}^3/\text{kg}$. The retarded velocity V_d is

$$V_d = \frac{V}{R_d} = \frac{188.273\ \frac{\text{m}}{\text{yr}}}{4} = 47.068\ \frac{\text{m}}{\text{yr}}$$

Since $V_d < V$, the cleanup time will be increased because of the slower contaminant movement toward the extraction wells. In addition, the tailing effect will be amplified due to desorption. This means that as the contaminant plume is reduced in concentration as a result of extracting water, the contaminant constituent will desorb from the soil and maintain the ratio of the partition coefficient.

Example 8-4

For an aquifer, the following values are given: $A = 18\text{ acres} = 18\left(43{,}560\text{ ft}^2\right) = 435{,}600\text{ ft}^2$, $b = 60\text{ ft}$, $n = 0.40$, $n_e = 0.30$, $\rho_b = 100\text{ lbm/ft}^3$, $\rho_w = 62.4\text{ lbm/ft}^3$, $K_d = 0.75\text{ mL/g}$, and $t_{ext} = 1\text{ yr}$. The flow in the aquifer is from one side of the contaminant zone to the other side with a volume discharged $Q = 100\text{ gpm}$. Determine the ground water velocity under sorbing conditions and compare it with the nonsorbing conditions case.

Solution

The ground water flux q (Darcy velocity) in the aquifer is

$$q = \frac{Q}{A} = \frac{\left(100\,\frac{\text{gal}}{\text{min}}\right)\left(1{,}440\,\frac{\text{min}}{\text{d}}\right)\left(\frac{\text{ft}^3}{7.48\text{ gal}}\right)\left(365\,\frac{\text{d}}{\text{yr}}\right)}{(660\text{ ft})\,(55\text{ ft})} = 193.574\,\frac{\text{ft}}{\text{yr}}$$

From Eq. (3-6), the ground water velocity is

$$V = \frac{q}{n_e} = \frac{193.574\,\frac{\text{m}}{\text{yr}}}{0.30} = 645.247\,\frac{\text{ft}}{\text{yr}}$$

From Eq. (4-68), the retardation factor is

$$R_d = 1 + \frac{\rho_b}{n}K_d = 1 + \left(\frac{100\,\frac{\text{lbm}}{\text{ft}^3}}{0.40}\right)\left(0.012014\,\frac{\text{ft}^3}{\text{lbm}}\right) = 4.0035$$

Notice that $K_d = 0.75\text{ mL/g} = 0.012014\text{ ft}^3/\text{lbm}$. The retarded velocity V_d is

$$V_d = \frac{V}{R_d} = \frac{645.247\,\frac{\text{ft}}{\text{yr}}}{4.0035} = 161.171\,\frac{\text{ft}}{\text{yr}}$$

Since $V_d < V$, the cleanup time will be increased because of the slower contaminant movement toward the extraction wells. In addition, the tailing effect will be amplified due to desorption. This means that as the contaminant plume is reduced in concentration as a result of extracting water, the contaminant constituent will desorb from the soil and maintain the ratio of the partition coefficient.

8.4.5 Description of Tailing and Rebound

Measured contaminant concentrations in ground water with a P&T system generate "tailing" and "rebound" effects on the concentration versus time curves shown in Figure 8-5. As shown in Figure 8-5, "tailing" refers to continuously declining contaminant concentrations with a slower rate in a P&T system. At many times, the asymptotic apparent residual contaminant concentration exceeds cleanup standards. Another potential problem is that dissolved contaminant concentrations may rebound if extraction of ground water is stopped after temporarily attaining a cleanup standard.

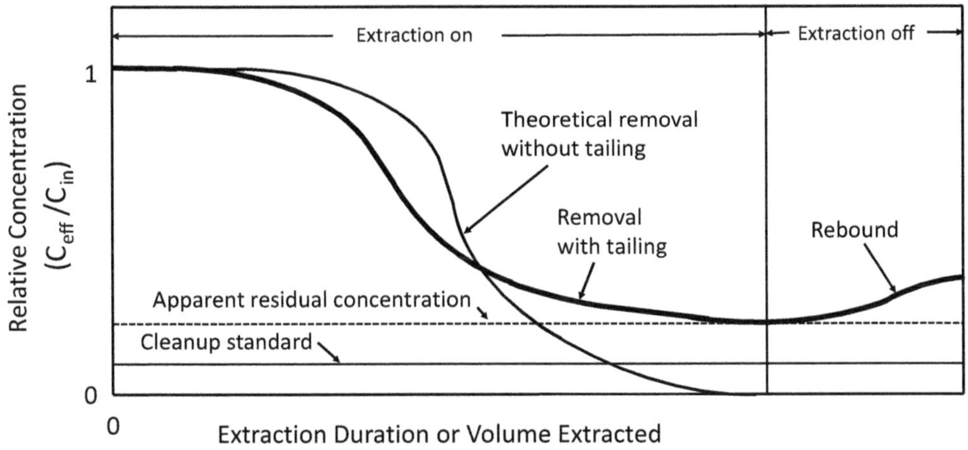

FIGURE 8-5
Concentration versus water extraction duration or volume showing tailing and rebound effects. Adapted from Cohen et al. (1994, p. 5, Figure 1.3; EPA/600/R-94/123).

8.4.6 Determination of Critical Inward Hydraulic Gradient

The mass of NAPLs can be contained by generating an inward hydraulic gradient. However, the adequacy of the inward hydraulic gradient must be evaluated using the method of Al-Niami and Rushton (1977), which is presented in Chapter 5.

8.5 Performance Evaluations of P&T Systems

Adequate understanding of the extent of contamination at a site may not be achieved unless the geologic, hydrogeologic, hydrologic, chemical, and biological complexities of the site are appropriately defined. For a selected remediation system for a particular site, (1) remedial action objectives and (2) monitoring program need to be described clearly. Remedial action objectives are the goals of the overall remediation process. To make sure that the remedial action objectives are fulfilled, appropriate monitoring of the site conditions must be carried out. If the results of the monitoring indicate that the objectives are not satisfied, changes for the remediation system as well as alternate objectives need to be specified. After that, their adequacy must be evaluated through the monitoring process. In the following sections, performance evaluations for P&T systems are described (Keely, 1989; Mercer et al., 1990).

8.5.1 Remedial Action Objectives

Numerous monitoring criteria and monitoring point locations at the site are used as parts of performance standards. Monitoring criteria can be divided into chemical, hydrodynamic, and administrative control. Chemical monitoring criteria are risk based (U.S. EPA, 1986) and include maximum contaminant levels (MCLs), alternate concentration limits (ACLs), detection limits, and natural water quality. Hydrodynamic compliance criteria may include to minimize infiltration through the vadose zone, maintaining an inward hydraulic gradient at the boundary of the contaminant plume or providing minimum flow to a surface water body. Administrative control monitoring criteria range from reporting requirements, such as frequency and character of operational and post-operational monitoring, to land-use restrictions, such as drilling bans and other access-limiting restrictions.

8.5.2 Monitoring

After the remedial action objectives are established and a remedial system is designed to meet the aforementioned remedial action objectives, the next step is to establish a monitoring program to evaluate the success of the remedial system.

Complex hydrogeologic conditions along with ground water flow patterns present significant technical difficulties for site characterization and evaluation of the migration pathways of contaminants. The location of monitoring wells is critical for a successful monitoring system of a contaminated site. For example, in Figure 8-6, water flowing along the flowline proceeding directly into and extraction well from upgradient moves most rapidly, whereas water at the lateral limits of the capture zone moves more slowly. One result is that certain parts of the contaminated aquifer flush well and the other parts poorly.

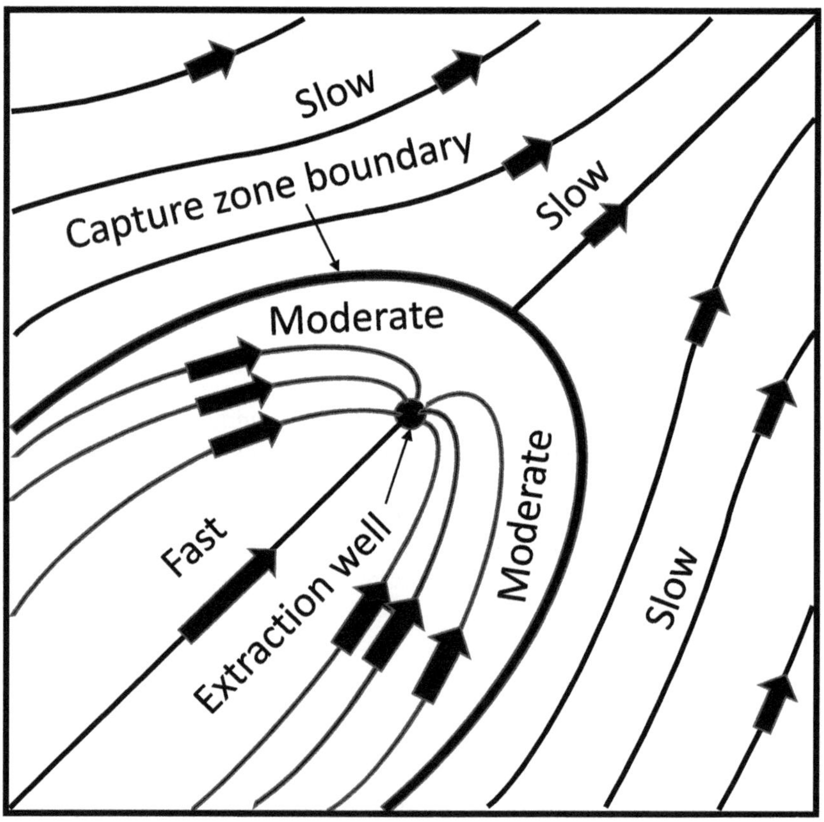

FIGURE 8-6
Flowline pattern generated by an extraction well. Adapted from Keely (1989, p. 2, Figure 1).

Another result of the water extraction is that previously uncontaminated portions of the aquifer at the outer boundary of the contaminant plume may become contaminated by the operation of an extraction well that is located too close to the plume boundary, because the flowline pattern extends downgradient of the well. Detailed field investigations are necessary during remediation to determine the locations of the various flow patterns generated by P&T operation. Consequently, there may be a need for more data to be generated during the site remediation, especially inside the contamination plume boundaries, than data generated during the site investigation. For most cases, it is likely that interpretations of the data that are collected during a P&T remediation may require the use of mathematical and statistical models to analyze those data.

A P&T system may require modification during the initial operational stages because of uncertainties in aquifer characterization. Modifications may result from improved estimates of horizontal and vertical hydraulic conductivities or more complete information on chemistry and loading to the treatment facility. Also, other modifications may be due to mechanical failures of pumps, wells, or surface plumbing equipment. Some key points regarding modifications are presented below (Keely, 1989; Mercer et al., 1990).

8.5.2.1 Effects of Low Hydraulic Conductivity Zones

Contaminated zones having low hydraulic conductivities in the contaminated volume of the aquifer may not be recovered by advection due to their hydrodynamic isolation. Movement of contaminants from these zones is similar to the movement in lower hydraulic conductivity zones. This situation can potentially be corrected by adjusting extraction rates and/or well locations.

8.5.2.2 Periodic Review

Periodic review and modification of the design, construction, maintenance, and operation of the P&T system may be necessary. Also the performance of the system should be evaluated annually or more frequently to determine if the goals and standards of the design criteria are being fulfilled. If it is not, adjustment or modification of the system may be necessary. Modifications may also be made as one part of the contaminant plume becomes clean or when portions are not showing the desired progress. Adjustments or modifications can include relocating or adding extraction wells or altering pumping rates.

8.5.2.3 Pulsed Extraction or Injection in P&T Remediations

One of the alternatives in P&T remediations to improve the efficiency of contaminant recovery is switching from continuous extraction to pulsed extraction. Pulsed extraction is the intermittent operation of a P&T system. In other words, pulsed operation of hydraulic systems is the cycling of extraction or injection wells on and off in active and resting phases (Figure 8-7). The resting phase of a pulsed-extraction operation can allow sufficient time for contaminants to diffuse out of low hydraulic zones and into adjacent high hydraulic conductivity zones until maximum concentrations are achieved in the higher hydraulic conductivity zones. For sorbed contaminants and NAPL residuals, sufficient time can be allowed for equilibrium concentrations to be reached in local ground water. Subsequent to each resting phase, the active phase of the cycle removes the minimum volume of contaminated ground water at the maximum possible concentrations for the most efficient

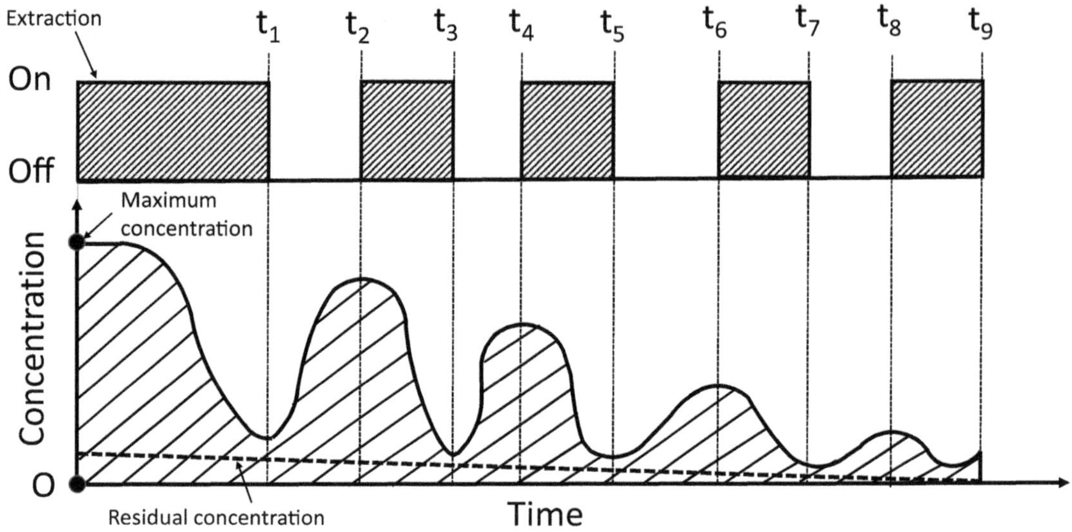

FIGURE 8-7
Remediation of residual contaminant mass by pulsed extraction. Adapted from Keely (1989, p. 7, Figure 12).

treatment. By occasionally cycling only select wells, stagnation zones may be brought into active flow paths and remediated. The durations of extraction and no-extraction periods (about 1–30 days) are site-specific and can only be optimized through trial-and-error operation.

8.6 Limitations of Pump-and-Treat (P&T) Systems

P&T cleanup systems may be the only option when deep ground water contamination exists. Well-designed and accurately located extraction wells may become effective for containing and/or remediating ground water contamination, but limitations exist. For many contaminants, reducing ground water concentrations below standards is a difficult task. Favorable and unfavorable conditions for the application of P&T technology are listed in Table 8-5.

Potentially, tailing and rebound may result from some physical and chemical processes that affect P&T remediation (e.g., Cohen et al., 1994, 1997). These are described below.

8.6.1 Limitations Due to Nonaqueous Phase Liquids (NAPLs)

The existence of NAPLs in subsurface flow systems can act as long-term ground water contamination due to their limited aqueous solubility, which may significantly exceed drinking water standards. If NAPLs are not removed by excavation or contained, tailing and rebound potentially may occur during and after P&T operation, respectively. NAPLs at residual saturation are one of the most difficult to clean up sources of ground water contamination. Among DNAPLS are substances such as halogenated aliphatic hydrocarbons,

TABLE 8-5

Favorable and Unfavorable Conditions for Pump-and-Treat Technologies

Favorable Conditions	**Unfavorable Conditions**
Source Term	
Source removed	NAPLs at residual saturation
Chemical Properties	
Mobile chemicals	Chemical sorbed or precipitated
Hydrogeology	
High hydraulic conductivity (e.g., $K > 10^{-5}$ cm/s)	Very low hydraulic conductivity (e.g., $K < 10^{-7}$ cm/s)
Homogeneous	Highly heterogeneous

Source: Adapted from USEPA: Mercer et al. (1990, p. 16, Table 5).

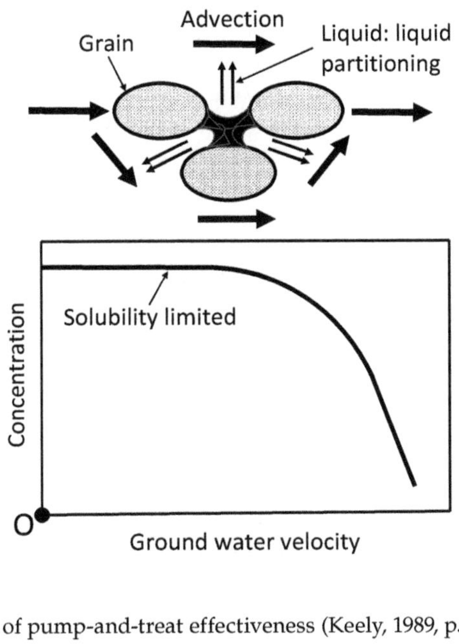

FIGURE 8-8
Liquid partitioning limitations of pump-and-treat effectiveness (Keely, 1989, p. 9, Figure 16; Mercer et al., 1990, p. 17, Figure 9).

halogenated benzenes, phthalate esters, and polychlorinated biphenyls for which there are particular difficulties for their cleanup.

When NAPLs are trapped in pores by interfacial tension, diffusive liquid–liquid partitioning controls dissolution. Flow rates during remediation may be too rapid to allow aqueous saturation levels of partitioned contaminants to be reached locally (see Figure 8-8). If insufficient contact time is allowed, the affected water may be advected away from the residual NAPLs before approaching chemical equilibrium and is replaced by water from upgradient. Because ground water extraction is not generally efficient at cleaning up this type of source, some other remedial action may be required.

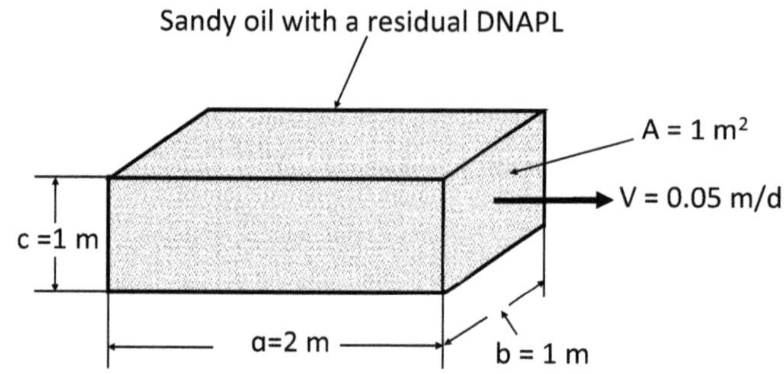

FIGURE 8-9
The geometry of Example 8-5.

Example 8-5

This example is adapted from Mercer et al. (1990, p. 16). The dimensions of a sandy residual DNAPL are $a = 2$ m, $b = 1$ m, and $c = 1$ m. The content of DNAPL is 60 L/m^3 (Figure 8-9). The ground water velocity is 0.05 m/d, and the effective porosity is 0.25. It is assumed that DNAPLs dissolve into ground water to 15% of their solubility. For trichloroethylene (TCE), the density is 1.47 g/cm^3 and solubility is 1,100 mg/L. Determine approximately how many years would be required to dissolve the DNAPLs.

Solution

The volume of the sandy oil:

$$V_{SO} = abc = (2\text{ m})(1\text{ m})(1\text{ m}) = 2\text{ m}^3$$

Residual DNAPLs content:

$$C_{R\text{-}DNAPL} = 60\,\frac{\text{L}}{\text{m}^3} = 60\,\frac{0.001\text{ m}^3}{\text{m}^3} = 0.06$$

Ground water velocity:

$$V = 0.05\,\frac{\text{m}}{\text{d}}$$

Effective porosity:

$$n_e = 0.25$$

Percentage of the DNAPLs solubility to be dissolved:

$$f_{TCE} = 15\%$$

TCE density:

$$\rho_{TCE} = 1.47\,\frac{\text{g}}{\text{cm}^3} = 1.47\,\frac{\text{g}}{(0.01\text{ m})^3} = 1{,}470{,}000\,\frac{\text{g}}{\text{m}^3}$$

Solubility of TCE:

$$C_{TCE-S} = 1{,}100 \frac{\text{mg}}{\text{L}}$$

Mass to be dissolved:

$$M_{TCE} = (C_{R-DNAPL})(V_{SO})(\rho_{TCE}) = (0.06)(2 \text{ m}^3)\left(1{,}470{,}000 \frac{\text{g}}{\text{m}^3}\right) = 176{,}400 \text{ g}$$

TCE solute concentration:

$$C_{TCE} = (f_{TCE})(C_{TCE-S}) = (15\%)\left(1{,}100 \frac{\text{mg}}{\text{L}}\right) = 165 \frac{\text{mg}}{\text{L}} = 165 \frac{\text{g}}{\text{m}^3}$$

Area perpendicular to the ground velocity:

$$A = bc = (1 \text{ m})(1 \text{ m}) = 1 \text{ m}^2$$

TCE mass flow rate through the area A:

$$\dot{M}_{TCE} = (V)(C_{TCE})(n_e)(A) = \left(0.05 \frac{\text{m}}{\text{d}}\right)\left(165 \frac{\text{g}}{\text{m}^3}\right)(0.25)(1 \text{ m}^2) = 2.0625 \frac{\text{g}}{\text{d}}$$

Time required to dissolve:

$$t_{dissolve} = \frac{M_{TCE}}{\dot{M}_{TCE}} = \frac{176{,}400 \text{ g}}{2.0625 \frac{\text{g}}{\text{d}}} = 85{,}527.273 \text{ d} = 234 \text{ yr}$$

8.6.2 Limitations Due to Contaminant Sorption

When NAPLs are trapped in pores by interfacial tension, diffusive liquid–liquid partitioning controls dissolution. Flow rates during remediation may be too rapid to allow aqueous saturation levels of partitioned contaminants to be reached locally (see Figure 8-10). If insufficient contact time is allowed, the affected water may be advected away from the residual NAPLs before approaching chemical equilibrium and is replaced by water from upgradient. Because ground water extraction is not generally efficient at cleaning up this type of source, some other remedial action may be required.

During P&T system operation, dissolved contaminant concentrations are reduced and contaminants sorbed to subsurface media desorb from the matrix into ground water. Sorption and desorption also cause tailing and concentration rebound. As a result, the ground water cleanup process becomes slower.

Equilibrium contaminant partitioning between porous media and ground water can be described by the Langmuir or Freundlich isotherms (Figure 8-11). For linear isotherm ($N = 1$) and limited ranges of the aqueous concentration C_w (particularly at low concentrations) where $N \neq 1$, the Freundlich constant K_f can be identified as a soil–water distribution coefficient K_d as

$$K_d = \frac{C_s}{C_w} \tag{8-3}$$

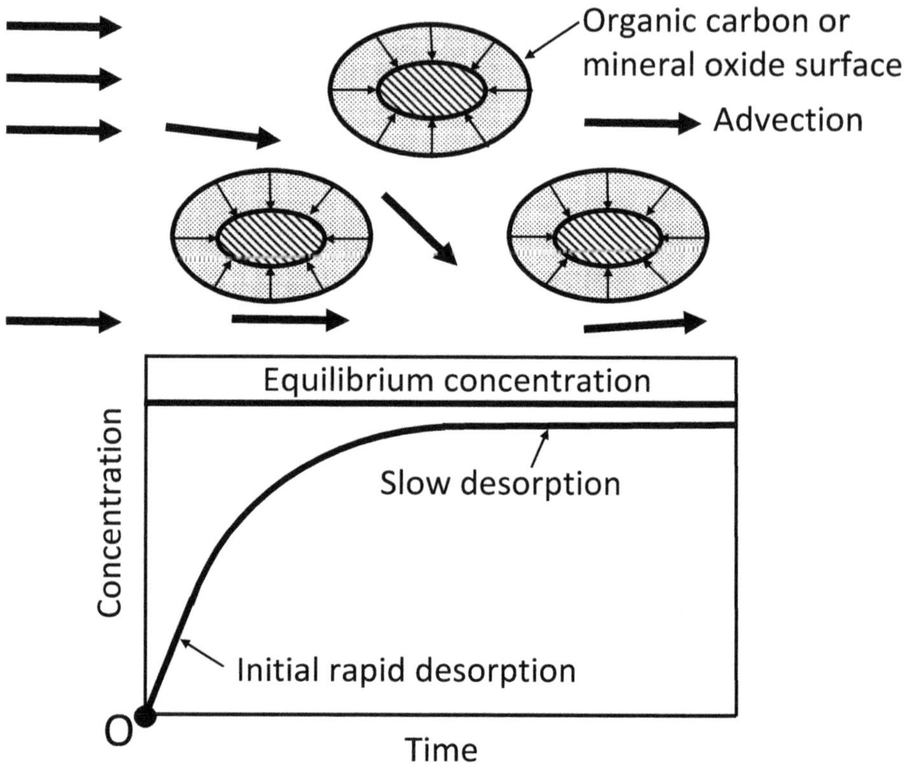

FIGURE 8-10
Sorption limitations to pump-and-treat effectiveness. Adapted from Keely (1989, p. 9, Figure 15; Mercer et al., 1990, p. 17, Figure 9).

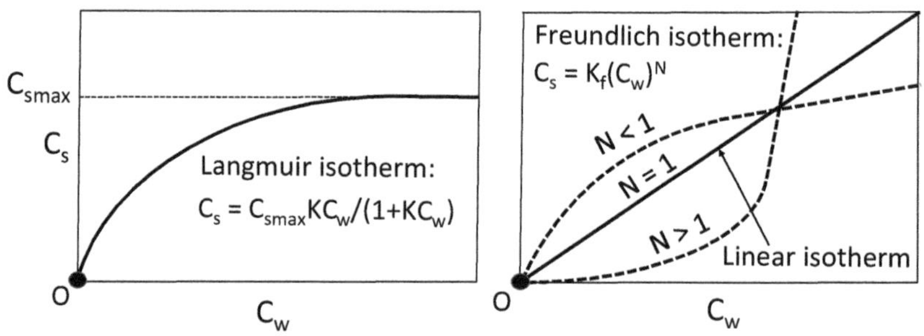

FIGURE 8-11
The Langmuir and Freundlich adsorption isotherms. Adapted from Palmer and Rich (1992).

in which C_s is the equilibrium contaminant concentration in soil and C_w is the equilibrium contaminant concentration in water.

To predict ground water restoration times by different P&T schemes or for natural ground water flushing, contaminant K_d values need to be determined. By assuming linear sorption, the K_d values can be used to estimate the retardation factor R_d given by Eq. (4-68) as

$$R_d = 1 + \frac{\rho_b}{n} K_d \tag{8-4}$$

The fraction of mass residing in the aqueous phase (f_w) is

$$f_w = \frac{C_w V_w}{(C_w V_w + C_s M_s)} = \frac{V_w}{V_w + \frac{C_s M_s}{C_w}} \tag{8-5}$$

in which V_w is the volume of water in the total volume and M_s is the mass of solids in the subject volume (Cohen et al., 1994, 1997).

Using Eq. (8-3) in Eq. (8-5),

$$f_w = \frac{V_w}{V_w + K_d M_s} \tag{8-6}$$

Sorption and retardation factor are dependent on the site characteristics. As indicated by Eq. (8-4), as K_d increases the retardation factor R_d increases as well. Desorption and retardation increase the volume of ground water that must be extracted to attain dissolved contaminant concentration reductions. Tailing and rebound effects will be increased significantly where desorption is slow with respect to the ground water flow and kinetic limitations prevent sustenance of equilibrium contaminant concentrations in ground water. Figure 8-12 illustrates the relationship between ground water velocity and the concentration of dissolved contaminants.

8.6.3 Limitations Due to Solids Dissolution

Solid dissolutions can also play a key role in limiting the P&T method for contaminated sites, and some key points are presented below (Cohen et al., 1997).

Important physicochemical processes affecting the solubility, reactivity, mobility, and toxicity of inorganic contaminants include (1) chemical speciation, (2) oxidation/reduction,

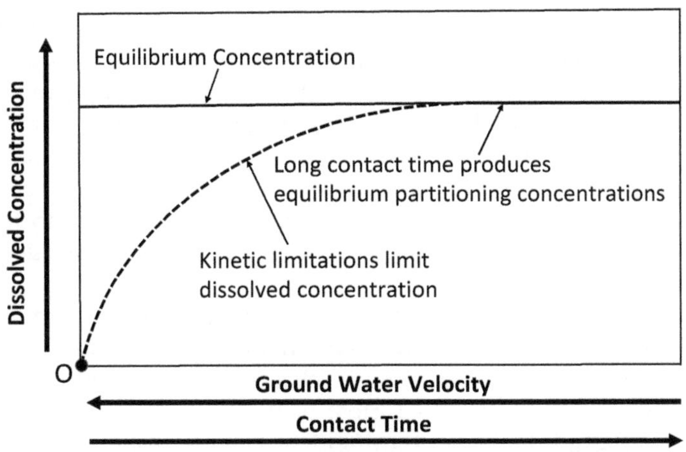

FIGURE 8-12
Relationship between ground water velocity and the concentration of dissolved contaminants. Adapted from Cohen et al. (1994).

(3) dissolution/precipitation, (4) ion exchange and sorption, and (5) particle transport (Palmer and Fish, 1991). Inorganic contaminants generally occur in various chemical forms or species. The known total concentration of an inorganic element in ground water or soil, which is commonly provided by laboratory analysis, may not have significant value alone to assess its subsurface behavior. Rather, it is often more important to determine contaminant speciation, which depends on several factors including pH, Eh, ion and gas concentrations, and temperature. For example, metal cations combine with different anions to form aqueous complexes that increase the solubility, mobility, and risk associated with potentially toxic metals such as chromium and arsenic.

There are publicly available computer codes to assess the solubility and geochemical behavior of inorganic species in ground water. Given the complex interaction between solid minerals, inorganics, and environmental factors (such as Eh–pH relations), these codes can be used to assess the solubility and geochemical behavior of inorganic species in ground water. Data requirements for these codes typically include field analysis of ground water samples for pH, temperature, Eh or dissolved oxygen, and alkalinity, and a complete inorganic chemical analysis for all major and minor ions and all metals and anions under investigation.

Apart from chemical analyses for speciation studies, investigations should be conducted in delineating inorganic contaminant plumes and estimating plume migration rates. Mineralogical characterization efforts can be used to identify solid phases (e.g., clay minerals and Fe and Mn oxyhydroxides) that control inorganic contaminant partitioning. Sorption–desorption and other tests can be conducted to assess inorganic contaminant partitioning, solubility, and mobility as a function of pH and other factors, and the potential for aquifer restoration.

8.6.4 Limitations Due to Hydraulic Conductivity and Ground water Velocity Variations

Localized variations of horizontal and vertical hydraulic conductivities and ground water velocity can also play key roles in limiting P&T remediations (Keely, 1989; Cohen et al., 1994, 1997).

Tailing and rebound, which are described in Section 8.4.5, also result from variable travel times associated with different flow paths taken by contaminants to extraction wells. Ground water at the edge of a capture zone travels a greater distance under a lower hydraulic gradient than ground water closer to the center of the capture zone. Travel times also vary as a function of initial contaminant distribution and hydraulic conductivity differences. If pumping is stopped, rebound will occur wherever the resulting flow path modification diminishes contaminant dilution. Horizontal and vertical hydraulic conductivities and contaminant distributions should be characterized to facilitate analysis of ground water stagnation and velocity variations that would be induced by alternative pumping schemes.

Localized variations in the rate of ground water flow velocities can exist in heterogeneous formations. When operating a P&T remediation system with extraction wells, the velocity variations potentially generate rapid cleansing of the sediments having higher hydraulic conductivities, which conduct most of the flow, and low hydraulic conductivity sediments act as though leaky contaminant reservoirs from which contaminants are removed very slowly by molecular diffusion (Figure 8-13).

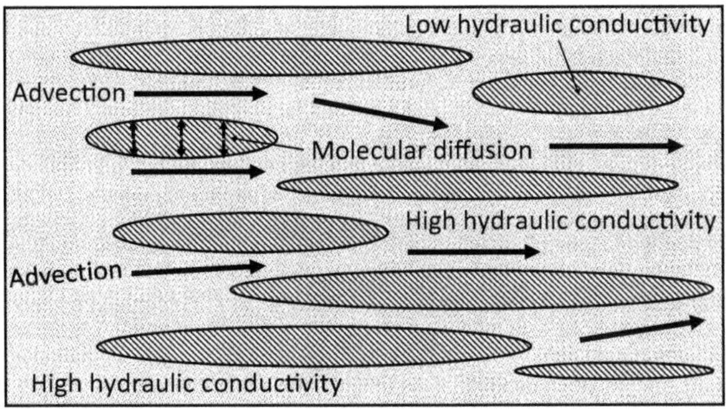

FIGURE 8-13
The effects of hydraulic conductivity and ground water velocity variations. Adapted from Keely (1989, p. 8, Figure 14).

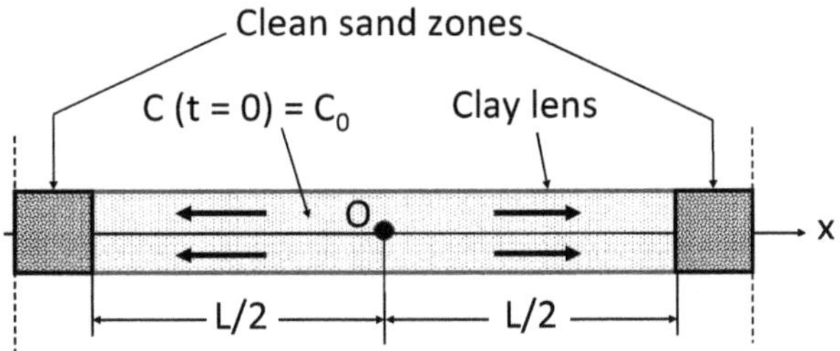

FIGURE 8-14
The conceptual model of the molecular diffusion contaminant migration from a single clay lens to clean sand zones.

8.6.5 Limitations Due to the Effects of Matrix Diffusion Contamination

In relatively less permeable media, the prime transport mechanism of dissolved contaminants is molecular diffusion. During a P&T operation, dissolved contaminant concentrations in the relatively permeable zones may be quickly reduced by advective flushing as compared with the less permeable zones. This slow migration by molecular diffusion can cause long-term tailing and rebound after the termination of water extraction. Matrix diffusion may increase the time necessary to complete the remediation process, especially in heterogeneous and fractured media in which contaminant transport through preferential pathways results in large concentration gradients (Gillham et al., 1984; McKay et al., 1993; Parker et al., 1994; Cohen et al., 1997).

8.6.5.1 Contaminant Migration from a Clay Lens to Clean Sand Zones Start

Consider a contaminated sandy aquifer with clay lenses. Advective transport induced by the extraction of water may quickly reduce contaminant concentrations in the sand. Concentrations in the clay lenses, however, will decrease slowly as contaminants slowly diffuse from the clay to the sand. If C_0 is a uniform initial contaminant concentration in a clay lens of thickness L (Figure 8-14) and that P&T maintains a very low concentration in the sand, then the time required for diffusion to reduce the average relative contaminant concentration C/C_0 in the clay lens can be estimated using the heat conduction solution in Carslaw and Jaeger [1959, p. 97, Eq. (8)] and given in Section 8.6.5.5.

8.6.5.2 Description of the Conceptual Model

As shown in Figure 8-14 of the conceptual model, the thickness of the clay lens is L and is located in a sandy aquifer. It is assumed that the contaminant migration by molecular diffusion is one-dimensional toward the clean sand zones. Initially, it is assumed that the contaminant concentration is uniform in the clay lens:

$$C(t=0)=C_0 \tag{8-7}$$

The problem is to determine the $C(x,t)$ concentration variation as a function of distance x and time t.

8.6.5.3 Governing Differential Equation

The one-dimensional differential equation under molecular diffusion conditions is

$$R_d\frac{\partial C}{\partial t}=D^*\frac{\partial^2 C}{\partial x^2} \tag{8-8}$$

in which C is the contaminant concentration, R_d is the retardation factor defined by Eq. (8-4), D^* is the effective molecular diffusion coefficient, x is the distance, and t is the time. Derivational details of Eq. (8-8) can be found in most books (e.g., Batu, 2006, pp. 12–33). Eq. (8-8) can also be written as

$$\frac{\partial C}{\partial t}=\frac{D^*}{R_d}\frac{\partial^2 C}{\partial x^2} \tag{8-9}$$

8.6.5.4 Boundary Conditions

The boundary conditions are

$$C=0 \quad x=-\frac{L}{2} \quad t=t \tag{8-10}$$

$$C=0 \quad x=+\frac{L}{2} \quad t=t \tag{8-11}$$

Equations (8-10) and (8-11) state that the contaminant concentrations at $x=-L/2$ and $x=+L/2$ are zero at the sand zones.

8.6.5.5 C(x, t) Solution in the Clay Lens

The problem described above is equivalent to the heat conduction problem described and solved in Carslaw and Jaeger (1959, pp. 93–97). Using the solution in Carslaw and Jaeger [1959, p. 97, Eq. (8)], the concentration solution can be written as

$$C(x,t)=\frac{4C_0}{\pi}\sum_{n=0}^{\infty}\frac{(-1)^n}{(2n+1)}e^{-\left(\frac{D^*}{R_d}\right)\left[\frac{(2n+1)^2\pi^2 t}{L^2}\right]}\cos\left[\frac{(2n+1)\pi x}{L}\right] \tag{8-12}$$

8.6.5.6 Average $C_{av}(t)$ Solution in the Clay Lens

From Eq.(8-12), the average concentration along the lens can be determined as

$$\frac{C_{av}(t)}{C_0}=\frac{8}{\pi^2}\sum_{n=0}^{\infty}\frac{1}{(2n+1)^2}e^{-\left(\frac{D^*}{R_d}\right)\left[\frac{(2n+1)^2\pi^2 t}{L^2}\right]} \tag{8-13}$$

which is equivalent to in Carslaw and Jaeger [1959, p. 97, Eq. (10)]. Derivation of Eq. (8-13) is given in Example 8-6. Using Eq. (4-30),

$$D^* = D_0\tau \quad 0<\tau<1.0 \tag{8-14}$$

in which D_0 is the molecular diffusion coefficient in pure water or diffusion coefficient in aqueous solution and τ is the dimensionless tortuosity factor and its value generally varies between 0.3 and 0.7 (e.g., van Genuchten and Wierenga, 1986, pp. 1026–1027). With Eq. (8-14), Eq. (8-13) takes the form

$$\frac{C_{av}(t)}{C_0}=\frac{8}{\pi^2}\sum_{n=0}^{\infty}\frac{1}{(2n+1)^2}e^{-\left(\frac{D_0\tau}{R_d}\right)\left[\frac{(2n+1)^2\pi^2 t}{L^2}\right]} \tag{8-15}$$

Equation (8-15) is exactly the same as Cohen et al. [1997, p. 35, Eq. (A.4)] with the exception of notation as well as a typo error [C_0 must not be at the right-hand side of Eq. (8-15) in Eq. (A-4)]. Beyond these, Cohen and others defined the tortuosity factor differently commonly encountered in the literature, i.e., their tortuosity factor is $\alpha = 1/\tau$.

Example 8-6

Derive Eq. (8-13) from Eq. (8-12).

Solution

The average of Eq. (8-12) along the lens can be written as

$$\frac{C_{av}(t)}{C_0}=\frac{4}{\pi}\sum_{n=0}^{\infty}\left[\frac{(-1)^n}{(2n+1)}e^{-\frac{D^*}{R_d}\frac{(2n+1)^2\pi^2 t}{L^2}}\right]I_{av} \tag{E8-6-1}$$

in which

$$I_{av}=\frac{\int_{-\frac{L}{2}}^{+\frac{L}{2}}\cos\left[\frac{(2n+1)\pi x}{L}\right]dx}{L} \tag{E8-6-2}$$

Performing the integral

$$I_{av} = \frac{L}{(2n+1)\pi}\left(\frac{1}{L}\right)\left|\sin\left[\frac{(2n+1)\pi x}{L}\right]\right|_{-\frac{L}{2}}^{+\frac{L}{2}}$$

$$= \frac{1}{(2n+1)\pi}\left\{\sin\left[\frac{(2n+1)\left(\frac{L}{2}\right)\pi}{L}\right] - \sin\left[\frac{(2n+1)\left(-\frac{L}{2}\right)\pi}{L}\right]\right\} \qquad \text{(E8-6-3)}$$

Since $\sin(-A) = -\sin(A)$, Eq. (E8-6-3) becomes

$$I_{av} = \frac{1}{(2n+1)\pi}\left\{\sin\left[\frac{(2n+1)\pi}{2}\right] + \sin\left[\frac{(2n+1)\pi}{2}\right]\right\}$$

$$= \frac{1}{(2n+1)\pi}2\sin\left[\frac{(2n+1)\pi}{2}\right] \qquad \text{(E8-6-4)}$$

$$= \frac{1}{(2n+1)\pi}2(-1)^n$$

Using this in Eq.(E8-6-1),

$$\frac{C_{av}(t)}{C_0} = \frac{4}{\pi}\sum_{n=0}^{\infty}\left[\frac{(-1)^n}{(2n+1)}e^{-\frac{D^*}{R_d}\frac{(2n+1)^2\pi^2 t}{L^2}}\right]\frac{1}{(2n+1)\pi}2(-1)^n$$

$$= \frac{8}{\pi^2}\sum_{n=0}^{\infty}\frac{(-1)^{2n}}{(2n+1)^2}e^{-\frac{D^*}{R_d}\frac{(2n+1)^2\pi^2 t}{L^2}} \qquad \text{(E8-6-5)}$$

Because $(-1)^{2n} = 1$ for $n = 0,1,2,\ldots$, it becomes

$$\frac{C_{av}(t)}{C_0} = \frac{8}{\pi^2}\sum_{n=0}^{\infty}\frac{1}{(2n+1)^2}e^{-\left(\frac{D^*}{R_d}\right)\left[\frac{(2n+1)^2\pi^2 t}{L^2}\right]} \qquad \text{(E8-6-6)}$$

which is exactly the same as Eq. (8-13).

8.6.5.7 Evaluation of the $C_{av}(t)/C_0$ Solution

The series in Eq. (8-13) rapidly converges and numerical evaluations by Excel showed that a couple of terms may be enough to generate stable results. However, for the results presented below, a total of $N = 17$ terms were used in order to generate very accurate $C_{av}(t)/C_0$ values.

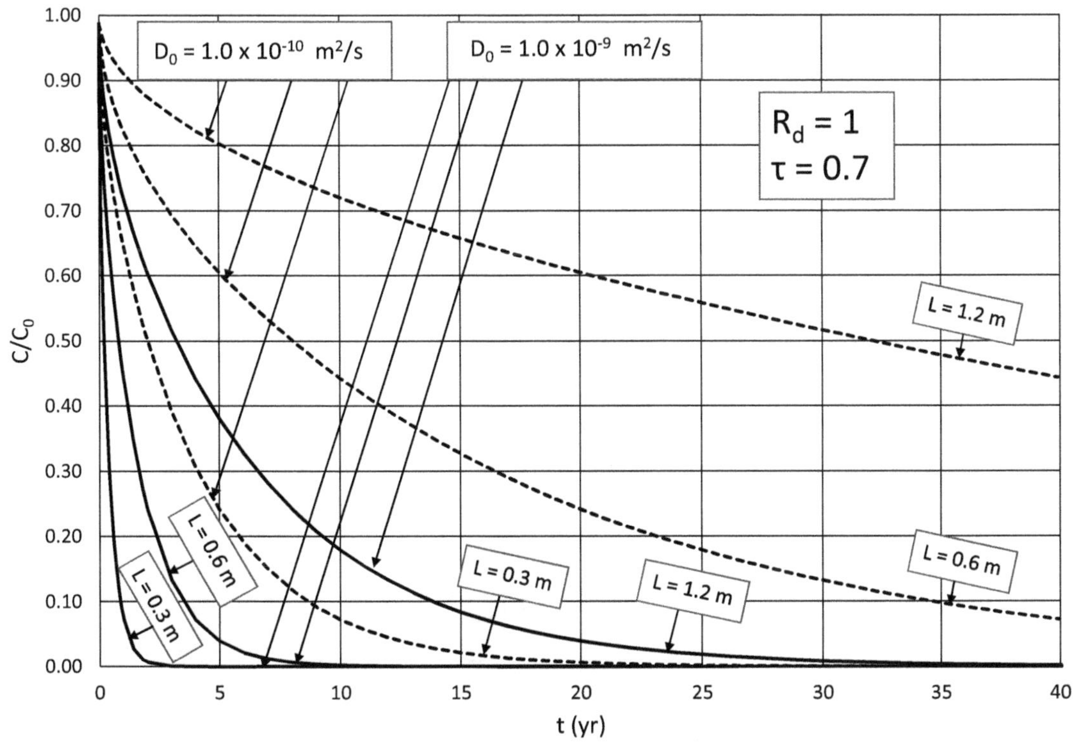

FIGURE 8-15
Normalized average concentration (C_{av}/C_0) versus time (t) in a clay lens of L thickness due to molecular diffusion to adjacent clean sand zones during P&T operation for $R_d = 1$ and $\tau = 0.7$ with two typical D_0 values. Adapted from Cohen et al. (1997, p. 35, Figure A.3).

Figure 8-15 displays $C_{av}(t)/C_0$ versus time for $R_d = 1$ and $\tau = 0.7$ with $L = 0.3$ m, 0.6 m, and 1.2 m clay lens thicknesses using typical diffusion coefficients $D_0 = 1.0 \times 10^{-10}$ m^2/s and 1.0×10^{-9} m^2/s. Figure 8-15 indicates that the molecular diffusion coefficient D_0 value can significantly affect the cleanup time of the aquifer. Figure 8-16 displays $C_{av}(t)/C_0$ versus time for $R_d = 5$ and the rest of values are the same as Figure 8-15. Comparison of Figures 8-15 and 8-16 shows that as the retardation factor increases, the contaminant concentrations in the clay lens increase as well, and this situation corresponds to increased cleanup times.

The potential for matrix diffusion contamination to cause tailing and rebound can be assessed based on (1) knowledge of the contaminant concentration history in the subsurface, (2) site stratigraphy, and (3) chemical analyses conducted on vertical core samples taken from low-permeability matrix material, (4) diffusion modeling, and (5) review of P&T monitoring data (Cohen et al., 1997, p. 35).

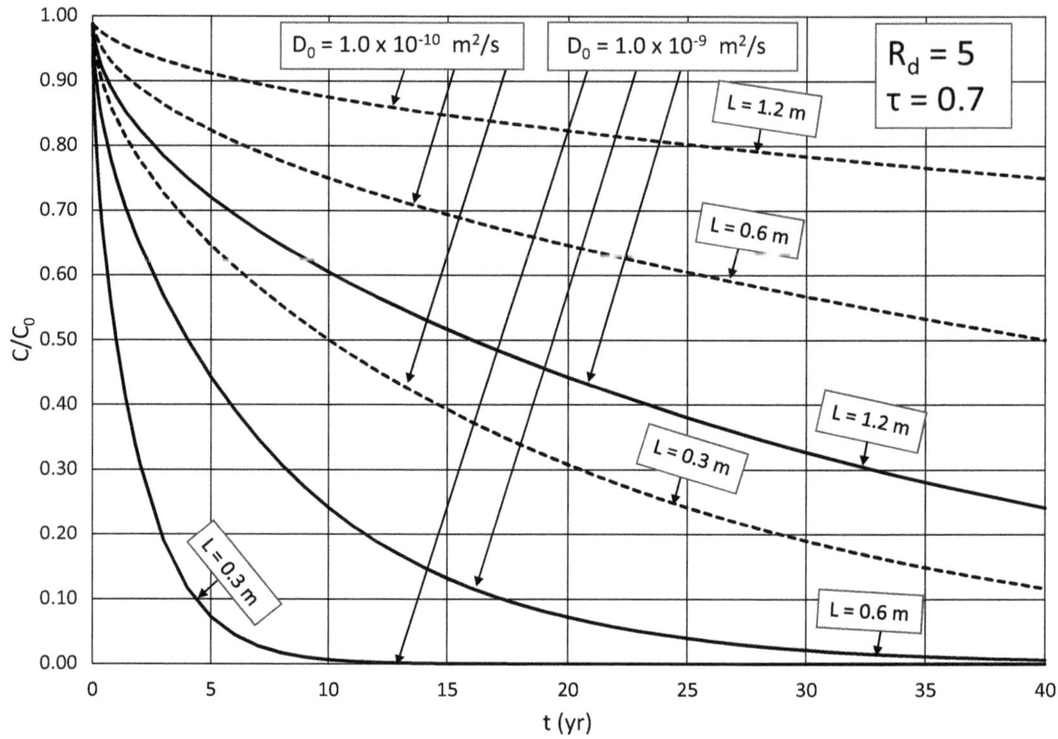

FIGURE 8-16
Normalized average concentration (C_{av} / C_0) versus time (t) in a clay lens of L thickness due to molecular diffusion to adjacent clean sand zones during P&T operation for $R_d = 5$ and $\tau = 0.7$ with two typical D_0 values. Adapted from Cohen et al. (1997, p. 35, Figure A.3).

Problems

8.1 For an aquifer, the following values are given: $A = 60,500\ \text{m}^2, b = 12.50\ \text{m}, n = 0.42$, and $t_{ext} = 1$ year. Determine the average extraction rate Q_{ext}.

8.2 For an aquifer, the following values are given: $A = 15$ acres, $b = 45$ ft, $n = 0.40$, and $t_{ext} = 1$ year. Determine the average extraction rate Q_{ext}.

8.3 For an aquifer, the following values are given: $A = (250\ \text{m})(250\ \text{m}) = 62,500\ \text{m}^2$, $b = 14$ m, $n = 0.44$, $n_e = 0.25$, $\rho_b = 1,500\ \text{kg/m}^3$, $\rho_w = 1,000\ \text{kg/m}^3$, $K_d = 0.65\ \text{mL/g}$, and $t_{ext} = 1$ year. The flow in the aquifer is from one side of the contaminant zone to the other side with a volume discharged $Q = 10\ \text{L/s}$. Determine the ground water velocity under sorbing conditions and compare it with the nonsorbing conditions case.

8.4 For a square aquifer, the following values are given: $A = 15$ acres $= 15(43,560\ \text{ft}^2) = 653,400\ \text{ft}^2\ (808.332\ \text{ft by } 808.332\ \text{ft})$, $b = 45$ ft, $n = 0.42$, $n_e = 0.28$, $\rho_b = 110\ \text{lbm/ft}^3$, $\rho_w = 62.4\ \text{lbm/ft}^3$, $K_d = 0.85\ \text{mL/g}$, and $t_{ext} = 1$ yr. The flow in the aquifer is from one side of the contaminant zone to the other side with a volume discharged $Q = 150$ gpm. Determine the ground water velocity under sorbing conditions and compare it with the nonsorbing conditions case.

8.5 The dimensions of a sandy residual DNAPL shown in Figure 8-9 are $a = 3$ m, $b = 2$ m, and $c = 1$ m. The content of DNAPL is 50 L/m^3. The ground water velocity is 0.10 m/d, and the effective porosity is 0.22. It is assumed that DNAPLs dissolve into ground water to 12% of their solubility. For TCE, the density is 1.47 g/cm^3 and solubility is 1,100 mg/L. Determine approximately how many years would be required to dissolve the DNAPLs.

References

Acker III, W.L., *Basic Procedures for Soil Sampling and Core Drilling*, Acker Drill Co., Inc., Scranton, Pennsylvania, 1974.

Al-Niami, A.N.S., and K.R. Rushton, "Analysis of Flow Against Dispersion in Porous Media," *Journal of Hydrology*, Vol. 33, pp. 87–97, 1977.

Al-Niami, A.N.S., and K.R. Rushton, "Finite Difference Solutions for One-Dimensional Dispersion Using an Improved Mesh Design," *Journal of Hydrology*, Vol. 39, pp. 301–310, 1978.

Aral, M.M., and Y. Tang, "Flow Against Dispersion in Two-Dimensional Regions," *Journal of Hydrology*, Vol. 140, pp. 261–277, 1992.

Baehr, A.L., and M.F. Hult, "Determination of the Air-Phase Permeability Tensor of an Unsaturated Zone at the Bemidji, Minnesota Research Site," U.S. Geological Survey Toxic Substances Hydrology Program, Proceedings of the Technical Meeting, Phoenix, Arizona, September 26–28, 1988, *U.S. Geological Survey Water-Resources Investigation Report 88-4220*, pp. 55–62, Reston, Virginia, 1989.

Baehr, A.L., and M.F. Hult, "Evaluation of Unsaturated Zone Air Permeability Through Pneumatic Tests," *Water Resources Research*, Vol. 27, No. 10, pp. 2605–2617, October, 1991.

Ball, J.W., E.A. Jenne, and D.K. Nordstrom, "WATEQ2: A Computerized Chemical Model for Trace and Major Element Speciation and Mineral Equilibria of Natural Waters", *ACS Symposium Series*, Vol. 83, pp. 815–835, 1979.

Barcelona, M.J., J.P. Gibb, J.A. Helfrich, and E.E. Garske, *Practical Guide for Ground-Water Sampling*, Illinois State Water Survey Contact Report No. 374, USEPA-RSKERL under Cooperative Agreement CR-809966-01, U.S. Environmental Protection Agency, Ada, Oklahoma, 1985.

Barcelona, M.J., J.P. Gibb, and R.A. Miller, *A Guide to the Selection of Materials for Monitoring Well Construction and Ground-Water Sampling*, Illinois State Water Survey Contract Report No. 327, USEPA-RSKERL, EPA-600/52-84/024, U.S. Environmental Protection Agency, 1983.

Batu, V., "A Generalized Two-Dimensional Analytical Solution for Hydrodynamic Dispersion in Bounded Media with the First-Type Boundary Condition at the Source," *Water Resources Research*, Vol. 25, No. 6, pp. 1125–1132, 1989.

Batu, V., "A Generalized Two-Dimensional Analytical Solute Transport Model in Bounded Media for Flux-Type Finite Multiple Source," *Water Resources Research*, Vol. 29, No. 8, pp. 2881–2892, 1993.

Batu, V., *Aquifer Hydraulics: A Comprehensive Guide to Hydrogeologic Data Analysis*, John Wiley & Sons, Inc., New York, 727 pp., 1998.

Batu, V., *Applied Flow and Solute Transport Modeling in Aquifers: Fundamental Principles and Analytical and Numerical Solutions*, CRC Press Taylor & Francis Group, Boca Raton, Florida, 667 pp., 2006.

Batu, V., "Estimation of Degradation Rates by Satisfying Mass Balance at the Inlet," *Ground Water*, Vol. 48, No. 4, pp. 560–568, July–August, 2010.

Batu, V., *Fluid Mechanics and Hydraulics: Illustrative Worked Examples of Surface and Subsurface Flows*, Taylor & Francis CRC Press, Boca Raton, Florida, 1,240, pp., 2024.

Batu, V., and M.T. van Genuchten, "First- and Third-Type Boundary Conditions in Two-Dimensional Solute Transport Modeling," *Water Resources Research*, Vol. 26, No. 2, pp. 339–350, 1990.

Batu, V., M.T. van Genuchten, and J.C. Parker, Reply to the comments of V. Gorokhovski and C.J. Neville, *Ground Water*, Vol. 51, No. 1, pp. 9–13, January–February, 2013.

Bear, J., "On the Tensor Form of Dispersion in Porous Media," *Journal of Geophysical Research*, Vol. 66, pp. 1185–1187, 1961.

Bear, J., *Dynamics of Fluids in Porous Media*, Dover Publications, Inc., New York, 1988, Originally published by American Elsevier Publishing Company, Inc., New York, 764 pp., 1972.

Bear, J., *Hydraulics of Ground water*, McGraw-Hill Book Company, New York, 569 pp., 1979.

Benjamin, J.R., and C.A. Cornell, *Probability Statistics and Decisions for Civil Engineers*, McGraw-Hill, New York, 1970.

Bögli, A., *Karst Hydrology and Physical Speleology*, Springer-Verlag, New York, 1980.
Burnett, R.D., and E.O. Frind, "An Alternating Direction Galerkin Technique for Simulation of Ground water Contaminant transport in Three Dimensions, 2, Dimensionality Effects," *Water Resources Research*, Vol. 23, No. 4, pp. 695–705, 1987.
Brigham, W.E., "Mixing Equations in Short Laboratory Columns," *Society of Petroleum Engineers Journal*, Vol. 14, pp. 91–99, 1974.
Buscheck, T.E., and C.M. Alcantar, "Regression Methods and Analytical Solutions to Demonstrate Intrinsic Bioremediation," In *Intrinsic Bioremediation*, Editors R.E. Hinchee, J.T. Wilson, and D.C. Downey, Batelle Press, Columbus, Ohio, pp. 109–116, 1995.
Butler, J.J., Jr., *The Design, Performance, and Analysis of Slug Tests* (2nd edition), Taylor & Francis CRC Press, Boca Raton, Florida, 266 pp., 2019.
Carslaw, H.S., and J.C. Jaeger, *Conduction of Heat in Solids* (2nd edition), Clarendon Press, London, United Kingdom, 510 pp., 1959.
Cartwright, K., and J.M. Shafer, "Selected Technical Considerations for Data Collection and Interpretation in Ground water", *National Water Quality Monitoring and Assessment*, Washington, DC, 1987.
Chapelle, H.C., P.M. Bradley, D.R. Lovley, and D.A. Vroblesky, "Measuring Rates of Biodegradation in a Contaminated Aquifer Using Field and Laboratory Methods," *Ground Water*, Vol. 34, No.4, pp. 691–698, 1995.
CLO (Chemical Libretexts Organization), Units of Concentration, Internet, January 29, 2023: https://chem.libretexts.org/Bookshelves/Physical_and_Theoretical_Chemistry_Textbook_Maps/Supplemental_Modules_(Physical_and_Theoretical_Chemistry)/Physical_Properties_of_Matter/Solutions_and_Mixtures/Solution_Basics/Units_Of_Concentration?readerView, 2023.
Cohen, R.M., A.H. Vincent, J.W. Mercer, C.R. Faust, and C.P. Spalding, *Methods for monitoring Pump-and-Treat Performance*, EPA/600/R-94/123, U.S. EPA, ORD, R.S. Kerr Environmental Research Laboratory, Ada, Oklahoma, 102 pp., 1994.
Cohen, R.M., J.W. Mercer, R.M. Greenwald, and M.S. Beljin, *Design Guidelines for Conventional Pump-and-Treat Systems*, EPA/540/S-97/504, U.S. EPA, ORD, R.S. Kerr Environmental Research Laboratory, Ada, Oklahoma, 38 pp., 1997.
Cooper, H.H., Jr., and C.E. Jacob, "A Generalized Graphical Method for Evaluation Formation Constants and Summarizing Well-Field History," *Transactions, American Geophysical Union*, Vol. 27, No. IV, pp. 526–534, August, 1946.
Dagan, G., "Time-Dependent Macrodispersion for Solute Transport in Anisotropic Heterogeneous Aquifers," *Water Resources Research*, Vol. 24, No. 9, pp. 1491–1500, September, 1988.
Dagan, G., *Flow and Transport in Porous Formations*, Springer-Verlag, Berlin, Heidelberg, Germany, 465 pp., 1989.
Dagan, G., and S. P. Neuman (Eds.), *Subsurface Flow and Transport: A Stochastic Approach*, Cambridge University Press, Cambridge, United Kingdom, 1997.
Darcy, H., *Les Fontaines Publiques de la Ville Dijon*, (in French) Dalmont, Paris, France, 1856.
Dettinger, M.D., and J.L. Wilson, "First-Order Analysis of Uncertainty in Numerical Models of Ground water Flow, I: Mathematical Development", *Water Resources Research*, Vol. 17, No. 1, pp. 149–161, 1981.
Domenico, P.A., and F.W. Schwartz, *Physical and Chemical Hydrogeology* (2nd edition), John Wiley & Sons, Inc., New York, 506 pp., 1998.
Ebach, F.A., and R.R. White, "Mixing of Fluids Flowing through Beds of Packed Solids," *Journal: American Institute of Chemical Engineering*, Vol. 4, pp. 161–169, 1958.
Eldho, T.I., "Flow Against Dispersion Analysis in Porous Media Using Boundary Elements," *Transactions on Ecology and the Environment*, Vol. 14, pp. 260–268, 1997.
Englund, E., and A. Sparks, *GEO-EAS (Geostatistical Environmental Assessment Software) User's Guide*, EPA/600/4-88/033a, U.S. Environmental Protection Agency, Las Vegas, Nevada, 1988.
Falta, R.W., "Analysis of Soil Gas Pump Tests," *Proceedings of the ER 93 Environmental Remediation Conference*, Augusta, Georgia, October 24–28, Vol. 1, pp. 441–447, 1993.

Falta, R.W., "A Program for Analyzing Transient and Steady-State Soil Gas Pump Tests," *Ground Water*, Vo. 34., No. 4, pp. 750–755, July–August, 1996.

Ferris, J.G., D.B. Knowles, R.H. Brown, and R.W. Stallman, "Theory of Aquifer Tests," *U.S. Geological Survey Water-Supply Paper 1536-E*, pp. 69–174, 1962.

Freeze, R.A., "A Stochastic-Conceptual Analysis of One-Dimensional Ground water Flow in Nonuniform Homogeneous Media," *Water Resources Research*, Vol. 11, No. 5, pp. 725–741, October, 1975.

Freeze, R.A., and J.A. Cherry, *Ground water*, Prentice-Hall, Ing., Englewood Cliffs, New Jersey, pp. 604, 1979.

Freeze, R.A., J. Massmann, L. Smith, T. Sperling, and B. James, "Hydrogeological Decision Analysis: 1. A Framework," *Ground Water*, Vol. 28, No. 5, pp. 738–766, 1990.

Freundlich, H., "Kapillarchemie, eine Darstellung der Chemie der Kolloide und verwandter Gebiete," Akademische Verlagsgesellschaft, 1909.

Garabedian, S. P., D. R. LeBlanc, L. W. Gelhar, and M. A. Celia, "Large-Scale Natural Gradient Tracer Test in Sand and Gravel, Cape Cod, Massachusetts: 2. Analysis of Spatial Moments for a Nonreactive Tracer," *Water Resources Research*, Vo. 27. No. 5, pp. 911–924, 1991.

Gelhar, L.W., *Stochastic Subsurface Hydrology*, Prentice-Hall, Inc., Englewood Cliffs, New Jersey, 390 pp., 1993.

Gelhar, L.W., and C.L. Axness, "Three-Dimensional Stochastic Analysis of Macrodispersion in Aquifers," *Water Resources Research*, Vol. 19, No. 1, pp. 161–180, February, 1983.

Gelhar, L.W., A. Mantoglou, C. Welty, and K.R. Rehfeldt, "A Review of Field-Scale Physical Solute Transport Processes in Saturated and Unsaturated Porous Media," *EPRI Report*, EA-4190, Electric Power Research Institute, Palo Alto, California, August, 1985.

Gelhar, L.W., C. Welty, and K.R. Rehfeldt, "A Critical Review of Data on Field-Scale Dispersion in Aquifers," *Water Resources Research*, Vol. 28, No.7, pp. 1955–1974, July, 1992.

Gershon, N.D., and A. Nir, "Effect of Boundary Conditions of Models on Tracer Distribution in Flow Through Porous Mediums," *Water Resources Research*, Vol. 5, No. 4, pp. 830–840, 1969.

Gillham, R.W., and J.A. Cherry, "Predictability of Solute Transport in Diffusion Controlled Hydrogeologic Regimes," *Proceedings of the Symposium on Low-Level Disposal: Facility Design, Construction and Operating Practices*, 1982a.

Gillham, R.W., and J.A. Cherry, "Contaminant Migration in Saturated Unconsolidated Geologic Deposits," *Recent Trends in Hydrogeology*, The Geological Society of America, Special Paper 189, Boulder, Colorado, pp. 31–62, 1982b.

Gillham, R.W., E.A. Sudicky, J.A. Cherry, and E.O. Frind, "An Advective-Diffusive Concept for Solute Transport in Heterogeneous Unconsolidated Geologic Deposits", *Water Resources Research*, Vol. 20, No. 3, pp. 369–378, 1984.

Glueckauf, E., K.H. Barker, and G.P. Kitt, "Theory of Chromatography, 8: The Separation of Lithium Isotopes by Ion Exchange and of Neon Isotopes by Low-Temperature Adsorption Columns," *Transactions of the Faraday Society*, Vol. 7, pp. 199–213, 1949.

Grisak, G.E., and J.F. Pickens, "Solute Transport through Fractured Media: 1. The Effect of Matrix Diffusion", *Water Resources Research*, Vol. 16, pp. 719–730, 1981.

Gröbner, W., and N. Hofreiter, *Integraltafel*, Springer, Vienna, Austria, Vol. 1, 1949.

Gröbner, W., and N. Hofreiter, *Integraltafel*, Springer, Vienna, Austria, Vol. 2, 1950.

Gutjahr, A.L., L.W. Gelhar, A.A. Bakr, and J.R. MacMillan, "Stochastic Analysis of Spatial Variability in Subsurface Flows, 2, Evaluation and Application," *Water Resources Research*, Vol. 14, No. 5, pp. 953–959, 1978.

Guvanasen, V., and R.E. Volker, "Experimental Investigations of Unconfined Aquifer Pollution from Recharge Basins," *Water Resources Research*, Vol. 19, pp. 707–717, 1983.

Hantush, M.S., "Hydraulics of Wells", In *Advances in Hydroscience*, Editor V.T. Chow, Academic Press, New York, pp. 281–442, 1964.

Hantush, M.S., and C.E. Jacob, "Non-Steady Radial Flow in an Infinite Leaky Aquifer," *Transactions, American Geophysical Union*, Vol. 36, No. 1, pp. 95–100, 1955.

Harleman, D.R.F., and R.R. Rumer, "Longitudinal and Lateral Dispersion in an Isotropic Porous Medium," *Journal of Fluid Mechanics*, Vol. 16, pp. 385–394, 1963.

IAHS (International Association of Hydrological Sciences), *Karst Hydrogeology and Karst Environment Protection*, IAHS Publication, Wallingford, Oxfordshire, 176, 1261 pp., 1988.

Iman, R.L., and M. Shortencarier, "A Fortran 77 Program and User's Guide for the Generation of Latin Hypercube and Random Samples for use with Computer Models," *Rep. NUREG/ CR-3624, SAND83-2365*, prepared for U.S. Nuclear Regulatory Commission by Sandia National Laboratory, Albuquerque, New Mexico, 1984.

Jacob, C.E., "Flow of Ground Water", In *Engineering Hydraulics*, Editor H. Rouse, John Wiley & Sons, Inc., New York, Chapter V, pp. 321–386, 1950.

Johnson, P.C., M.W. Kemblowski, and J.D. Colthart, "Quantitative Analysis for the Cleanup of Hydrocarbon-Contaminated Soils by In-Situ Soil, Venting", *Ground Water*, Vol. 28, No. 3, pp. 413–429, 1990.

Journel, A.G., and C.J. Huijbregts, *Mining Geostatistics*, Technology & Engineering, 600 pp., 2003.

Kalos, M.H., and P.A. Whitlock, *Monte Carlo Methods Volume I: Basics*, John Wiley & Sons, New York, 186 pp., 1986.

Keely, J.F., *Performance Evaluation of Pump-and-Treat Remediations*, EPA/540/4-89/005, U.S. EPA, ORD, R.S. Kerr Environmental Research Laboratory, Ada, Oklahoma, 1989.

Kemblowski, M.W., J.P. Salanitro, G.M. Deeley, and C.C. Stanley, "Fate and Transport of Residual Hydrocarbon in Ground water – A Case Study," *Proceedings, Petroleum Hydrocarbons and Organic Chemicals in Ground Water: Prevention, Detection, and Restoration*, National Water Well Association/API, Houston, Texas, pp. 207–231, 1987.

Klinkenberg, L.J., "The Permeability of Porous Media to Liquids and Gases," *Drilling and Production Practices*, American Petroleum Institute, pp. 200–213, 1941.

Knox, R.C., L.W. Canter, D.F. Kincannon, E.L. Stover, and C.H. Ward, *Aquifer Restoration: State of the Art*, Noyes Publications, Park Ridge, New Jersey, 1986.

Kreft, A., and A. Zuber, "On the Physical Meaning of the Dispersion Equation and Its Solution for Different Initial and Boundary Conditions," *Chemical Engineering Science*, Vol. 33, pp. 1471–1480, 1978.

Kreft, A., and A. Zuber, "On the Use of the Dispersion Model of Fluid Flow," *The International Journal of Applied Radiation and Isotopes*, Vol. 30, pp. 705–708, 1979.

Kreft, A., and A. Zuber, "Comments on 'Flux-Averaged and Volume-Averaged Concentrations in Continuum Approaches to Solute Transport," *Water Resources Research*, Vol. 22, pp. 1157–1158, 1986.

Lambe, T.W., and R.V. Whitman, *Soil Mechanics*, John Wiley & Sons, New York, 553 pp., 1969.

Langevin, C.D., A.M. Provost, S. Panday, and J.D. Hughes, "Documentation for the MODFLOW 6 Ground water Transport Model," U.S. Geological Survey Techniques and Methods, Book 6, Chapter A61, 197 pp., 2022.

Lapidus, L., and N.R. Amundson, "Mathematics of Adsorption in Beds: IV. The Effect of Longitudinal Diffusion in Ion Exchange Chromatographic Columns," *Journal of Physical. Chemistry*, Vol. 56, pp. 984–988, 1952.

Lattman, L.H., and R.R. Parizek, "Relationship Between Fracture Traces and the Occurrence of Ground Water in Carbonate Rocks, *Journal of Hydrology*, Vol. 22, pp. 73–91, 1964.

Lindstrom, F.T., R. Haque, V.H. Freed, and L. Boersma, "Theory of Movement of Some Herbicides in Soils: Linear Diffusion and Convection of Chemicals in Soils," *Journal of Environmental Science and Technology*," Vol. 1, pp. 561–565, 1967.

Lohman, S.W., *Ground-Water Hydraulics*, Professional Paper 708, U.S. Geological Survey, Washington, DC, 70 pp., 1972.

Marino, M.A., "Distribution of Contaminants in Porous Media," *Water Resources Research*, Vol. 10, pp. 1013–1018, 1974.

Marino, M.A., "Flow Against Dispersion in Nonadsorbing Porous Media," *Journal of Hydrology*, Vol. 37, pp. 149–158, 1978a.

Marino, M.A., "Flow Against Dispersion in Adsorbing Porous Media," *Journal of Hydrology*, Vol. 38, pp. 197–205, 1978b.

Massmann, J.W., "Applying Ground water Flow Models in Vapor Extraction System Design," *Journal of Environmental Engineering*, American Society of Civil Engineers, Vol. 115, No. 1, pp. 129–149, 1989.

Mayer, S.W., and E.R. Tompkins, "Theoretical Analysis of the Column Separation Process," *Journal of American Chemical Society*, Vol. 69, pp. 2866–2874, 1947.

McKay, L.D., R.W. Gillham, and J.A. Cherry, "Field Experiments in a Fractured Clay Till: 2. Solute and Colloid Transport," *Water Resources Research*, Vol. 29, pp. 3879–3890, 1993.

McNab, W.W., and B.P., Dooher, "A Critique of a Steady Analytical Method for Estimating Degradation Rates," *Ground Water*, Vol. 36, No. 6, pp. 983–987, November–December, 1998.

McWhorter, D.B., "Unsteady Radial Flow of Gas in the Vadose Zone," *Journal of Contaminant Hydrology*, Vol. 5, pp. 297–314, 1990.

McWhorter, D.B., and D.K. Sunada, *Ground water Hydrology and Hydraulics*, Water Resources Publications, Littleton, Colorado, 256 pp., 1977.

Mercer, J.W., D.C. Skipp, and D. Griffin, *Basics of Pump-and-Treat Ground-Water Remediation Technology*, EPA/600/8-90/003, U.S. EPA, ORD, R.S. Kerr Environmental Research Laboratory, Ada, Oklahoma, 31 pp., 1990.

Mercer, J.W., S.D. Thomas, and B. Ross, "Parameters and Variables Appearing in Repository Models", *U.S. Nuclear Regulatory Commission NUREG/CR-3066*, Washington, DC, 1982.

Moltyaner, G. L., M. H. Klukas, C. A. Wills, and R. W. D. Killey (1993), "Numerical Simulations of Twin Lake Natural-Gradient Tracer Tests: A Comparison of Methods," *Water Resources Research*, Vol. 29, No. 10, pp. 3433–3452, 1993.

Morel, F., and J. Morgan, "A Numerical Method for Computing Equilibria in Aqueous Chemical Systems", *Environmental Science Technology*, Vol. 6, pp. 58–67, 1972.

Morris, D.A., and A.I. Johnson, "Summary of Hydrologic and Physical Properties of Rocks and Soil Materials as Analyzed by the Hydrologic Laboratory of the U.S. Geological Survey", *U.S. Geological Survey Water Supply Paper 1389-D*, 1967.

Neuman, S.P., "Universal Scaling of Hydraulic Conductivities and Dispersivities in Geologic Media," *Water Resources Research*, Vo. 26, No. 8, pp. 1749–1758, August, 1990.

Neuman, S.P., "Comment on "Is Unique Scaling of Aquifer Macrodispersivity Supported by Field Data," by A. Zech et al.," *Water Resources Research*, Vol. 52, pp. 4199–4202, 2016.

Neuman, S.P., C.L. Winter, and C.M. Newman, "Stochastic Theory of Field-Scale Fickian Dispersion in Anisotropic Porous Media," *Water Resources Research*, Vol. 23, No. 3, pp. 453–466, March, 1987.

Neuman, S.P., and P.A. Witherspoon, "Theory of Flow in a Confined Two Aquifer System," *Water Resources Research*, Vol. 5, No. 4, pp. 803–816, August, 1969.

Ogata, A., "Theory of Dispersion in a Granular Media," U.S. Geological Survey, *Professional Paper 411-I*, 134 pp., 1970.

Ogata, A., and R.B. Banks, "A Solution of the Differential Equation of Longitudinal Dispersion in Porous Media," *Geological Survey Professional Paper 411-A*, United States Government Printing Office, Washington, DC, 7 pp., 1961.

Palmer, C.D., and W. Fish, *Physicochemical Processes: Inorganic Contaminants, in Site Characterization for Subsurface Remediation*, U.S. EPA Seminar Publication, EPA/925/4-91/026, U.S. EPA, ORD, R.S. Kerr Environmental Research Laboratory, Ada, Oklahoma, pp. 179–192, 1991.

Palmer, C.D., and W. Fish, *Chemical Enhancements to Pump-and-Treat Remediation*, EPA/540/S-92/001, U.S. EPA, ORD, R.S. Kerr Environmental Research Laboratory, Ada, Oklahoma,1992.

Parker, B.L., R.W. Gillham, and J.A. Cherry, "Diffusive Disappearance of Immiscible Phase Organic Liquids in Fractured Geologic Media", *Ground Water*, Vol. 32. No. 5, pp. 805–820, 1994.

Parker, J.C., "Analysis of Solute Transport in Column Tracer Tests," *Soil Science Society of America Journal*, Vol. 48, pp. 719–724, 1984.

Parker, J.C., and M.T. Van Genuchten, "Flux-Averaged and Volume-Averaged Concentrations in Continuum Approaches to Solute Transport," *Water Resources Research*, Vol. 20, No. 7, pp. 866–872, July, 1984.

Pickens, J.F., and G.E. Grisak, "Scale-Dependent Dispersion in a Stratified Granular Aquifer," *Water Resources Research*, Vol. 17, No. 4, pp. 1191–1211, August, 1981.
Polubarinova-Kochina, P.Y., *Theory of Ground Water Movement*, translated from the Russian by R.J.M. De Wiest, Princeton University Press, Princeton, New Jersey, 613 pp., 1962.
Ptak, T., and G. Teutsch, "Forced and Natural Gradient Tracer Tests in a Highly Heterogeneous Porous Aquifer: Instrumentation and Measurements," *Journal of Hydrology*, Vol. 159, pp. 79–104, 1994.
Quinlan, J.F., and R.O. Ewers, "Ground Water Flow in Limestone Terrains: Strategy Rationale and Procedure for Reliable, Efficient Monitoring of Ground Water Quality in Karst areas", *Proceedings of the National Symposium and Exposition on Aquifer Restoration and Ground Water Monitoring* (5th), National Water Well Association, Dublin, Ohio, pp. 197–234, 1985.
Raimondi, P., G.H.F. Gardner, and C.B. Petrick, "Effect of Pore Structure and Molecular Diffusion on the Mixing of Miscible Liquids," *A.I.Ch.E. Joint Symposium on Fundamental Concepts of Miscible Fluid Displacement*, Part II, Preprint 43, San Francisco, California, December 6–9, 1963.
Robinson, R.A., and R.H. Stokes, *Electrolyte Solutions* (2nd edition), Butterworth, London, Great Britain, 1965.
Rosenberg, J.L., *Theory and Problems of College Chemistry*, Shaum's Outline Series (6th edition), McGraw-Hill Book Company, New York, 314 pp., 1980.
Rubin, Y., *Applied Stochastic Hydrogeology*, Oxford University Press, Oxford, New York, 391 pp., 2003.
Rumer, R.R., "Longitudinal Dispersion in Steady and Unsteady Flow," *Journal of Hydraulics Division, Proceedings of the American Society of Civil Engineers*, Vol. 88, pp. 147–172, July, 1962.
Sauty, J. P., "Contribution a L'identification des Paramètres de Dispersion Dans les Aquifères par Interpretation des Experiments de Traçage," Ph.D. Thesis, Univ. Scientifique et Medicale et Institut Natl. Polytechnique de Grenoble, Grenoble, France, 1977.
Scheidegger, A.E., "General Theory of Dispersion in Porous Media," *Journal of Geophysical Research*, Vol. 66, pp. 3273–3278, 1961.
Shan, C., R.W. Falta, and I. Javandel, "Analytical Solutions for Steady State Gas Flow to a Soil Vapor Extraction Well," *Water Resources Research*, Vol. 28, No.4, pp. 1105–1120, April, 1992.
Spiegel, M.R., *Theory and Problems of Laplace Transforms*, Shaum's Outline Series, McGraw-Hill Book Company, New York, 261 pp., 1965.
Spiegel, M.R., *Probability and Statistics*, Shaum's Outline Series, McGraw-Hill Book Company, New York, 372 pp., 1975.
Spiegel, M.R., *Mathematical Handbook of Formulas and Tables*, Shaum's Outline Series, McGraw-Hill Book Company, New York, 271 pp., 1968.
Streltsova, T.D., *Well Testing in Heterogeneous Formations*, John Wiley & Sons, New York, 413 pp., 1988.
Theis, C.V., "The Relation Between the Lowering of the Piezometric Surface and the Rate and Duration of Discharge of a Well Using Ground-Water Storage," *Transactions*, American Geophysical Union, Vol. 16, pp. 519–524, August, 1935.
U.S. Army Corps of Engineers, *Soil Vapor Extraction and Bioventing*, EM 1110-1-40001, Engineer Manual, Washington, DC, 3 June, 2002.
U.S. Environmental Protection Agency, *Superfund Public Health Evaluation Manual*, EPA/540/1-86/060, U.S. Environmental Protection Agency, Washington, DC, 1986.
U.S. Environmental Protection Agency, *Handbook of Ground Water*, EPA/625/6-87/016, U.S. Environmental Protection Agency, Cincinnati, Ohio, 212 pp., 1987.
van Genuchten, M.Th., "Mass Transfer in Porous Media," Ph.D. dissertation, New Mexico State University, Las Cruces, New Mexico, December, 148 pp., 1974.
van Genuchten, M.Th., "Analytical Solutions for Chemical Transport with Simultaneous Adsorption, Zero-Order Production and First-Order Decay," *Journal of Hydrology*, Vol. 49, pp. 213–233, 1981.
van Genuchten, M.Th., and J.C. Parker, "Boundary Conditions for Displacement through Short Laboratory Soil Columns," *Soil Science Society of America Journal*, Vol. 48, pp. 703–748, 1984.

van Genuchten, M.Th., and P.J. Wierenga, "Solute Dispersion Coefficients and Retardation Factors", *Methods of Soil Analysis, Part I: Physical and Mineralogical Methods*, Agronomy Monographs, No. 9, Second Edition, American Society of Agronomy – Soil Science Society of America, Madison, Wisconsin, pp. 1025–1054, 1986.

Vennard, J.K., and R.L. Street, *Elementary Fluid Mechanics*, John Wiley & Sons, Sixth edition, 689 pp., New York, 1982.

Vereecken, H., U. Döring, H. Hardelauf, U. Jaekel, U. Hashagen, O. Neuendorf, H. Schwarze, and R. Seidemann, "Analysis of Solute Transport in a Heterogeneous Aquifer: The Krauthausen Field Experiment, *Journal of Contaminant Hydrology*, Vol. 45, pp. 329–358, 2000.

Vermeulen, T., and N.K. Hiester, "Ion exchange Chromatography of Trace Components," *Industrial & Engineering Chemistry Research*, Vol. 44, pp. 636–651, 1952.

Verruijt, A., *Theory of Ground water Flow*, Grand and Breach Science Publishers, New York, 190 pp., 1970.

Wolery, T.J., "Calculation of Chemical Equilibrium between Aqueous Solution and Minerals: The EQ3/6 Software Package, UCRL-52658", Lawrence Livermore Laboratories, Livermore, California, 1979.

Xu, M., and Y. Eckstein, "Use of Weighted Least-Squares Method in Evaluation of the Relationship Between Dispersivity and Field Scale," *Ground Water*, Vol. 33, No. 6, pp. 905–908, November–December, 1995.

Xue, L., X. Guo, and H. Chen, *Fluid Dynamics in Porous Media: Fundamentals and Applications*, World Scientific Pub Co Inc, Hackensack, New Jersey. 408 pp., 2020.

Zech, A., S. Attinger, A. Bellin, V. Cvetkovic, G. Dagan, P. Dietrich, A. Fiori, and G. Teutsch, "Evidence Based Estimation of Macrodispersivity for Ground water Transport Applications", *Ground water*, Vol. 61, No. 3, pp. 346–362, doi: 10.1111/gwat.13252, May–June, 2023.

Zech, A., S. Attinger, A. Bellin, V. Cvetkovic, P. Dietrich, A. Fiori, G. Teutsch, and G. Dagan, "A Critical Analysis of Transverse Dispersivity Field Data", *Ground water*, Vol. 57, No. 4, pp. 632–639, doi: 10.1111/gwat.12838, July–August, 2019.

Zech, A., S. Attinger, V. Cvetkovic, G. Dagan, P. Dietrich, A. Fiori, Y. Rubin, and G. Teutsch, "Is Unique Scaling of Aquifer Macrodispersivity Supported by Field Data?", *Water Resources Research*, Vol. 51, pp. 7662–7579, doi: 10.1002/2015WRO17220, 2015.

Zech, A., S. Attinger, V. Cvetkovic, G. Dagan, P. Dietrich, A. Fiori, Y. Rubin, and G. Teutsch, "Reply to Comment by S. Neuman On 'Is Unique Scaling of Aquifer Macrodispersivity Supported by Field Data?'", *Water Resources Research*, Vol. 52, pp. 4203–4205, doi: 10.1002/2016WRO18812, 2016.

Zhang, Y.-K., and R.C. Heathcote, "An Improved Method for Estimation of Biodegradation Rate with Field Data," *Ground Water Monitoring & Remediation*, Vol. 23, No. 3, pp. 112–116, 2003.

Zheng, C., and P.P. Wang, "MT3DMS: A Modular Three-Dimensional Multispecies Transport Model for Simulation of Advection, Dispersion, and Chemical Reactions of Contaminants in Ground water Systems; Documentation and User's Guide," US Army Corps of Engineers, December, 219 pp., 1999.

Index

adiabatic gases, pressure-density relationship 13
adiabatic exponent 13
adiabatic process 13
advective-dispersive solute transport equations, general case 50, 52
advective-dispersive solute flux for flux-averaged concentration 134
advective-dispersive solute flux for volume-averaged concentration 137
air-filled porosity 154
air-filled porosity determination 168
air permeability 154
air permeability determination 168
air permeability determination from field data 178
aquifer 23
aquifer, confined 24
aquiclude 24
aquitard 24
aquifuge 24
aquifer, homogeneous and anisotropic 51
artesian aquifer 24; *see also* confined aquifer

Bernoulli equation 28
biodegradation 64
bioventing 1, 2, 4, 152

capillary fringe 2
capillarity 18
capillary drop 20
capillary rise 20
cleanup time estimation for nonreactive solutes 209–211
cleanup time estimation for reactive solutes 211–212
coefficient of permeability 31
compressibility of fluids 12
compressible fluids 12
concentration types 44
concentration units 37
concentration units, based on moles 41
concentration units, dilute mass 39
concentration, flux-averaged 45
concentration unit, mass/mass 38
concentration units, mass/volume 39
concentration unit, parts per billion (ppb) 40
concentration unit, parts per million (ppm) 39
concentration unit, parts per trillion (ppt) 41
concentration units, relative 38
concentration units, volume/volume 38
concentration, volume-averaged 45
confined aquifer 24
confining layer 24
converting actual vapor flowrates to equivalent standard vapor flowrates 192
Cooper and Jacob method for soil gas flow, adapted 168
correlation lengths 78
correlation scales 78
covariance function 78
critical inward hydraulic gradient determination 213

Darcy experiment 27–29
Darcy flux 29, 133, 135
Darcy's law 27, 33
Darcy's law validity 30
Darcy velocity components 52, 153
decay parameter of the autocorrelation function 79
degradation 63
degradation parameter 133, 146
degradation parameter estimation method 134
degradation parameter estimation methods, comparative evaluation 146
degradation parameter estimation method, curve fitting 148–151
degradation parameter estimation method evaluation based on first-type source solution 143–145
degradation rate constant 3
dense nonaqueous phase liquids (DNAPLs) 198
density 6, 62
diffusive transport 46
dimensions of units 5
discharge velocity 26, 27, 29
dispersion coefficient, longitudinal 50
dispersion coefficient, mechanical 49, 66–71
dispersion coefficients, under uniform flow conditions 71
dispersion, homogeneous and isotropic, with three constant velocity components 57

dispersion, homogeneous and isotropic, with uniform flow, one-dimensional case 58
dispersion, homogeneous and isotropic, with uniform flow, three-dimensional case 57
dispersion, homogeneous and isotropic, with uniform flow, two-dimensional case 58
dispersion, hydrodynamic 50
dispersive flux, mechanical 49
dispersivities, estimation methods in aquifers 72
dispersivity, intrinsic 67
dispersion tensor, hydrodynamic 51, 72
distribution coefficient 61
drawdown 25
Dupuit-Forchheimer's seepage equation 27
dynamic viscosity 15, 16

English FSS (foot-slug-second) units system 5, 6, 13

Fickian approach 72
Fick's first law 37, 46
Fick's first law for porous media 47
Fick's first law for pure water 46
Fick's second law 48
first-type and third-type source analytical solutions relationship 137
first-type source solution 133
first-type steady state source solution 133, 140
flow and dispersion in opposite directions 101
fluids 5
flux-averaged concentration 45, 133, 134, 138

gas pressure distribution equation, steady state 186
gradient of piezometric head 26
ground water velocity 26

half-life 63
Hantush's steady state drawdown equation for soil gas flow, partially-penetrating wells, leaky aquifers, adapted 178
Hantush's steady state drawdown equation for soil gas flow, partially-penetrating wells, nonleaky aquifers, adapted 177
Hantush's transient drawdown equation for soil gas flow, partially-penetrating wells, leaky aquifers, adapted 175–178
Hantush's transient drawdown equation for soil gas flow, partially-penetrating wells, nonleaky aquifers, adapted 172–175

hydraulic conductivity 30
hydraulic gradient 28, 29, 52
hydraulic head 25

incompressible fluids 12
isotherm, Freundlich nonlinear sorption 60
isotherms, linear sorption 60

Joule (J) 10

K_d approach 61
Kelvin degree 10
kinematic viscosity 15, 16, 32
Klinkenberg's effects 154

leaky aquifer 24
log hydraulic conductivity variance (σ_f^2) 3, 73
longitudinal correlation scale (λ_f) 3, 73, 82
longitudinal dispersivity, asymptotic 81–83
longitudinal dispersivity (α_L) 3, 69, 71, 72, 133
longitudinal dispersivity (α_L) estimation, Neuman method 74, 75, 85
longitudinal dispersivity (α_L) estimation, Pickens and Grisak method 73, 75, 85
longitudinal dispersivity (α_L) estimation, Xu and Eckstein method 74, 75, 85
longitudinal dispersivity (α_L) estimation, Zech et al. method 76, 81, 84–90

macrodispersivity 3
mass in slugs 6
modulus of elasticity 13
molality 43
molar concentration 41
mole fraction 41
molecular diffusion coefficient in pure water 48
molecular diffusion coefficient, effective 48, 102, 133
monitored natural attenuation 64
monitoring of the sources of the monitoring system 214

natural attenuation 64
Navier-Stokes equations 30
Newton (N) 6
Newtonian fluids 17, 18
Newton's second law 6
Newton's viscosity law 14
non-newtonial fluids 17, 18

perfect gas 9
permeability 30
permeability units 32

permeability versus coefficient of permeability 31
porosity, effective 25, 26, 48, 133–135
porosity, total 25, 62
porous materials, manmade 37
porous materials, natural 37
porous media 37
potentiometric head surface 24
pressure aquifer 24; *see also* confined aquifer
pressure-density relationship for vapor flow 156, 157
pump-and-treat (P&T) method 1, 2, 4, 198, 199
pump-and-treat (P&T) systems, components 200
pump-and-treat (P&T) systems, conceptual design 208–213
pump-and-treat (P&T) systems, data collection 204–206
pump-and-treat (P&T) systems, data interpretation 207–208
pump-and-treat (P&T) systems, data requirements 200–204
pump-and-treat (P&T) systems, data uncertainties 208
pump-and-treat (P&T) systems, general description 198
pump-and-treat (P&T) systems, limitations 216–227
pump-and-treat (P&T) systems, limitations due to hydraulic conductivity and ground water velocity variations 222
pump-and-treat (P&T) systems, limitations due to NAPLs 216–219
pump-and-treat (P&T) systems, limitations due to solid dissolution 221–222
pump-and-treat (P&T) systems, limitations due to sorption 219–221
pump-and-treat (P&T) systems, limitations due to the effects of matrix diffusion 223–227
pump-and-treat (P&T) systems, main limitations 200
pump-and-treat (P&T) systems, performance evaluation 213
pump-and-treat (P&T) systems, pulsed extraction or injection 215

radioactive decay 63, 132
radioactive decay equation 63
radius of influence determination 187
Rankine degree in the English FSS units 10, 11
remedial action objectives 213
remediation methods 1
retardation factor 60–62, 133

saturated hydraulic conductivity 30
saturated zone 2, 23
seepage velocity 27
separation length 78
shear stress 15
SI unit system (Système International d'Unités) 5, 6, 9, 13
slug 6, 9
soil gas flow to fully-penetrating wells 158–167
soil gas flow to fully-penetrating wells in unsaturated zone, steady state 182–187
soil vapor extraction 1, 2, 4, 152
soil venting 1, 4, 152
solid matrix 23
solute 37
solute flux equation due to advective transport 48
solute flux equation due to diffusive transport 46
solute flux equation due to dispersive transport 49
solute flux equation in unsaturated porous media for constant water content 50
solute flux equations for homogeneous and anisotropic porous media 51
solute flux equations for homogeneous and isotropic porous media 53
solute transport differential equations in saturated porous media, homogeneous and anisotropic dispersion, generalized 64
solute transport differential equations in saturated porous media, homogeneous and isotropic dispersion, special cases 65
solute transport differential equations in saturated porous media, homogeneous and isotropic dispersion, uniform flow field 66
solute transport differential equations in saturated porous media, homogeneous and isotropic dispersion, uniform flow field, three-dimensional case 66
solute transport differential equations in saturated porous media, homogeneous and isotropic dispersion, uniform flow field, two-dimensional case 66
solute transport differential equations in saturated porous media, homogeneous and isotropic dispersion, uniform flow field, one-dimensional case 66
solute transport equations under degradation conditions 63

solute transport first-and third-type source solutions, comparative evaluation 138–142
solute transport first-type one-dimensional steady state analytical continuous source solution 95
solute transport first-type one-dimensional transient analytical continuous source solution 93–95
solute transport in saturated porous media with sorption and decay 58
solute transport in saturated porous media without sorption and decay 54
solute transport in saturated porous media without sorption and decay, one-dimensional case 46
solute transport one-dimensional analytical solutions 93
solute transport third-type one-dimensional steady state analytical continuous source solution 98–99
solute transport third-type one-dimensional transient analytical continuous source solution 96–98
solute transport with sorption and desorption, one-dimensional case 59
solvent 37
source containment 199
source contamination cleanup 209
source removal, contaminated 199
sources of ground water contamination 199
specific discharge 29
specific gas constant 9, 10
specific gravity 9
specific heat ratio 13
specific volume 7, 9
specific weight 6
standard deviation 79
statistically homogeneous 77
statistically isotropic 77
steady state reaching time estimation 179–182
stochastic subsurface hydrology, some preliminaries 77
stress-strain diagram for liquids and gases 12, 13
subsurface water 23
surface tension 18

tailing and rebound description 212
Theis equation for soil gas flow, adapted 162
Theis transient drawdown equation for soil gas flow, fully-penetrating wells, adapted 158–167
Theis type-curve method for soil gas flow 168
Third-type source solution 135, 140
tortuosity factor 48
transmissivity 30, 33
transverse dispersivities estimation 90, 102
transverse dispersivity 69
transverse horizontal dispersivity (α_{TH}) 3, 71, 72, 90
transverse vertical dispersivity (α_{TV}) 3, 71, 72, 90

unconfined aquifer 24
unsaturated zone 23

vadose zone 23
vapor flow 156, 157
vapor flow rate prediction 193
vapor flow velocity equation 188
variance 79, 82, 84
viscosity dimensions in the English FSS system 16
viscosity dimensions in the SI unit system (Système International d'Unités) 15
viscosity of fluids 14
viscosity, definition 14
volatile organic compounds 1, 4, 152
volume-averaged concentration 45, 135, 136, 138
volumetric air flow rate 190

water table 23
water table aquifer 24
water table upwelling 193

For Product Safety Concerns and Information please contact our EU
representative GPSR@taylorandfrancis.com
Taylor & Francis Verlag GmbH, Kaufingerstraße 24, 80331 München, Germany

www.ingramcontent.com/pod-product-compliance
Ingram Content Group UK Ltd.
Pitfield, Milton Keynes, MK11 3LW, UK
UKHW052052230926
475702UK00025B/1123
9781041239222